Lecture Notes in Physics

Edited by J. Ehlers, München K. Hepp, Zürich
R. Kippenhahn, München H. A. Weidenmüller, Heidelberg
and J. Zittartz, Köln

155

Quantum Optics

Proceedings of the South African Summer School in Theoretical Physics. Held at Cathedral Peak, Natal Drakensberg, South Africa, January 19–30, 1981

Edited by C.A. Engelbrecht

Springer-Verlag
Berlin Heidelberg New York 1982

Editor
C.A. Engelbrecht
The Merensky Institute of Physics, University of Stellenbosch
Stellenbosch 7600, South Africa

ISBN 3-540-11498-X Springer-Verlag Berlin Heidelberg New York
ISBN 0-387-11498-X Springer-Verlag New York Heidelberg Berlin

Printed in Germany

Printing and binding: Beltz Offsetdruck, Hemsbach/Bergstr.
2153/3140-543210

TABLE OF CONTENTS

LECTURERS

H Haken, Institut für Theoretische Physik, University of Stuttgart

F A Hopf, Optical Sciences Center, University of Arizona, Tucson

L A Lugiato, Institute di Fisica, University of Milano

A Schenzle, Physics Department, University of Essen

ORGANIZING COMMITTEE

C A Engelbrecht (CHAIRMAN), University of Stellenbosch
J J Henning, S Afr Atomic Energy Board, Pelindaba
R H Lemmer, University of the Witwatersrand, Johannesburg
T I Salamon, NPRL, CSIR, Pretoria
T B Scheffler, University of Pretoria

PARTICIPANTS

D Bedford, University of Natal, Durban
H K Bouwer, NPRL, CSIR, Pretoria
J H Brink, S Afr Atomic Energy Board, Pelindaba
J D Comins, University of the Witwatersrand, Johannesburg
J A de Wet, Mount Marlow, P O Witmos 5825
E F du Plooy, University of Stellenbosch
E E Erasmus, University of Stellenbosch
H Fiedeldey, University of South Africa, Pretoria
W E Frahn, University of Cape Town
M Gering, University of the Witwatersrand, Johannesburg
H B Geyer, S Afr Atomic Energy Board, Pelindaba
I Gledhill, University of Natal, Durban
F J W Hahne,
P J Harper, NPRL, CSIR, Pretoria
W D Heiss, NRIMS, CSIR, Pretoria
J D Hey, University of Cape Town
E G Jones, NPRL, CSIR, Pretoria
D P Joubert, University of Stellenbosch
S P Klevansky, University of the Witwatersrand, Johannesburg
F J Kok, University of Pretoria
H U Kranold, S Afr Atomic Energy Board, Pelindaba

P Krumm, University of Natal, Durban
P E Lourens, S Afr Atomic Energy Board, Pelindaba
R E Raab, University of Natal, Pietermaritzburg
D E Roberts, S Afr Atomic Energy Board, Pelindaba
D Schmieder, NPRL, CSIR, Pretoria
P du T van der Merwe, S Afr Atomic Energy Board, Pelindaba
W S Verwoerd, University of South Africa, Pretoria
J du P Viljoen, S Afr Atomic Energy Board, Pelindaba
H M von Bergmann, NPRL, CSIR, Pretoria

PREFACE

South Africa shares with other countries far from the North Atlantic scientific community the drawbacks of distance. In a very small number of fields the number of local physicists actively engaged in research is large enough to provide students and scientists with the opportunity to make contact with the conceptual framework and the latest developments. In most fields this is not the case.

It was realized that a partial solution to this problem lies in the establishment of a school along the lines of a summer school, where the participants are immersed in a concentrated course on a specific topic with lectures delivered by a group of experts. During the seventies the theoretical physicists organized themselves and worked, through the South African Institute of Physics (SAIP), to achieve this goal. Success came at last when the Council for Scientific and Industrial Research (CSIR) agreed to provide the all-important financial support for this venture.

A theme was sought as the topic of the first course which is of current interest and combines intriguing conceptual structures with useful applications. Besides meeting these specifications, the theory of quantum optics is also a field in which very little research has been done locally and which could thus profit much from the stimulation provided by such a course.

The first school was held at the Cathedral Peak Hotel in the Natal Drakensberg from 19 to 30 January 1981. These lecture notes consist of the typed manuscripts as supplied by the authors, or, where necessary, prepared from notes taken by participants. They deal with related aspects of quantum optics and present a very readable review of the current ideas in this field.

I would like to use this opportunity to thank the SAIP for its support, the Cathedral Peak Hotel for the use of its facilities, the other members of the organizing committee for their assistance, and the participants for the enthusiasm with which they joined in the discussions. The CSIR bore the brunt of the financial burden and provided invaluable technical and organizational aid. The four lecturers from abroad presented us with inspired lectures, which will always be remembered. We would especially like to thank Hermann Haken, who also acted as unofficial godfather with his advice on the organization of the course.

Finally, we are grateful to the editors of Lecture Notes in Physics for their willingness to publish these proceedings.

C.A. Engelbrecht

THE THEORY OF LASERS AND LASER LIGHT

A course of lectures by:

Hermann Haken

Institut für Theoretische Physik der Universität Stuttgart

Manuscript compiled, from lecture notes taken by them, by:

J J Henning
Physics Division
S.A. Atomic Energy Board

T B Scheffler
Physics Department
University of Pretoria

CONTENTS

1. INTRODUCTION

These lectures treat the theory of the laser and the quantum theory of coherence. Most of the material can also be found in Haken (1966), (1979a, b) or (1981).

Laser physics started in the microwave region. "Microwave Amplification by Stimulated Emission of Radiation" was abbreviated to MASER. In re-deriving Planck's radiation formula, Einstein (1917) postulated that the interaction between matter and light takes 3 forms: in addition to absorption and spontaneous emission of photons occur. A photon impinging on an atom or molecule in an excited state can cause the atom to emit an additional photon, thereby transmitting its excitation energy to the light field. The stimulated photon is exactly in phase with the stimulating photon - they are described by exactly the same wave function.

In order for stimulated emission to predominate over absorption, there must be more atoms in an upper than in a lower energy state. This condition - a so-called population inversion - can never hold in thermal equilibrium. In an ammonia molecule, the ground and first excited states (symmetric and antisymmetric with respect to the position of the nitrogen atom relative to the plane of hydrogen atoms) are very close together, but well separated from other states. As $\Delta E \equiv E_1 - E_0 << kT$, both are equally populated at normal temperatures. By sending a beam of NH_3 molecules through an electric field, Gordon, Zeiger and Townes (1954) separated upper and lower energy molecules into separate beams. The upper state beam was then sent through a cavity with natural (mode) frequency tuned to be in resonance with the Bohr frequency $\Delta E/h$.

First, spontaneous decays will occur. As they are random, and independent of each other, they lead to unco-ordinated chaotic waves with random wave vectors and with the whole range of frequencies corresponding to the molecular natural linewidth. A spontaneously emitted (or stimulated) photon with a frequency and wave vector corresponding to a cavity mode will, however, probably traverse the cavity many times to and fro before escaping. It thus has a good chance to stimulate a molecule to emit a photon with an identical associated electromagnetic wave. Each of the resultant 2 photons may repeat the process, and so forth, so that initially this cascade of coherent photons (all with the same EM wave) will grow exponentially. This growth will be prolonged and lead to a strong coherent wave dominating all spontaneous emission, if the particular wave is a low loss cavity mode.

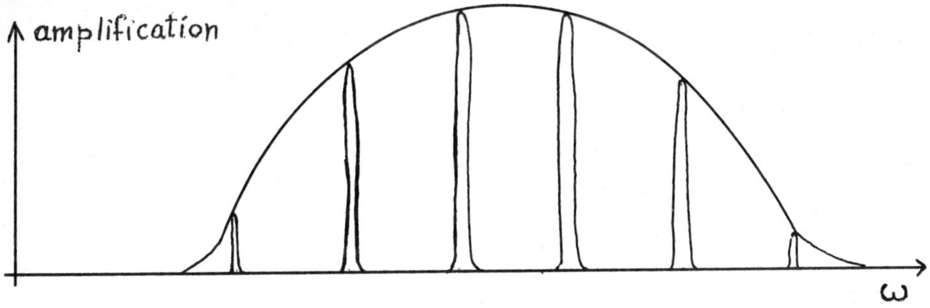

The spacing between cavity lines is usually smaller than the atomic linewidth (envelope), so that amplification by stimulated emission may occur on several cavity modes at the same time. However, in the completely homogeneous case, competition between modes for upper state atoms (more accurately - for inverted population) to "feed" on eliminates all but the strongest mode - a laser equivalent of "survival of the fittest. This can only happen in a ring laser, where $I \propto |\mathcal{E}|^2 \propto |e^{ikx}|^2$ is independent of position. In other cases, the wave has nodes. No stimulated emission "eats" the population inversion at a node, where it can thus feed another mode which has an antinode at this position - a variety of environments permit "peaceful coexistence" of several modes.

As cavity lifetimes T_c may exceed natural lifetimes T_n by many orders of magnitude, the associated cavity linewidth ($\Delta\omega_c = 1/T_c$) may be far sharper than the natural linewidth ($\Delta\omega_n = 1/T_n$). Nonlinear interactions in the laser can further reduce the remaining frequency spread by many orders of magnitude. Laser light can thus be highly monochromatic, with a spectral width of 10^6 to 10^{10} times smaller than that of the best spectral lamp.

In an essentially closed cavity, as is normally used in masers, the mode structure is three dimensional. Since $\omega = kc$ depends only on the magnitude of $\bar{k}$, all modes which (in $\bar{k}$-space) corresponding to a spherical shell of volume $d^3k = 4\pi k^2 dk$ are degenerate, within the same frequency range $d\omega = c dk$.

The number of modes is proportional to the volume in phase space

$$dn = \frac{V\, d^3k}{(2\pi)^3} = \frac{V\, k^2 dk}{2\,\pi^2} = \frac{4\pi V p^2 dp}{(2\pi\hbar)^3}$$

For a cubic cavity, with k/k kept constant (at say 10^{-6}), dn will increase with the cube of the parameter $n_o \equiv \frac{kL}{\pi} = \frac{2L}{\lambda}$. For a microwave laser this is no problem, as the wavelength and cavity dimensions are about the same, so that n_o is a small integer. In the optical region, n_o is typically $\frac{2 \times 0{,}25\ \mathrm{m}}{0{,}5 \times 10^{-6}\ \mathrm{m}} = 10^6$, so that about $10^{-6}.\ n_o^3 = 10^{12}$ modes will compete.

Schawlow and Townes (1960) drastically reduced this number by using a cavity open in the Y and Z directions, so that the number of modes reduces to that of the one-dimensional case. In practice, one to a few hundred modes may compete. With such a cavity, Maiman (1960) obtained laser action in a ruby rod. As the ruby pump scheme is of type A below, it was not considered a likely candidate for successful laser action. Thus Phys. Rev. Letters rejected the first report of laser action, which then appeared in the New York Times after a press conference!

When more than one mode appears with frequency inside the atomic (or molecular) line profile, "survival of the fittest" may assert itself, so that only the mode with the strongest gain (nearest the centre of the atomic line) will survive.

The initial exponential growth of the coherent radiation cascade will slow down when the number of photons, and hence the number of stimulated emissions becomes so large that the population of excited state atoms is significantly depleted. (The number of lower state atoms simultaneously increases, which leads to absorption). The "gain" of the medium thereby decreases, and the exponential growth slows down and stops. (All this typically takes a few nanoseconds). A continuous steady output may be achieved if upper state atoms are continuously supplied, and lower state ones removed. In the ammonia maser this happens by a molecular beam passing through the cavity. In most lasers this is, however, achieved by an optical, electrical, chemical, gasdynamic or other "pump" mechanism which transfers a given atom or molecule from the lower to the upper state via one or more other quantum states.

Sometimes, a pulsed laser output is desired. This may be achieved by a procedure called Q-switching or Q-spoiling: The Q value of the cavity is degraded (the cavity feedback mechanism spoiled) for a sufficient time to allow the "pump" procedures to build up a very high population inversion, storing a large amount of energy. During this period laser action cannot reduce the population inversion, as the positive feedback provided by to and fro reflections in the cavity is prevented by Q spoiling. One form of Q-switching is to rapidly rotate one cavity mirror, thereby preventing laser action except during the nanosecond or so when the mirrors are substantially parallel. During this brief period the laser cascade develops, as a so-called giant pulse that even in a tiny laser may reach a megawatt. In large neodymium-glass or CO_2 systems, many terawatts may be obtained.

Passive Q-switching (without mechanically moving parts) may be achieved by inserting inside the cavity a saturable (bleachable) dye to prevent cavity feedback. When the population inversion exceeds a sufficiently high value, sufficient spontaneously emitted photons are absorbed in the dye to raise half the dye molecules to an upper state. In the dye, spontaneous emission now matches absorption, so that it becomes transparent to light at the laser frequency, and no longer spoils the cavity. The laser cascade develops in the next nanosecond in a giant pulse.

Mode-locking, in which a nonlinear element (either in the laser medium itself, or in a separate cell) establishes a phase relation between different modes, yields a train of even briefer pulses (of picosecond or shorter duration) separated from each other by the laser round-trip time $t = 2L/c$. By selecting one pulse from such a train, and amplifying it by passing it once through a laser medium without mirrors, the highest intensities, as used for achieving fusion, can be reached.

We have seen that laser light is extra-ordinarily monochromatic. Features such as coherence time $t_c = 1/\Delta\omega$ and coherence length $\ell_c = ct_c$ are related to this, as is the possibility to focus laser light to a spot with a diameter of about a wavelength. High intensities are achievable with pulsed focused lasers (up to 10^{25} watt/m^2 on a small spot). The correspondingly strong electric fields can exceed that felt be an electron in the lowest Bohr orbit in hydrogen, and can hence "instantaneously" ionize atoms, even when the energy per photon is far less than the ionization energy. However, these properties do not uniquely characterize laser light. With interference filters, extremely narrowband (monochromatic) light can be prepared. And high intensities are not an essential feature of laser radiation, as typical helium-neon lasers deliver 1 milliwatt of light.

Laser light is, however, uniquely characterized by other coherence, or statistical properties. An example is the 2 photon correlation function, which is measured in a Hanbury Brown-Twiss experiment. To explain the essential difference between laser and other light, we consider a component of the electric field

$$\varepsilon(xt) = E(t) \sin kx\, e^{i\omega t},$$

and focus our attention on the comparatively slowly varying amplitude $E = E(t)$ - the rapid fluctuations being contained in the factor $e^{i\omega t}$. Thermal light has a Gaussian distribution function $\exp[-(\Delta\omega\cdot t)^2]$, whereas light from an interference filter (or a laser below threshold) has a Lorentzian form $1/[1 + (\Delta\omega\cdot t)^2]$. In all these cases the statistical distribution of the amplitude E is around a maximum and average at $E = 0$.

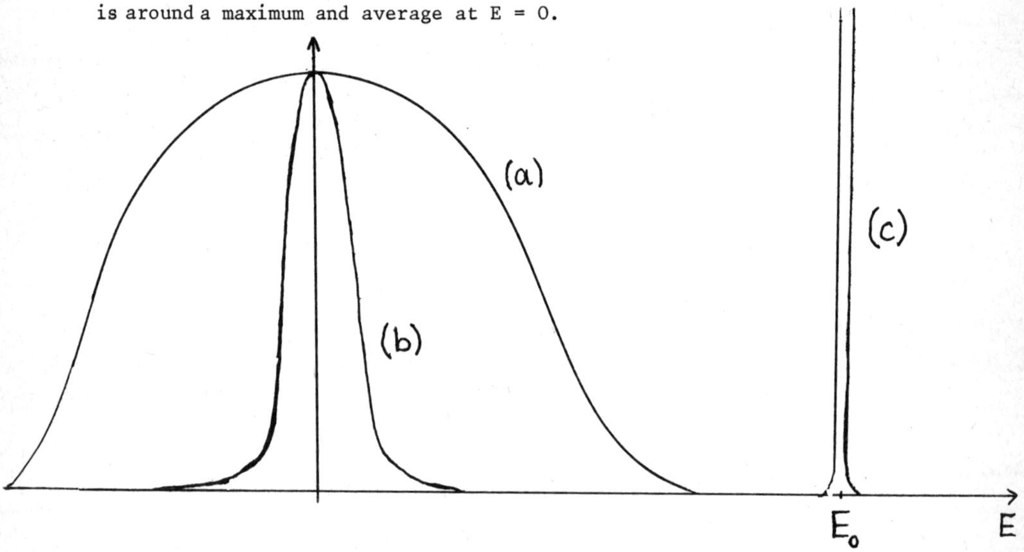

Statistical distribution of the amplitude E, for (a) thermal (Gaussian) light; (b) light from a spectral source or interference filter (Lorentzian); and (c) light from a laser above threshold.

As E may contain a slowly varying phase factor, it is complex. Hence a more correct representation would be obtained by rotating the diagram around the vertical axis, to generate a horizontal E plane.

For a laser above threshold, the picture changes totally. The amplitude E is now well stabilized around a nonzero average value E_o. It shows extremely small fluctuations around E_o, due to quantum fluctuations (spontaneous emission) and interactions with the pumping and other "heatbath" variables which include a reservoir to which atoms can lose energy. There is an effective restoring force which maintains $|E|$ near E_o. The phase of E will however execute a rondom walk, as it is not subject to any "restoring force".

To see how this comes about, we give a classical interpretation to the Heisenberg equation

$$\dot{E} = (G-K)E - CE^3 + F(t).$$

Here G (which increases with increasing pump power) represents the gain to the light field due to stimulated emission, K the absorption due to cavity losses, C is positive, and F is the random complex fluctuations. This equation of motion is similar to that for the displacement q(t) of a particle with small mass m in damped motion subject to a potential $V(q) = \frac{1}{2}(G-K)q^2 + \frac{1}{4}c\,q^4$ and rondomly fluctuating forces F(t):

$$m\ddot{q} + \dot{q} = -\frac{\partial V}{\partial q} + F(t).$$

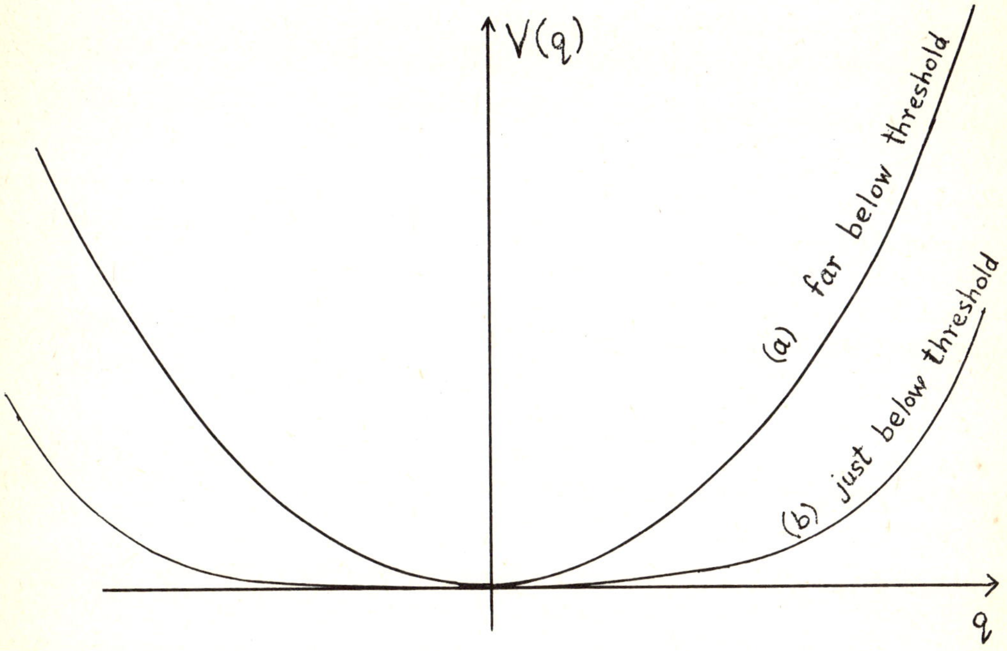

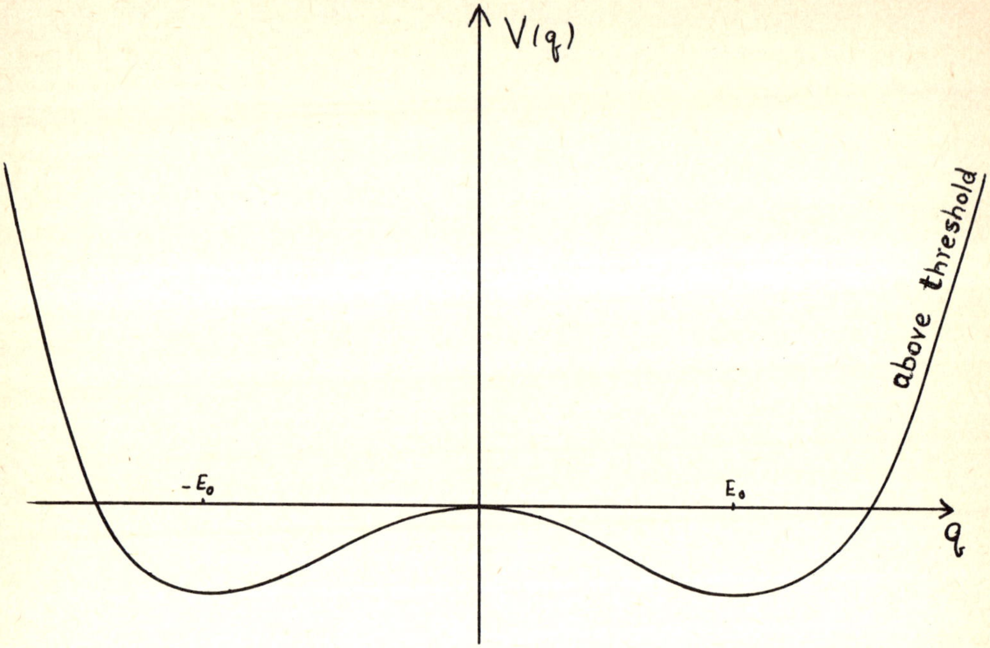

The potential V(E) is sketched for the cases (a) far below threshold ($G \ll K$), (b) still below, but nearer to threshold, and (c) above threshold ($G > K$). The phase of E is not represented in these two-dimonsional diagrams: a more correct, three-dimonsional diagram would be obtained by rotating the given ones about the vertical (V) axis. The other 2 axes would then represent the real and imaginary parts of E. The restoring force in the cases below threshold is towards $E = 0$, and decreases as the gain increases towards threshold. Consequently the time-relaxations become slower, and the linewidth narrower.

Above threshold these is a restoring force for $|E|$ towards a value E_o. Both E_o and the restoring force increase with increasing gain, i.e. as the laser moves further above threshold. The phase of E experiences no restoring force in the rotationally symmetric well.

The transition from below to above threshold resembles a phase transition. The system is self organizing in that stimulated emission (which is dominant above threshold) "synchronizes" the emissions of the individual atoms.

2. RATE EQUATIONS

Laser theory is a rich field and many questions can be asked:

1. What is the laser condition? What is the intensity?

2. What is the time-behaviour of the electric field? In a laser we have a number of modes, each, say, with index λ, with n_λ the number of photons in a mode λ. Can modes exist together? Do they compete? All these questions can be answered by the so called rate equations. The quantities we are working with therefore are n_λ, $N_{\mu,2}$, (number of atoms in upper level), $N_{\mu,1}$ (number in lower level).

But we can improve on this description by also describing the system in terms of the light frequency and the phases. The semiclassical equations yield this description.

2.1 The Laser condition

This was first derived in a paper by Schawlow and Townes[(2)]. That was the beginning of the rat race to make the first laser work. The condition is not so difficult. What we want to have is the process of stimulated emission dominating the whole process.

We have a system of two level atoms and we have in the beginning all in the ground state. Then we excite the system from the outside. Now we look for the rate at which light is produced. We have first of all stimulated emission. Now we go back to Einstein, who in his 1917-paper has introduced the rate for stimulated emission. This rate is as follows. Let us have n photons (in a single mode). The intensity of this mode is proportional to the number of photons n in this mode.

The rate of stimulated emission is proportional to $N_2 nW$, $W \equiv$ transition probability. To look what this probability means I remind you of spontaneous emission. The rate of spontaneous emission is proportional to $N_2 W$. Now we can relate W experimentally to the optical lifetime

$$W = \frac{1}{\tau}.$$

This τ means lifetime with respect to emission into all possible modes. The excited atom decays into the lower state and it may emit all sorts of radiation which is compatible with selection rules and with the energy conservation. However, in the laser case we are interested only in a *specific mode,* because we have a cavity and we want to know what the transition rate is in the specific mode. We have therefore still to divide by p, the number of modes within the spontaneous linewidth $\Delta\nu$ of the atom. How many modes are in the optical linewidth?

$$p = \frac{8\pi\nu^2\Delta\nu V}{c^3} \quad = \text{ number of modes in the interval } \Delta\nu$$

So $\quad W = \frac{1}{\tau p}$.

The virtue of the Schawlow-Townes calculation is that all quantities are experimentally known.

We also have absorption. The absorption rate is $N_1 nW$ (the W is the same).

Now we not finished, because we have a cavity and the photons may escape and the loss rate is proportional to the number of photons present and proportional to the inverse of the mean lifetime $t_1 = \frac{L}{c(1-R)}$.(R = reflectivity). The remaining losses are due to diffraction. Thus we finally obtain

$$\frac{dn}{dt} = (N_2 - N_1)\,Wn - \frac{n}{t_1}$$

We have neglected spontaneous decay and retained only the coherent parts. If spontaneous decay is included, we would get a term N_2W

$$\frac{dn}{dt} = (N_2 - N_1)Wn + N_2W - \frac{n}{t_1}$$

This term (N_2W) represents an incoherent contribution.

$\frac{n}{t_1}$ → losses

$(N_2 - N_1)Wn$ → correspond to coherent processes.

Since spontaneous decay is so much smaller than stimulated emission, the laser condition $\frac{dn}{dt} \geqslant 0$ reduces to

$$n(N_2 - N_1)W \geqslant \frac{n}{t_1}$$

i.e. $$(N_2 - N_1)W \geqslant \frac{1}{t_1} \qquad \text{if } n \neq 0$$

or $$\frac{N_2 - N_1}{V\,\tau\, 8\pi\nu^2\Delta\nu/c^3} \geqslant \frac{1}{t_1}\ .$$

If R = 90%, L = 30 cm, then $t_1 = 10^{-8}$ seconds. In order to satisfy the laser condition $\Delta\nu$ must be small and the frequency ν should not be too big, also $N_2 - N_1$ must be high enough. For $\nu = 10^{15}\ \text{sec}^{-1}$, $c = 3 \times 10^{10}$ m/sec,

$$\tau = 10^{-8}, \qquad \Delta\nu = 10^{10}, \quad \text{then}$$

$$\frac{N_2 - N_1}{V} = 10^{10}\ \text{cm}^{-3}$$

which is a good figure.

2.2 Pump schemes

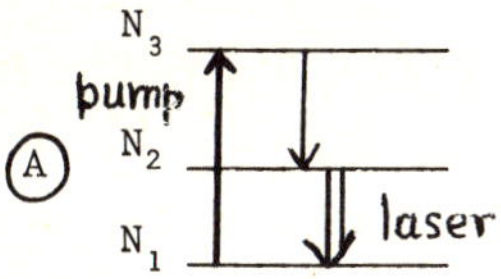

Pumping takes place from 1 to 3. From 3 to 2 we then have recombination and the lasing action is between 2 and 1.

The disadvantage of this scheme is that more than 50% of the lower atoms must be excited before anything can be achieved. The ruby laser is an example using this scheme. Note for example that in the Ruby laser (Al_2O_3) the density of the doped chrome-ions Cr^{3+} is 10^{19} cm^{-3}. Maiman did it with sufficient power.

A better system is

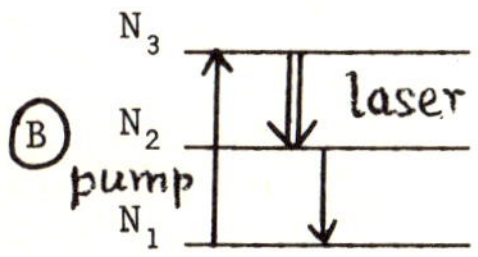

where the pumping is from 1 to 3 and the lasing is from 3 to 2, provided of course that the recombination rate from 2 to 1 is big enough. Lastly one gets (in a 4 level system) a combination of these:

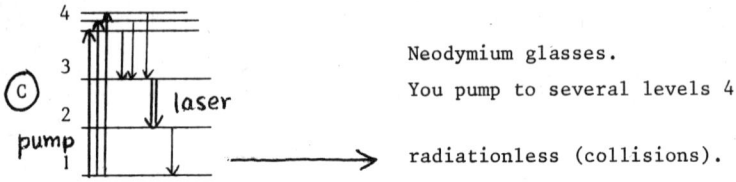

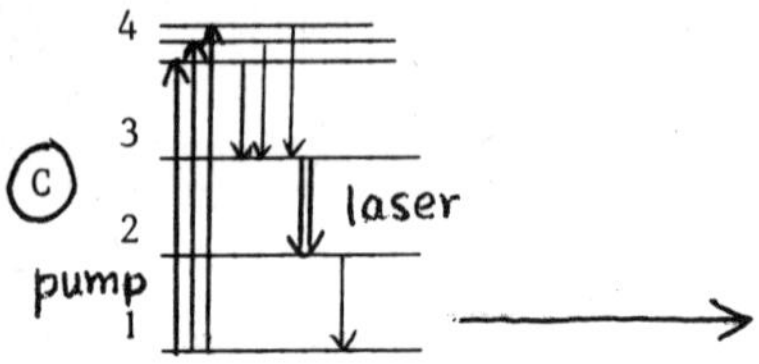

Neodymium glasses.
You pump to several levels 4.

radiationless (collisions).

These are the pump schemes you can have in principle. You can realise these pump schemes by many solids, gases, etc.

2.3 Matter equations

For simplicity we take one kind of photon only (one mode) and two-level systems of atoms, where all atoms are equivalent. The creation rate for photons is given by the field equation

$$\frac{dn}{dt} = N_2W + (N_2 - N_1)nW - 2kn \tag{1}$$

where the term N_2W due to spontaneous emission will be neglected and where $2k \equiv \frac{1}{\tau}$, is the cavity loss probability.

Now, if we introduce a transition rate W_{21}[from 1 → 2, determined by the external pump] and a rate W_{21} determined by radiationless processes like collisions as well as by spontaneous emission, we obtain for the atoms the two equations

$$\frac{dN_2}{dt} = -(N_2 - N_1)nW + N_1W_{21} - N_2W_{12}$$

$$\frac{dN_1}{dt} = (N_2 - N_1)nW - N_1W_{21} + N_2W_{12}$$

The total number of atoms $N = N_1 + N_2$ is constant, whereas the difference

$$N_2 - N_1 = D$$

satisfies the equation

$$\frac{dD}{dt} = \frac{D_o - D}{T} - 2DWn \tag{2}$$

where the constants are defined by

$$\frac{1}{T} = W_{12} + W_{21}$$

$$D_o = \frac{W_{12} - W_{21}}{W_{12} + W_{21}} N$$

Equation(2) says that if 2DWn is neglected (photon number = 0, no optical processes) then D will relax to D_o at a rate determined by the time constant T.

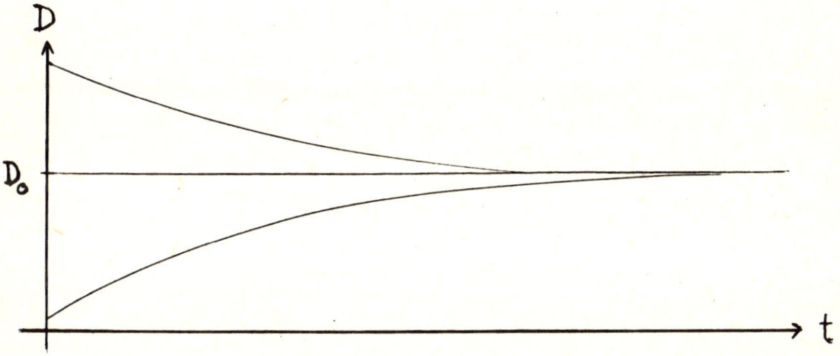

Relaxation of the population difference D in the case of "no coherent optical processes", n = 0.

2.4 Steady state

Let us look at the steady state case of (1) and (2). In the steady state the inversion is given by

$$D = \frac{2K}{W}, \quad n \neq 0$$

and from $\frac{dD}{dt} = 0$ we obtain

$$D = \frac{D_o}{1+2TWn}$$

So in the steady state, the inversion is (naturally) lower than in the unsaturated case. The solution for n in the steady state is

$$n = \frac{WD_o - 2K}{4KTW} \qquad (3)$$

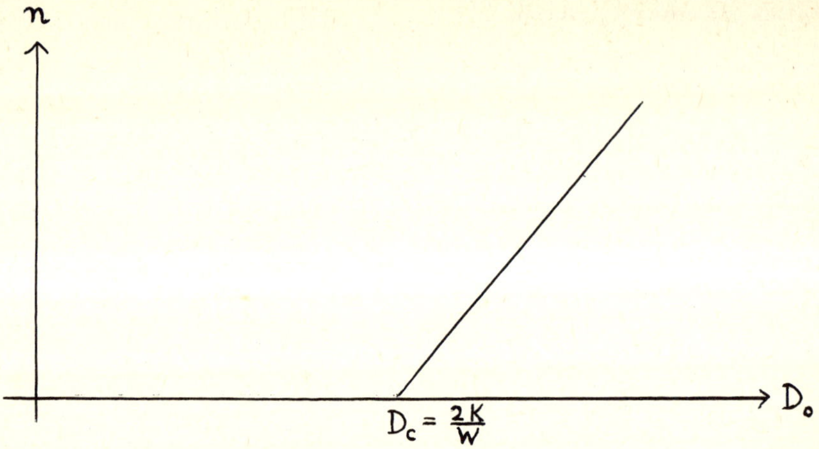

Photon number as function of pump-rate $\sim D_o$

D_o contains the pump rate, $\sim D_o \sim W_{21} - W_{12}$. Now we see there is a critical pump rate D_c. As the pumping increases, at first a certain amount of pump-energy is eaten up by incoherent processes and then any surplus pump energy we provide the system with, is converted into the laser light.

There is a problem with rate equations. The photon number n has to be nonnegative even if the laser condition

$$WD_o > 2K$$

is not fulfulled.

2.5 The adiabatic approximation

We assume that D relaxes very quickly in the two equations

$$\dot{n} = DWn - 2Kn \tag{4}$$

$$\dot{D} = \frac{D_o - D}{T} - 2DWn \tag{5}$$

Then $\dot{D} = 0$ and

$$D = \frac{D_o}{1+2TWn}, \tag{6}$$

the value of D to which it relaxes. For a small photon number, i.e. not too far above threshold

$$D = D_o(1-2TWn).$$

Substitution into (4) gives

$$\frac{dn}{dt} = (WD_s - 2K)n - Cn^2$$

The stationary solution n_s is known to us. Let us assume n is small - then we have an exponential increase of n. If n is large it starts decreasing and reaches the stationary value n_s.

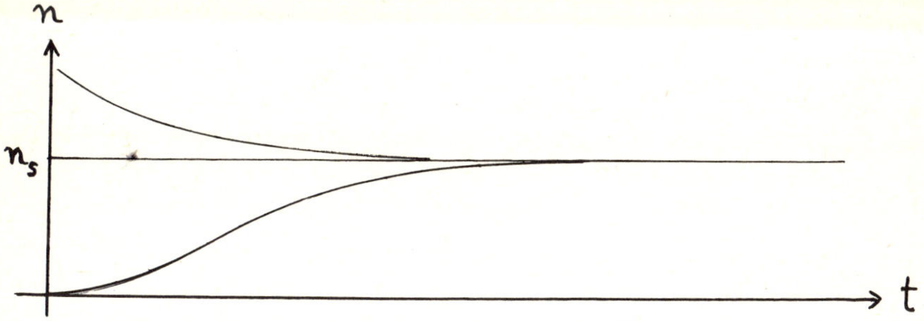

If you have a large deviation then you cannot use this approach. A possible appreach then is one of linearization:

$$n = n_s + \delta n, \qquad \delta n = a\,e^{\alpha t}$$

$$D = D_s + \delta D$$

We obtain two linear equations for δn, δD and obtain from the determinant = 0 condition some value

$$\alpha = -\Gamma + i\omega.$$

Usually $\Gamma > 0$, which means that the system is stable. The $i\omega$ does not appear in 2-level systems, only in three level systems. The relaxation oscillates in 3-level systems. Sometimes Γ negative is observed - then the rate equations are too restrictive - we then need phases too.

2.6 The giant pulse (Q switch)

The rate equations can be used in a semi-stationary way to describe the functioning of the Q-switch. With mirrors not aligned, or the dye not yet bleached, n = 0 and the equation $D = \frac{D_o}{1+2TWn}$ yields $D = D_o$. Here we have a very high inversion $D = D_o$. When the mirrors are aligned, or the dye gets bleached, $\frac{dn}{dt}$ and n increases (initially exponentially) according to

$$\frac{dn}{dt} = (WD_o - 2K)n - Cn^2$$

and D decreases according to (6). The resulting behaviour is illustrated below.

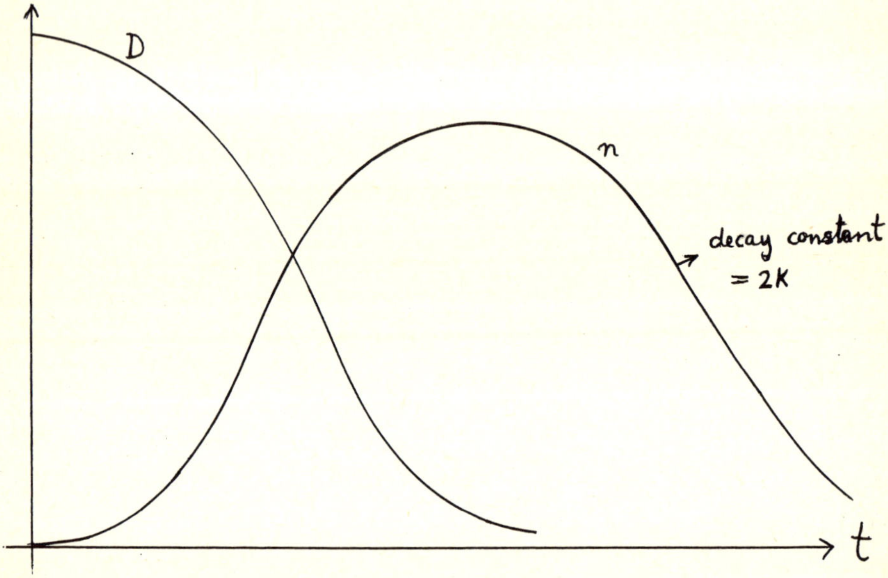

Approximate behaviour of D and n for Q-switch according to rate equations.

2.7 Multilevel atoms

Now we consider three levels, with 3 → 2 the lasing transition.

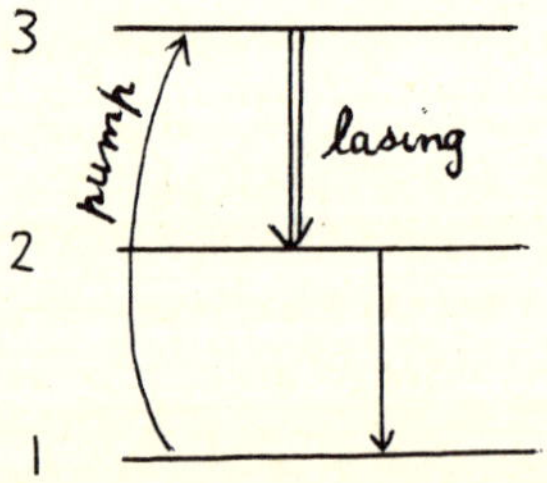

With just one kind of photon (single-mode case) the photon rate equation is

$$\frac{dn}{dt} = (N_3 - N_2)Wn - 2Kn$$

whereas the matter equations become

$$\frac{dN_3}{dt} = -(N_3 - N_2)Wn + W_{31}N_1 + W_{32}N_2 - N_3(W_{23} + W_{13})$$

$$\frac{dN_2}{dt} = -(N_2 - N_3)Wn + W_{21}N_1 + W_{23}N_3 - W_{12}N_2 - W_{32}N_2$$

$$\frac{dN_1}{dt} = -N_1(W_{31} + W_{21}) + N_2W_{12} + N_3W_{13}$$

In reality some terms are always small or zero. For example in the 3-level systems above, $W_{13} = 0$ because the pumping conditions just do not cause a transition $3 \rightarrow 1$ and $W_{32} = 0$ due to the same reason.

Also $W_{21} = 0$ and $W_{13} = 0$. Under these conditions the system simplifies to

$$\frac{dn}{dt} = (N_3 - N_2)Wn - 2Kn$$

$$\frac{dN_3}{dt} = -(N_3 - N_2)Wn + W_{31}N_1 - W_{23}N_3$$

$$\frac{dN_2}{dt} = -(N_2 - N_3)Wn + W_{21}N_1 + W_{23}N_3 - W_{12}N_2$$

$$\frac{dN_1}{dt} = -N_1W_{31} + N_2W_{12}$$

If the pump rates W_{31}, W_{21} are much smaller than the decay rates W_{23} and W_{12}, most of the atoms are in their ground states, and $N_1 \sim N$. Then

$$\frac{dN_3}{dt} = W_{31}N - W_{23}N_3 - W(N_3 - N_2)n$$

$$\frac{dN_2}{dt} = W_{21}N + W_{23}N_3 - W_{12}N_2 + W(N_3 - N_2)n$$

$$\approx W_{21}N - W_{12}N_2 + W(N_3 - N_2)n$$

where we have also neglected $W_{23}N_3$. Without laser action N_3 and N_2 will arrive independently at their equilibrium values

$$N_1 = \frac{W_{31}}{W_{23}} N \quad \text{and}$$

$$N_2 = \frac{W_{21}}{W_{12}}$$

respectively.

2.8 Multimode cases

We know that a number of modes fit under the atomic line profile - see the diagram on page 2. We assume that each of these modes can be occupied by a number of photons - n_λ for mode λ. Now we also want to extend the approach with respect to the atoms:

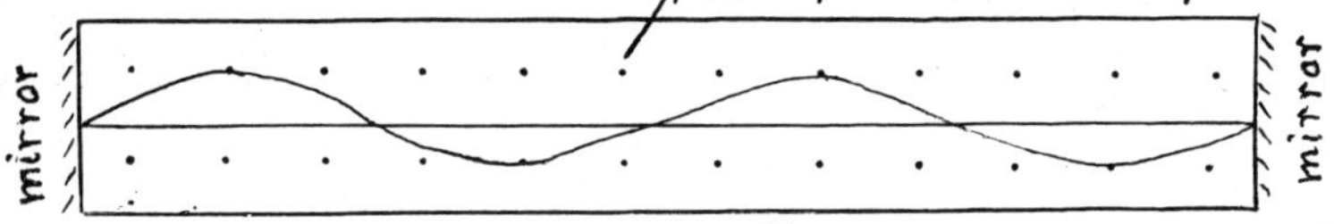

We distinguish these atoms by the label μ, and let $\underline{x}_\mu$ be the position within the cavity.

The field equation, for the number of photons in mode λ (ignoring spontaneous emission) is

$$\frac{dn_\lambda}{dt} = n_\lambda \sum_\mu W_{\lambda\mu} d_\mu - 2K_\lambda n_\lambda \qquad (7)$$

which replaces $\frac{dn}{dt} = DWn - 2Kn.$

The matter equation for atom μ describes the "local inversion"

$$d_\mu = N_{2\mu} - N_{1\mu} = \rho_{22\mu} - \rho_{11\mu} = |a|^2 - |b|^2 \qquad (8)$$

where $\psi(t) = a(t)\theta_u + b(t)\theta_L$, with θ_u and θ_L the upper and lower state atomic wave functions for atom μ. The matter equation is

$$\frac{d\,d_\mu}{dt} = \frac{d_o - d_\mu}{T} - \sum_\lambda 2n_\lambda W_{\lambda\mu} d_\mu \qquad (9)$$

to replace $\frac{dD}{dt} = \frac{d_o - d_\mu}{T} - 2WDn.$

The rate coefficient for stimulated emission is

$$W_{\lambda\mu} = \frac{2\gamma_\mu}{\gamma_\mu^2 + (\nu_\mu - \omega_\lambda)^2} \; |g_{\mu\lambda}|^2 \tag{10}$$

$$\text{with} \quad g_{\mu\lambda} = ie < \phi_\mu |\underline{x}| \phi_L >_\mu \cdot \underline{e}_\lambda \; u_\lambda(\underline{x}_\mu) \cdot \left(\frac{2\pi\omega_\lambda}{\hbar}\right)^{\frac{1}{2}} . \tag{11}$$

Here $\underline{e}_\lambda$ and u_λ are the mode polarization and wavefunction.

Very interesting consequences follow from the multimode formulation. Firstly we consider mode competition, and prove that only certain modes can survive.

2.8.1. Mode selection in the completely homogeneous case

Where the laser active atoms are inbedded in a matrix (as the Cr^{++} in $A\ell_2O_3$ for ruby, or the Nd^{+3} in glass), individual atoms at non-equivalent sites may suffer different degrees of shifting or individual broadening of the lasing transition. In (19), ν_μ and/or γ_μ would then differ from atom to atom. In a gas discharge, different atoms have different Doppler shifts, so that their line centres ν_μ differ. Such processes lead to inhomogeneous broadening of a spectral transition. In a laser, yet another, additional, source of inhomogeneity may arise: near the nodes of the field $\mathcal{E}_\lambda$ for a given node, the probability for stimulated emission (and absorption) is much less than elsewhere, so that different spatial positions x_μ are not equivalent.

In the completely homogeneous case, all laser-active atoms are equivalent with respect to the lasing transition: $W_{\lambda\mu}$ of (10) must be the same for all atoms. This can only happen in a ring laser where $\mathcal{E}_\lambda = E_o U_\lambda(x_\mu) = E_o e^{ikx_\mu}$ so that the intensity $|\mathcal{E}|^2$ is the same for all atoms.

In such a case

$$W_{\lambda\mu} = W_\lambda , \qquad \sum_\mu d_\mu = D,$$

and for a steady state ($\dot{n}_\lambda = 0$), equation (7) yields that

$$(DW_\lambda - 2k_\lambda) = 0.$$

It follows that for all modes which partake in the laser action (i.e., for which $n_\lambda \neq 0$), the ratio $\frac{W_\lambda}{K_\lambda}$ of gain to loss must be the same (= 2/D). What happens is that only the strongest mode (or degenerate modes) in terms of $\frac{W_\lambda}{K_\lambda}$ will survive in a completely homogeneous case. In terms of the diagram on page 2 for an optical laser (where K_λ is the same for neighbouring longitudinal modes), this means the mode corresponding to the spike nearest the peak of the atomic transition.

2.8.2 Coexistence of modes due to spatial inhomogeneities

In a steady ($\dot{d}_\mu = 0$) state, (9) gives that

$$d_\mu = \frac{d_o}{1+2T\sum_\lambda n_\lambda W_{\lambda\mu}} \approx d_o(1-2T\sum_\lambda n_\lambda W_{\lambda\mu}). \qquad (12)$$

When the frequency dependence of $W_{\lambda\mu}$ (the spectral line profile) is homogeneous, the spatial dependence

$$W_{\lambda\mu} \sim |U_\lambda(x_\mu)|^2 = |\sin k_\lambda x_\mu|^2$$

implies [by (12)] a corresponding variation in the local population inversion.

As explained below the diagram on page 2, these spatial inhomogeneities (spatial hole burning) of the mode functions for standing waves, provide a "variety of habits" which permit the coexistence of several longitudinal mode "species".

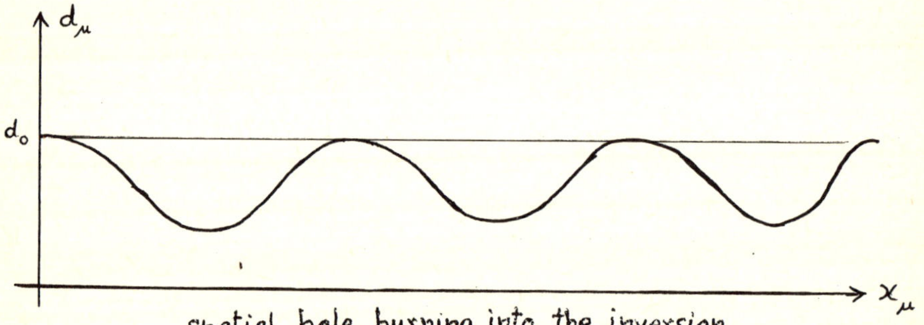

spatial hole burning into the inversion

2.8.3 The Lamb dip: holes in a Doppler or other inhomogeneously broadened line

In gas lasers, the effects of inhomogeneous Doppler broadening of the spectral line profile imply that in (10), the line centre ν_λ for an individual atom depends on the velocity v of the atom. If we consider (12) for a single mode, this implies the burning of a hole corresponding to the mode frequency into the inhomogeneously broadened atomic spectral profile (the envelope on page 2) - the so-called Lamb dip.

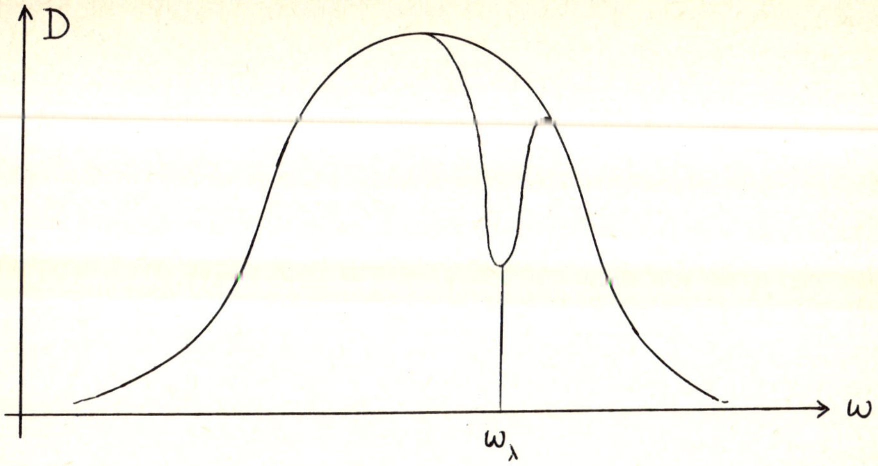

The original inversion (outer curve) has the shape of the atomic spectral profile. Lasing on the mode ω_λ removes from the inverted state atoms with frequencies ν_μ within a distance γ_μ from the mode frequency profile, to yield the inner curve.

3. SEMICLASSICAL THEORY

The Maxwell equations for a dielectric with polarization $\underline{P}$ and conductivity σ are:

CGS units:

$$\text{curl } \underline{E} = -\frac{1}{c} \dot{\underline{B}}$$

$$\text{curl } \underline{H} = \frac{4\pi}{c} \underline{j} + \frac{1}{c} \dot{\underline{D}}$$

$$\underline{D} = \underline{E} + 4\pi \underline{P}$$

$$\underline{j} = \sigma \underline{E}$$

CGS:

$$\underline{H} = \underline{B}$$

$$\text{div } \underline{E} = 0$$

SI:

$$\text{curl } \underline{E} = -\frac{\partial \underline{B}}{\partial t}$$

$$\text{curl } \underline{H} = \underline{J} + \frac{\partial \underline{D}}{\partial t}$$

$$\underline{D} = \epsilon \underline{E} + \underline{P}$$

$$\underline{H} = \mu^{-1} \underline{B}$$

By eliminating H we obtain

$$(\nabla^2 - \frac{1}{c^2}) \ddot{\underline{E}} = \frac{4\pi}{c^2} \sigma \dot{\underline{E}} + \frac{4\pi}{c^2} \ddot{\underline{P}} \tag{13}$$

under the assumption that $\text{div}\,\underline{P} = 0$. For a discussion of the atoms we
consider the interaction of the electromagnetic field $\underline{E}$ with a single
atom. The interaction energy is $H^P = e\underline{\xi}\cdot\underline{E}(\underline{x}_\mu, t)$, and the total hamiltonian is

$$H = H_o + H^P \qquad\qquad \underline{\xi} \equiv \underline{x} - \underline{x}_\mu$$

with H_o representing the unperturbed atom.

We want to reduce the problem to a two-level problem. Let the two
states σ_1, ϕ_2 be known eigenstates of H_o:

$$H_o \phi = \hbar \epsilon \phi(\underline{\xi})$$

$$\psi(\underline{\xi}, t) = c_1 e^{-i\epsilon_1 t} \phi_1(\underline{\xi}) + c_2 e^{-i\epsilon_2 t} \phi_2(\underline{\xi})$$

Let $\epsilon_2 - \epsilon_1 \equiv \nu$.

It is then easily shown that the Schrödinger equation

$$H\psi = i\hbar \frac{\partial \psi}{\partial t}$$

is equivalent to

$$\left.\begin{aligned} i\hbar\,\dot{c}_1(t) &= \underline{E}(t,x_\mu)e^{-i\nu t}\cdot\underline{\theta}_{12}\cdot c_2(t) \\ i\hbar\,\dot{c}_2(t) &= \underline{E}(t,x_\mu)e^{+i\nu t}\cdot\underline{\theta}_{21}\,c_1(t) \end{aligned}\right\} \quad (13a)$$

where $\underline{\theta}_{ij} = \langle\phi_i|e\underline{x}|\phi_j\rangle$.

The states ϕ_i and ϕ_j normally possess definite parity, in which case $\underline{\theta}_{ii} = 0$ and the polarization $\underline{p}_\mu = \langle\psi|e\underline{\xi}|\psi\rangle$ for atom μ reduces to

$$p_\mu = \underline{\theta}_{12}c_1^*c_2e^{-i\nu t} + \underline{\theta}_{21}c_1c_2^*e^{i\nu t},$$

i.e. $$p_\mu = \alpha_\mu\underline{\theta}_{12} + \alpha_\mu^*\underline{\theta}_{21} = \alpha_\mu\underline{\theta}_{12} + \text{c.c.} \quad (14)$$

where $\alpha_\mu \equiv c_1^*c_2e^{-i\nu t}$ is a density matrix element ρ_{21} in Heisenberg form.

From (13a) and (14), both α and the atomic polarization (and hence the light field) are zero when the atom is in either the upper or the lower state. Absorption and stimulated emission only takes place when both c_1 and c_2, and hence the polarization and α, are nonzero. From (13a), the field equation (for α_μ) is

$$\dot{\alpha}_\mu = -i(\nu_\mu+\gamma)\alpha_\mu + \frac{i}{\hbar}\,\underline{E}(\underline{x}_\mu)\cdot\underline{\theta}_{21}d_\mu \quad (15)$$

where $d_\mu = |c_2|^2 - |c_1|^2$ is the population inversion for atom μ. The matter equations are

$$\dot{d}_\mu = \frac{2i}{\hbar}\,\underline{E}(\underline{x}_\mu)\cdot(\underline{\theta}_{12}\alpha_\mu - \text{c.c.}) + \frac{d_o - d_\mu}{T} \quad (16)$$

The first term follows from (13a). It describes the interaction of the inversion d_μ with the field α_μ, i.e. absorption and stimulated emission, the last term describes incoherent (pump and relaxation) processes. For a homogeneous medium, the index μ may of course be dropped.

We write the total polarization as a sum over the atomic contributions

$$\underline{P}(\underline{x},t) = \sum_\mu \delta(\underline{x} - \underline{x}_\mu)\underline{p}_\mu$$

3.1. Mode decomposition

The modes have been treated elsewhere— in the lecture by D.J. Brink. We assume a complete orthonormal set of modes satisfying $(\nabla^2 + k_\lambda^2)\underline{u}_\lambda(\underline{x}) = 0$ and appropriate boundary conditions on the mirrors, and exp nd the field in terms of these modes:

$$\underline{E}(\underline{x},t) = \sum_\lambda E_\lambda(t)\underline{U}_\lambda(\underline{x}), \qquad \omega_\lambda = k_\lambda c,$$

and similarly for $\underline{P}(\underline{x},t)$.

The equation (13) for $\underline{E}$ then becomes

$$\omega_\lambda^2 E_\lambda + \ddot{E}_\lambda + \sum_{\lambda'} 4\pi\sigma_{\lambda\lambda'}\dot{E}_{\lambda'} = -4\pi \sum_\mu \underline{u}_\lambda(\underline{x}_\mu)\cdot\ddot{\underline{p}}_\mu \tag{17}$$

where $\sigma_{\lambda\lambda'} = (u_\lambda, \sigma u_{\lambda'}) \approx \delta_{\chi\lambda}\sigma_\lambda$ if σ does not vary rapidly over the mode volume (laser cavity). In the mode picture, (15) and (16) become

$$\dot{\alpha}_\mu = (-i\nu_\mu - \gamma)\alpha_\mu + \frac{i}{\hbar} d_\mu \sum_\lambda E_\lambda(t)\underline{u}_\lambda(\underline{x}_\mu)\cdot\underline{\theta}_{21} \tag{18}$$

and

$$\dot{d}_\mu = \frac{d_o - d_\mu}{T} + \frac{2i}{\hbar}(\underline{\theta}_{12}\alpha_\mu - c.c.)\cdot \sum_\lambda E_\lambda(t)\underline{U}_\lambda(\underline{x}_\mu) \tag{19}$$

$$= \frac{d_o - d_\mu}{T} + \frac{2\underline{p}_\mu}{\hbar\nu_\mu}\cdot \sum_\lambda E_\lambda \underline{U}_\lambda(\underline{x}_\mu) \tag{19a}$$

3.2. The slowly varying amplitude and rotating wave approximations

We illustrate these approximations for the mode picture of the previous section [they may also be directly applied to the original $\underline{E}(\underline{x}t)$ and $\underline{P}(\underline{x}t)$] . The fast variations, at the mode (laser) frequency ω_λ, which (see p2) is very close to the atomic (Bohr) frequency ν_μ, is factored out in the slowly varying amplitude approximation. Hence the amplitudes

$\tilde{E}^{\pm}$ and $\tilde{p}^{\pm}$ will vary slowly compared to $e^{\pm i\omega t}$ and $e^{\pm i\nu t}$, so that, when differentiating, $\dot{\tilde{E}}$ and $\dot{\tilde{p}}$ are neglected conpared to $\omega\tilde{E}$ and $\nu\tilde{p}$ in this approximation:

$$E_\lambda(t) = e^{-i\omega_\lambda t}\,\tilde{E}^{+}_\lambda(t) + e^{+i\omega_\lambda t}\,\tilde{E}^{-}_\lambda(t) \equiv E^{+}_\lambda(t) + E^{-}_\lambda(t) = E^{+}_\lambda + \text{c.c.} \qquad \text{(20a)}$$

$$\underline{p}_\mu(t) = e^{-i\nu_\lambda t}\tilde{\underline{p}}^{+}_\mu(t) + e^{i\nu_\mu t}\tilde{\underline{p}}^{-}_\mu(t) \equiv \underline{p}^{+}_\mu(t) + \underline{p}^{-}_\mu(t) = \underline{p}^{+}_\mu + \text{c.c.} \qquad \ldots \text{(20b)}$$

With the positive and negative frequency parts $E^{\mp}$ and $\underline{p}^{\mp}$ thus separated [by (14), we may set

$$\underline{p}^{+}_\mu(t) = \alpha_\mu\,\underline{\theta}_{12} \qquad \text{and} \qquad \underline{p}^{-}_\mu(t) = \underline{\alpha}^{*}\underline{\theta}_{21}] ,$$

the essence of the rotating wave approximation is to keep, in a product such as

$$E_\lambda(t)\ \underline{p}_\mu(t) = (E^{+} + E^{-})(p^{+} + p^{-}) = E^{+}p^{+} + E^{-}p^{-} + E^{+}p^{-} + E^{-}p^{+} \qquad \ldots \text{(21)}$$

only slowly varying term such as $E^{+}p^{-}$ and $E^{-}p^{+}$. The other terms $\sim e^{\pm 2i\omega t}$ oscillate at twice the Bohr (and laser) frequency, will rapidly average to almost zero over any short time covering many Bohr (optical) periods. Products similar to (19d) occur in most basic equations, such as (18),(19) and (19a) above.

3.3 Dimensionless quantities b_λ.

$$\text{Let}\quad E^{+}_\lambda(t) = i\sqrt{(2\pi\hbar\omega_\lambda)}\cdot b_\lambda , \qquad E^{-}_\lambda(t) = -i\sqrt{(2\pi\hbar\omega_\lambda)}\,b^{*}_\lambda \qquad \ldots \text{(22)}$$

With this definition, and the slowly varying amplitude and rotating wave approximations, the basic equations (17)-(19) become

$$\dot{b}_\lambda = (-i\omega_\lambda - k_\lambda)b_\lambda - i\sum_\mu g_{\mu\lambda}\,\alpha_\mu \qquad \ldots \text{(23)}$$

$$\dot{\alpha}_\mu = (-i\nu_\mu - \gamma)\alpha_\mu + \sum_\lambda g_{\mu\lambda} b_\lambda d_\mu \qquad \ldots \text{(24)}$$

$$\dot{d}_\mu = \frac{d_o - d_\mu}{T} + 2i\left(\sum_\lambda g^{*}_{\mu\lambda}\,\alpha_\mu b^{*}_\lambda - \text{c.c.}\right) \qquad \ldots \text{(25)}$$

The first describes variations in the field; the last 2 are the matter equations. Here an effective conductivity σ, related to κ_λ was used to incorporate both material and cavity losses. The term γ and the term

$\frac{d_o - d_n}{T}$ also describe incoherent relaxation processes.

3.4 Single mode operation

With a homogeneously broadened spectral line, and single mode operation (see § 2.8.1), the equations (22) simplify: indices λ disappear, and $\nu_\mu \to \nu$. For a steady state, $\dot{d}_\mu = 0$.

The ansatz of a coherent field coherently interacting with the atoms

$$b = B\,e^{-i\omega t}, \qquad \alpha_\mu = A_\mu e^{-i\Omega t} \qquad \ldots (26)$$

with B, A_μ and Ω constant, leads to

$$A_\mu(-i\Omega + i\nu + \gamma) = i\,g_{\mu\lambda}\,B\,d_\mu\,, \qquad A_\mu = \frac{i\,g_{\mu\lambda}\,d_\mu B_\lambda}{i(\nu-\Omega) + \gamma} \qquad \ldots (24a)$$

$$0 = \frac{d_o - d_\mu}{T} + 2i(g^*AB^* - c.c.) = \frac{d_o - d_\mu}{T} - 2d_\mu|B|^2 W_{\lambda\mu} \qquad \ldots (25a)$$

$$\text{with } W_{\lambda\mu} = |g_{\mu\lambda}|^2 \frac{2\gamma}{(\nu-\Omega)^2 + \gamma^2}\,. \qquad \ldots (27)$$

We set the dimensionless quantity

$$|b_\lambda|^2 \equiv |b|^2 = |B|^2 = n. \qquad \ldots (27a)$$

Then n is proportional to the intensity I $|E^+|^2$. Then we have

$$0 = \frac{d_o - d_\mu}{T} - 2d_\mu n\, W_{\lambda\mu}$$

and, from (26) into (23):

$$B(-i\Omega + i\omega - K) = B \sum_\mu |g_{\mu\lambda}|^2 d_\mu \frac{\gamma - i(\nu-\Omega)}{(\nu-\Omega)^2 + \gamma^2}$$

This implies that either B = 0 (mode not lasing) or the dispersion relation obtained by cancelling B. Real and imaginary parts give (see (27)).

$$2\kappa = \sum_{\mu} W_{\mu\lambda} d_{\mu} \qquad \text{and} \qquad 2(\omega - \Omega) = \sum_{\mu} W_{\mu\lambda} d_{\mu} \frac{(\nu-\Omega)}{\gamma} = -2\kappa \frac{(\nu-\Omega)}{\gamma}$$

which gives $\Omega = \frac{\nu\kappa + \omega\gamma}{\kappa + \gamma} = \frac{\nu T_2 + \omega t_1}{T_2 + t_1}$ ... (27a)

in terms of the cavity lifetime $t_1 = \frac{1}{2\kappa}$ and natural lifetime $T_2 = \frac{1}{2\gamma}$.

3.5 Recovering the rate equations

By averaging over phases, we now rederive the rate equations. Into (23 we substitute for α_μ from (26) and (24a):

$$\frac{d\,b_\lambda}{dt} = (-i\omega - \kappa_\lambda) b_\lambda - i \sum_{\mu,\lambda'} g_{\mu\lambda} g_{\mu\lambda'} \frac{b_\lambda\, d_\mu}{i(\nu_\mu - \Omega_{\lambda'}) + \gamma}$$

From this equation we may calculate $\frac{db_\lambda}{dt} b^*_\lambda + b_\lambda \frac{bd^*_\lambda}{dt} = \frac{d}{dt}(b_\lambda b^*_\lambda)$. We also use the fact that the phase average (see also (27a))

$$\overline{b^*_{\lambda'} b_\lambda} = \delta_{\lambda\lambda'}\, n_\lambda,$$

i.e. that the phases of different modes are independent (as mode frequencies differ and - in the absence of mode-locking - are not commensurable). Then it follows directly that the photon rate equation is

$$\frac{d}{dt}(b^*_\lambda b_\lambda) = \frac{dn_\lambda}{dt} = -2\kappa_\lambda n_\lambda + \sum_\mu n_\lambda W_{\lambda\mu} d_\mu,$$

as given earlier.

3.6 Adiabatic elimination. The nonlinear laser equation.

In order to investigate the time dependent nonlinear behaviour of the light field, we eliminated the matter variables from (23) - (25) according to the following scheme, which we illustrate for a

3.6.1 Single mode resonant two-level case with homogeneous broadening.

As in § 3.4, indices λ disappear, and $\nu_\mu \to \nu$. As a first step in our iteration, we take $d = d_o$ (the value without coherent optical processes), and on resonance ($\nu \approx \omega$) we self-consistently put

$$b = B\,e^{-i\omega t}$$

in (24), which becomes a linear first order differential equation, with solution

$$\alpha_\mu^{(1)} = i\,g_\mu\,d_o B\,e^{-i\omega t}/\gamma.$$

Inserting this into (25), another linear d.e . is obtained, with solution

$$d_\mu^{(1)} = d_o[1 - 4T|g_\mu B|^2/\gamma].$$

Re-substituting this improved inversion into (24), we obtain an improved dipole moment

$$\alpha_\mu^{(2)} = \frac{i}{\gamma}\,g_\mu\,d_o\,b(t)[1-4T|g_\mu b|^2/\gamma]. \qquad \ldots \text{(a28)}$$

We now substitute this into (23) to eliminate the matter variable, and obtain a self-consistent equation for the field alone:

$$\dot{b} = (-i\omega-\kappa)b + \frac{g^2Nd_o}{\gamma}\,b - \frac{4NTd_og^4}{\gamma^2}\,|b|^2b, \qquad \ldots \text{(28)}$$

as with a homogeneously broadened transition, $\sum^N g_\mu^2 = Ng^2$ with N atoms present. With

$$G = g^2Nd_o/\gamma\,, \qquad C = 4N\,T\,d_og^4/\gamma^2,$$

we are below threshold (see p) if $G < \kappa$; and above threshold (coherent processes predominating when the "gain coefficient" G exceeds the loss coefficient κ.

Substitute $b = B\,e^{-i\omega t}$ into (28), then

$$\kappa = G - C|b|^2.$$

The nonlinearity of (28) enables us to find the photon number $n = \frac{G-\kappa}{C}$ for a given pump rate (which implies a given unsatured inversion d_o and a given gain coefficient). Without the nonlinear term, no steady state would be possible above or below threshold (for $G \neq \kappa$).

3.6.2 Effect of detuning the mode from the atomic line centre ($\omega_\mu \neq \nu_\lambda$).

In the most general (multimode, off-resonance, but still steady state) case, the above elimination procedure is modified by starting with

$$b_\lambda = B_\lambda e^{-i\Omega_\lambda t}, \qquad d = d_{o\mu}.$$

Equations (a28) and (28) are replaced by lengthy ones [involving summations over 3 different λ indices in the nonlinear (cubic) term] - see e.g. Haken (1970a) or p229 of Haken (1970b).

In the single mode inhomogeneously broadened case, we find, in addition to the above results, that a hole is burned into the gain profile, as in §2.8.3. Also, one finds a frequency shift

$$\Omega - \omega = (\Delta\Omega)_\mu + n(\Delta\Omega)_s$$

away from the cavity frequency. The first contribution is just the dispersion due to the atomic transition. The other contribution is proportional to the number of photons (light intensity) and describes the change of the dispersion du to the adjustment of the atomic occupation to the intensity of the light field.

3.6.3 Multimode case. Frequency and phase (mode) locking

With 2 modes, we find in addition to the above dispersion effects, a mutual influence of the individual mode dispersions on each other. As expected, two holes are burned into the inhomogeneous gain profile or atomic inversion spectrum. These influence each other when $\Omega_1 - \Omega_2 \lesssim \gamma$.

In the multimode case, for an inhomogeneously broadened laser transition, an important phenomenon occurs at very high inversion (Q switch), or when the cavity is so tuned that the frequency spacing $\Omega_{\lambda+1} - \Omega_\lambda$ between adjacent pairs of modes become nearly the same: As the inversion is increased, or the tuning improved, the frequencies suddenly jump to values where the spacings $\Delta\Omega$ are all equal. The nonlinearity also establishes a phase relation between the modes, which therefore beat, and produce a train of ultrashort light pulses.

The interval from one pulse to the next is $\tau = \frac{1}{\Delta\Omega} \approx \frac{1}{\Delta\omega} = \frac{2L}{c}$, the "round trip" time for light to traverse the cavity forth and back. If n modes (all under the gain profile) are thus phase locked together, the duration of a pulse becomes τ/n - less than 10^{-12} seconds in a Nd^{+3} glass laser.

4. THE QUANTUM THEORY OF THE LASER

This is the most complete theory of the laser, and can yield all the results of the semiclassical theory (and therefore also of the rate equations). In addition, it gives the effect of fluctuations (noise) due to spontaneous emussion, and of fluctuations due to interactions with the pump and other heatbath variables [lossy cavity and all incoherent (spontaneous) processes - see diagram below].

As is well known (and evident from equation (23)), when the electro-magnetic field in a cavity is decomposed into modes (as was done in the semiclassical theory), then the mode amplitudes $E_\lambda(t)$ and $b_\lambda(t)$ behave as harmonic oscillators. In the fully quantum mechanical treatment, these field oscillators must now be quantized. The dimensionless complex amplitudes (positive and negative frequency parts) b_λ and b_λ^* now are replaced by nonhermitian boson (in the present case, photon) operators b_λ and $b_\lambda^\dagger$, which satisfy commutation relations:

$$[b_\lambda, b_{\lambda'}] = [b_\lambda^\dagger, b_{\lambda'}^\dagger] = 0, \qquad [b_\lambda, b_{\lambda'}^\dagger] = \delta_{\lambda\lambda'} 1. \tag{30}$$

The number operator $n_\lambda = b_\lambda^\dagger b$ represents the number of photons in mode λ.

Matter equations. The atom μ is again described in terms of eigenfunctions ϕ_i of the atomic hamiltonian: with $H_i = \hbar\phi_i = \hbar\epsilon_i\phi_i$

$$\psi_\mu(\underline{\xi}) = a_{1\mu}\phi_1(\xi), + a_{2\mu}\phi_2(\xi), \qquad \xi = \underline{x} - \underline{x}_\mu.$$

The $a_{i\mu}$ describe electrons, and may be regarded as fermion operators, satisfying anticommutation relations

$$\{a_{i\mu}, a_{j\mu'}\} = \{a_{i\mu}^\dagger, a_{j\mu'}^\dagger\} = 0, \qquad \{a_{i\mu}, a_{j\mu}^\dagger\} = \delta_{ij}\,\delta_{\mu\mu'} \tag{31}$$

We now illustrate schematically the interactions relevant for a laser:

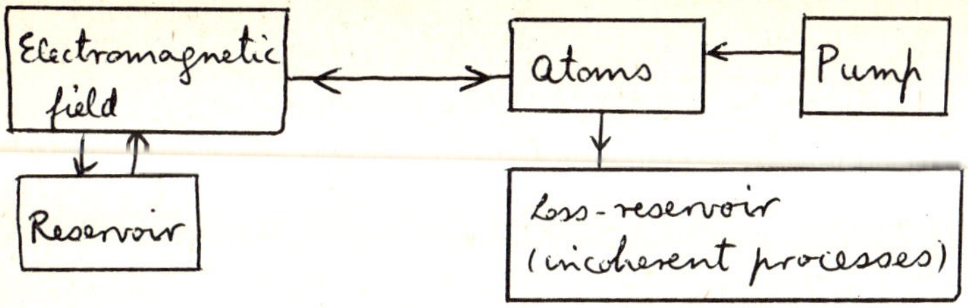

The hamiltonian for this scheme is

$$H = \underbrace{H_{field} + H_{atoms} + H_{field-atoms}}_{H_{system}} + H_{field-reservoir} + H_{atoms-reservoir} \tag{32}$$

$$\left.\begin{array}{ll} \text{where } H_{field} = \hbar\omega b^{\dagger} b , & H_{atoms} = \sum_{\mu,j} \hbar\epsilon_j (a^{\dagger}_{j\mu}\, a_{j\mu} + h.c.) \\ \text{and } H_{field-atoms} = \sum_{\mu} g_{\mu}\, \hbar (b^{\dagger}\alpha_{\mu} + h.c.) . & \end{array}\right\} \tag{33}$$

There are several equivalent ways to introduce the field-reservoir interaction. We shall use the Heisenberg picture, as this makes the physics clear. Then

$$\alpha_{\mu} = a^{\dagger}_{1\mu}\, a_{2\mu}$$

is again, as in (14), a Heisenberg density matrix element. The Heisenberg equation

$$\frac{dA}{dt} = \frac{1}{i\hbar}[A,H]$$

used successively for A = b, α_{μ} and d_{μ} give ($\nu_{\mu} \equiv \epsilon_{2\mu} - \epsilon_{1\mu}$)

$$\dot{b}_{\mu} = i\omega b - i\Sigma_{\mu} g_{\mu}\alpha_{\mu} \tag{34}$$

$$\dot{\alpha}_{\mu} = i\nu_{\mu}\alpha_{\mu} + ibg_{\mu}d_{\mu} \tag{35}$$

$$\dot{d}_{\mu} = 2ig_{\mu}(\alpha_{\mu}b^{\dagger} - h.c.) \tag{36}$$

So far we have included $H_{field} + H_{atoms} + H_{field-atoms}$ - the interactions with the heatbaths (reservoirs) would give additional terms, and make these equations more similar to (23) - (25), and also introduce statistical fluctuations (noise).

In order to incorporate pumping and damping terms into these, let us first consider the interaction of the field modes with a heat bath, according to the following model Hamiltonian with a bi-boson interaction

$$H = \hbar\omega_o\, b^\dagger b + \Sigma\, \hbar\omega B^\dagger_\omega B_\omega + \sum_\omega G_\omega\, bB^\dagger_\omega + \Sigma\, G^*_\mu\, b^\dagger\, B_\omega$$

(Remark: Our system is sufficiently large so that the Poincaré recurrence time is so large that it does not matter)

The equations of motion for this sub-system are

$$\dot{b} = -i\omega_o b - i\sum_\omega G_\omega B_\omega$$

$$\dot{B}_\omega = -i\omega B_\omega - i\, G^*_\omega\, b$$

and the time-integrated form of the last equation is

$$B_\omega(t) = B_\omega(o)e^{-i\omega t} - i\int_o^t e^{-i\omega(t-\tau)} G^*_\omega(\tau)b(\tau)d\tau\,.$$

The idea is to eliminate the reservoir coordinates $B_\omega(t)$ because they are not of interest:

$$\dot{b} = -i\omega_o b - \sum_\omega \int_o^t d\tau |G_\omega|^2\, e^{-i\omega(t-\tau)} b(\tau)$$

$$-i\sum_\omega G_\omega B_\omega(o)\, e^{-i\omega t} \qquad (37)$$

For a broad spectrum of the reservoir the sum

$$\sum_\omega e^{-i\omega(t-\tau)} |G_\omega|^2$$

can be approximated by $2\,\kappa\,\delta(t-\tau)$ with the result that (57) assumes the form

$$\dot{b}(t) = -i\omega_o b - \kappa\, b(t) + F(t)$$

The operator F(t) varies in a fluctuating manner, since we can consider $B_\omega(o)$ as an element of an ensemble. Classically this term will be absent, $\dot{b} = (-i\omega_0-\kappa)b$, but then we find that the commutation relations, assumed valid for t=0, are not satisfied for all times. In order to remedy this, it is therefore necessary to modify the damping equation with the term F(t).

The operator F(t) has the following properties

$$\langle F(t)\rangle = 0$$

$$\langle F^+(t)\, F(t')\rangle = \delta(t-t')2\kappa\bar{n}\ .$$

This result is another statement of the so-called fluctuation dissipation theorem – the system has a short "memory". The expectation values are defined in terms of, say, a thermal distribution with temperature T:

$$\langle\ \Omega\ \rangle = \mathrm{tr}(\rho\Omega) = \sum_m \langle m|\Omega|m\rangle\ e^{-m\hbar\omega/\kappa T}$$

$$\langle\ F(t)F^+(t')\ \rangle = 2\kappa(\bar{n}+1)\delta(t-t').$$

If we should now modify our five equations with dissipative and pumping terms similar to those just derived, we obtain

$$\dot{b} = i\omega b - ig\sum_\mu \alpha_\mu - \kappa b + F(t) \tag{38}$$

$$\dot{\alpha}_\mu = i\nu\,\alpha_\mu + ibgd_\mu - \gamma\alpha_\mu + \Gamma_\mu(t) \tag{39}$$

$$\dot{d}_\mu = 2ig(\alpha_\mu b^\dagger - bd_\mu^\dagger) + \frac{d_o - d_\mu}{T} + \Gamma_{\mu 1}d(t) \tag{40}$$

The analogy with semiclassical equations has a nice feature. We have learnt lessons there, among other things the process of adiabatic elimination. We want to solve the quantum mechanical nonlinear stochastic equations. [Remark: Should we average over the heat bath variables we would have obtained

$$\langle F\rangle = 0, \qquad \langle\Gamma_\mu\rangle = 0, \qquad \langle\Gamma_\mu, d\rangle = 0,$$

and this would amount to the transition to the semiclassical case.]

As in the semiclassical approach, adiabatic elimination would lead to

$$\dot{b} = (-i\omega + \frac{g^2D}{\gamma} - \kappa)b - C\,b^\dagger bb + F_{total} \tag{41}$$

where $C = \frac{4g^2 \Gamma D_o}{\gamma^2}$, $D_o = Nd_o$

and $F_{total} = F(t) - \frac{ig}{\gamma} \sum_\mu \Gamma_\mu$.

The light field is driven by two noise sources (1) firstly the loss-mechanism due to the mirrors and other causes, (2) secondly due to fluctuations in the dipole moments, or noise due to the atoms (spantaneous emission).

The fact that the quantum mechanical average $< F_{total} >$ is zero, helps us to interpret (41) as a classical equation.
Firstly, let

$$b = q e^{-i\omega t} .$$

Then the classical variable q will satisfy

$$\dot{q} = Gq - Cq^3 + F_{tot} e^{i\omega t} \tag{42}$$

The constant $G = \frac{g^2 D_o}{\gamma} - \kappa \equiv G' - \kappa$ is called the nett gain. Eq (42) is a nonlinear stochastic equation, and a simple mechanical interpretation makes life much easier. Rewrite (42) as an equation of motion

$$m\ddot{q} + q = - \frac{\delta V}{\delta q} + F_{tot} e^{i\omega t} \tag{43}$$

for a particle with a very small mass m and with a potential

$$V = \tfrac{1}{2}Gq^2 + \tfrac{1}{4}Cq^4$$

There are two cases of interest, namely $G > 0$ and $G < 0$.
The case when $G < 0$ is called the case below threhold, whereas the case $G > 0$ is called the case above threshold. A simple graphical representation of V(q) (see pages 6-7) will greatly help us in understanding the nature of these two cases and will replace a complicated stability analysis.

When V(q) becomes smaller, the amplitude of q becomes bigger. These then become "critical fluctuations". (We do not take $G = 0$ exactly and the q^4-part of the potential has not been taken into account).

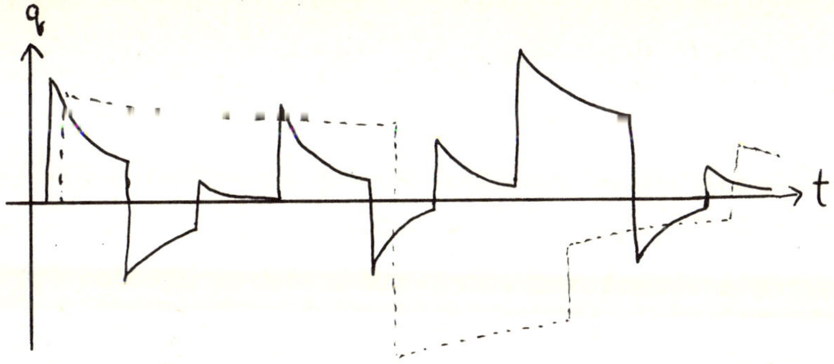

The position of the "particle" as a function of time moving under the influence of the fluctuating forces (a random superposition of decaying wave tracts). If G increases (becomes closer to zero) the damping is smaller - see the dotted curve. The slower decaying wave implies that we have a smaller linewidth.

For positive G we have the behaviour sketched on p.7.

$$b(t) = qe^{-i\omega t}$$

$$= (r_o + \rho(t))e^{i\phi(t)}\, e^{-i\omega t} \qquad (44)$$

We have, in (44), taken the quantity q not as a real quantity, but as a complex quantity with a phase $\phi(t)$. Both ρ and ϕ are stochastic variables. The stochastic force causes the particle to oscillate radially. There is no restoring force for the motion in the ϕ-direction. This implies a broken symmetry.

Above threshold we have an average field b_o, but below threshold we have no average field, just chaotic light. We shall later treat the case G = 0 when we introduce the Fokker-Planck equation in stead of the Langevin equations.

Below threshold

$$\dot{b} = Gb + \tilde{F}_{tot}$$

$$= -|G|b + \tilde{F}_{tot} , \qquad (45)$$

where the cubic terms have been neglected. The formal solution of (45) is

$$b(t) = \int_o^t e^{-|G|(t-\tau)}\tilde{F}_{tot}(\tau)d\tau + b(o)e^{-|G|t} \qquad (46)$$

In order to obtain information from (46) we shall use it in the calculation of correlation functions in a later stage. To complete the analysis, we now also solve for the Langevin equation above threshold. Substitute

$$q(t) = (r_o + \rho(t))e^{i\phi(t)} \qquad \text{into (42):}$$

$$(r_o + \rho)i\dot{\phi}e^{i\phi} = -\dot{\rho}e^{i\phi} + G(r_o + \rho)e^{i\phi} - (r_o + \rho)^3 e^{i\phi} + F_{tot}e^{i\omega t}$$

or

$$(r_o + \rho)i\dot{\phi} + \dot{\rho} = G(r_o + \rho) - C(r_o + \rho)^3 + F_{tot}e^{i\omega t}\, e^{-i\phi(t)}$$

For large G we have

$$\dot{\phi} = \frac{1}{r_o} \operatorname{Im} F_{tot}\, e^{i\omega t}\, e^{-i\phi(t)}$$

$$\dot{\rho} = Gr_o - Cr_o^3 + G\rho - 3Cr_o^2\rho + \operatorname{Re} \tilde{\tilde{F}}_{tot}$$

or, if $\dot{\rho} = 0$, $r_o^2 = \frac{G}{C}$

Now r_o^2 is related to the photon-number-density.
If G grows, r_o^2 grows, but simultaneously $\dot{\phi}$ becomes smaller, and we have a smaller linewidth because the more constant $\phi(t)$ is, the sharper is the linewidth. The time to go around the V-avis is 1/linewidth. The fluctuating ϕ actually gives rise to the linewidth. To summarise: Above threshold $G > 0$, r_o exists and as G grows the sharper the linewidth of the light becomes. Below threshold $G < 0$ and $r_o = 0$, i.e. no average light field exists. We have excluded the case $G = 0$. This case will be dealt with when the Fokker-Planck equation is discussed.

5. CLASSICAL AND QUANTUM THEORIES OF COHERENCE

In this chapter we shall treat the classical and especially the quantum theory of coherence. One wishes to know the difference between the coherence properties of light from thermal sources and light from lasers, but at the same time, we have to solve the problem of what to measure. We have introduced concepts like creation operators but we must also know what to answer an experimentalist if we are asked what can be measured. We have decomposed the electric field

$$E \Rightarrow E_{operator}(x,t) = C(b^\dagger - b)\sin \kappa x$$

We would have obtained Heisenberg equations for $b(t)$, $b^{\dagger}(t)$. It remains to devise procedures to construct measurable quantities from these operators. It is useful to start with the concepts of the classical theory of coherence. So let me remind you of Young's double slit experiment, by which Young proved the wave nature of light.

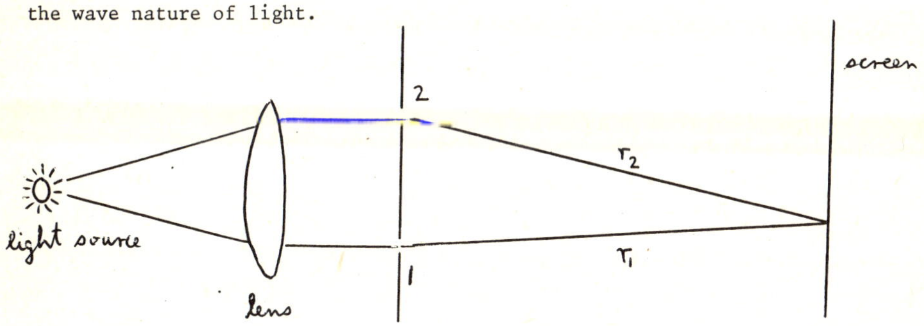

What does the field look like? The positive frequency part of the classical field emanating from slit i is ($\omega = kc$)

$$E_i^+ = E_o \exp[jk\cdot z - \omega t)]$$

$$E_i^+ = E_o \exp[ik(r_i - ct)] = (E_i^-)^* \, . \tag{47}$$

The instantaneous intensity I_I due to field is (proportional to)

$$I_I = \tfrac{1}{2} E^2 = \tfrac{1}{2}(E^+ + E^-)^2 = \tfrac{1}{2}(E^+)^2 + \tfrac{1}{2}(E^-)^2 + E^+E^-$$

The response times T of the human eye, and of conventional light detection systems equal very many millions of periods of the light field. Thus what is actually observed is the "moving average" intensity

$$I(t) = \langle I_I \rangle = \frac{1}{T} \int_{t_-}^{t_+} I_I(t')dt', \text{ with } t_\pm = t \pm \tfrac{1}{2}T. \tag{48}$$

With $\tfrac{1}{2}(E^\pm)^2 \sim e^{\pm 2i\,\omega t}$, their contributions are smaller by a factor $\frac{1}{4\omega T} <<< 10^{-6}$ than that of $E^+E^- = |E^+|^2$, which is thus the only term to survive the averaging process:

$$I = \langle \tfrac{1}{2}E^2 \rangle = \langle E^+E^- \rangle = |E^+|^2 = \langle(E_1^+ + E_2^+)(E_1^- + E_2^-)\rangle = G_{11} + G_{22} + G_{12} + G_{21} \tag{49}$$

where the last step follows from the superposition principle. In Young's experiment the distance to the screen very much exceeds that between the slits. Thus, as in (47), the "amplitude factor" E_o is practically the same for both slits. Hence, with

$$G_{ij} = \langle E_i^+ E_j^- \rangle = G_{ji}^*, \text{ so that } G_{ii} \text{ is real,} \tag{50}$$

it follows from (49) that

$$G_{11} = G_{22} = E_0^2 \quad \text{and} \quad G_{12} = G_{21} = E_0^2\, e^{i\phi}, \qquad \phi = k(r_1 - r_2),$$

so that from (49),

$$I = 2E_0^2\,(1 + \cos\phi) \tag{51}$$

–see illustration for case $A = B = E_0^2$.

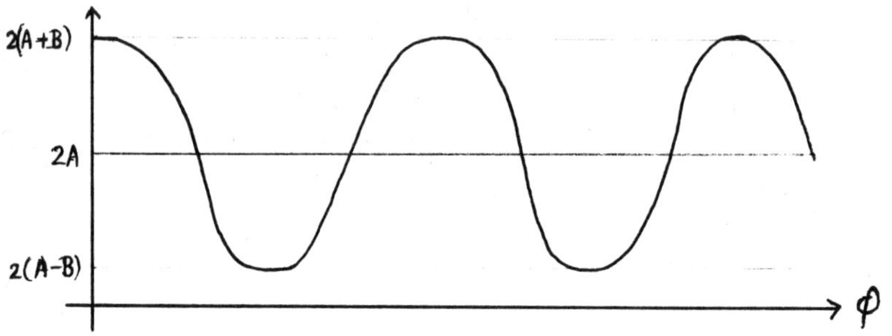

Now (49) implies perfect coherence, as both E_1 and E_2 have perfectly well-defined frequencies, wave-numbers, amplitudes and phases - each represents a train of waves that extends infinitely over all of space and time. With the fringe intensity minima equal to zero in (51), the fringe visibility

$$V \equiv \left(\frac{I_{max} - I_{min}}{I_{max} + I_{min}}\right)_{\text{neighbouring maximum and minimum}} \tag{52}$$

and degree of coherence is one (hundred percent).

With less than perfectly monochromatic light, $I_{min} > 0$ and $V < 1$. It is clear from the above that (49) is still valid, but not (47). With the usual Young's experiment, each slit separately (with the other one closed) yields the same intensity ($I_1 = I_2$), so that

$$\left.\begin{aligned} G_{11} &= G_{22} \equiv A \\ \text{and } G_{12} &= G_{21}^{*} \equiv B\,e^{i\phi}, \end{aligned}\right\} \tag{53}$$

with ϕ once more the phase difference between the waves from slits 1 and 2. Substituting (53) into (49), we now obtain

$$I = 2(A + B\cos\phi)\,. \tag{54}$$

Now $I_{max} = 2(A+B)$, $I_{min} = 2(A-B)$, and the fringe visibility (and the degree of coherence $|\gamma|$ between the light from slits 1 and 2) becomes

$$V = |\gamma| = \frac{B}{A}. \tag{50a}$$

As I_{min} cannot be negative, it is clear than $B \leq A$ in all cases.*

More generally, we define the complex degree of mutual coherence as

$$\gamma_{12}(\tau) = \langle E_1^+(t+\tau)E^-(t)\rangle\langle|E_1^+|^2\rangle^{-\frac{1}{2}} \langle|E_2^+|^2\rangle^{-\frac{1}{2}}$$

$$= G_{12}(\tau)/\sqrt{G_{11}\, G_{22}} \tag{55a}$$

It is a measure of the "sharpness" of the fringes in an interference experiment. The link between experiment and theory is given by G_{12} - the correlation or mutual coherence function, "mutual" because it refers to two different beams. [G_{12} - cross-correlation]

Eventually we want to replace the time average by an ensemble average, and also for this reason T has to be long, otherwise the ergodic property mentioned cannot be used.

Now let us come to quantum mechanics. The classical expression

$$G(1,2) = \langle E^+(x,t)\, E^-(x',t')\rangle \tag{55}$$

can be generalised to a quantum-mechanical expression. We shall see how a qusntum mechanical analysis of a detector experiment leads to a natural quantum mechanical generalisation of (55). Consider the following interacting systems:

[Source] ⟷ [field] ⟷ [detector]

$$H_s + H_{s-f} + H_f + H_{f-d} + H_d = H$$

$$H_F = H_s + H_{s-f} + H_f$$

$$H_o = H_F + H_d$$

$$H_{int} = H_{f-d}$$

---

*This also follows from the definition (50) for G_{ij}. With $i = j$, the integrand in (48) is always positive, and permits no cancellation due to varying phase, as happens when $i \neq j$. Hence with $I_1 = I_2(|E_1^+|^2 = |E_2^+|^2)$, $|G_{12}| = |G_{21}|$ is always smaller than $G_{11} = G_{22}$.

The eigenstates of H_F are called the dressed field modes, for the source-field system. The Schrödinger equation

$$i\hbar\dot{\Phi} = (H_o + H_{int})\,\Phi$$

can be solved in the interaction representation $\tilde{\Phi}$

where $\Phi = e^{-iH_o t/\hbar}\tilde{\Phi}$ and the Schrödinger equation assumes the form

$$\tilde{H}_{int}\,\tilde{\Phi} = i\hbar\dot{\tilde{\Phi}} \qquad (56)$$

in which $H_{int} = e^{iH_o t/\hbar}\, H_{int}\, e^{-iH_o t/\hbar}$.

The simplest detector is a two level atom. How does this work?

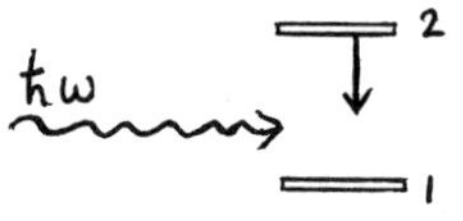

$$H_d = \hbar\omega a_2^+ a_2$$

$$H_{int} = \int dv\, \psi^*(-e/m)\, \underline{A}\cdot\frac{\hbar}{i}\,\underline{\nabla}\,\psi$$

If $\psi = a_1\phi_1 + a_2\phi_2$

we obtain $H_{int} = a_2^*\,a_1 \int \phi_2^*(x)(-e/m)\frac{\hbar}{i}\, A^+(x)\, \underline{\nabla}\phi_1\, dV$

Since H_d commutes with H_F, we have

$$\tilde{H}_{int} = e^{iH_d t/\hbar}\, a_2^* a_1 e^{-iH_d t/\hbar} \int dv\, \phi_2^*\left(-\frac{e}{m}\right)\frac{\hbar}{i}\, e^{iH_F t/\hbar}\, \underline{A}^+\, e^{-iH_F t/\hbar}\, \underline{\nabla}\phi_1$$

$$\equiv e^{i\nu t/\hbar} a_2^* a_1 \int dv\, \phi_2^*\left(-\frac{e}{m}\right)\frac{\hbar}{i}\, A^+(\underline{x},t)\underline{\nabla}\phi_1$$

and the wave equation contains an operator

$$\underline{A}^+(x,t) \equiv e^{iH_F t/\hbar}\, \underline{A}^+(\underline{x}) e^{-iH_F t/\hbar}$$

It must look as if the possibility to calculate $\underline{A}^+(\underline{x},t)$ is a hopeless one. Fortunately this is just the Heisenberg operator $\underline{A}^+(\underline{x},t)$ corresponding to the Schrödinger operator $\underline{A}^+(\underline{x})$ at $t = 0$. We shall assume that the interaction is weak in (56) and use perturbation theory to calculate the detection probability

$$P(t) = \langle \tilde{\Phi}(t) | \tilde{a}_2^+ \tilde{a}_2^+ | \tilde{\Phi}(t) \rangle$$

$$= \langle \tilde{\Phi}(t) | a_2^+ a_2 | \tilde{\Phi}(t) \rangle \qquad (57)$$

which is the probability that at time t the upper state is occupied. The first order form for $\tilde{\Phi}(t)$ is

$$\tilde{\Phi}(t) = \frac{1}{i\hbar} \int_0^t \tilde{H}_{int} \, \tilde{\Phi}(o) dt' + \tilde{\Phi}(o) \qquad (58)$$

where

$$\tilde{\Phi}(o) = \Phi_F \times \Phi_d(o)$$

$$= \Phi_F \times a_2^+ \bar{\Phi}_o$$

is the initial, uncoupled state of field and atom.

Substituting (58) into (57) we obtain

$$P(t) = \frac{\hbar^2 e^2}{m^2} \int_0^t d\tau \int_0^t d\tau^1 \langle \Phi_F | \underline{A}^-(x,\tau) \underline{A}^+(x^1 \tau^1) | \Phi_F \rangle$$

$$\cdot \phi_2 \, \underline{\bar{v}} \, \phi_1^*(x) \; . \; \phi_2^* \, \underline{\bar{v}} \, \phi_1(x^1) \, d^3 x \;\; d^2 x' \qquad (59)$$

In reality we must replace the P by a new P that contains an average over thermal states. In reality one has a set of atoms and we then have a sum over the various ν of the atoms. If the frequencies in the light are sharp, compared to the broad spectrum $-\alpha < \nu < \alpha$ of the atoms, we can replace the sum $\sum_\nu e^{-i\nu(\tau-\tau')}$ by $2\pi \, \delta(\tau-\tau')$. This is an ensemble of atoms - called a broad-band detector. Then in (59) we replace τ' by τ. One can also think ideally of one atom only (no statistical arguments) with a large linewidth. If the wavelength of light is larger than the dimensions of the atoms we could replace $x \to x_o$, $x' \to x_o$ in the integral and we thus would obtain for the absorption rate

$$\frac{dP}{dt} = |P_{21}|^2 \langle \phi_F | A^-(x_o,\tau) A^+(x_o,\tau) | \phi_F \rangle. \qquad (60)$$

It looks as if this result is purely classical in the sense that, by the ergodic theory, time averages and ensemble averages are equal, so that (60) could have resulted as a classical ensemble average from a time average.

But this is not quite true. The role of the quantum theory is to prescribe the sequence $A^{-}(\ )A^{+}(\ )$ in which the operators occur.

Thus in a double-slit experiment, to check the coherence properties of a field A, we look at

$$\underline{A} = \underline{A}(\underline{x},t) + \underline{A}(\underline{x}',t')$$

where $A(\underline{x}',t')$ is, say, a time-delayed and/or space-delayed beam prepared from $A(\underline{x},t)$ by some delay or splitting mechanism. (A delay can be realised by using a semitransparent mirror.) By recombining the two A's we can look at things like

$$G(1,2) = \langle \phi_F | A^{-}(1)\ A^{+}(2) |\ \phi_F \rangle$$

We have here a one-photon process. We could also take n-photon processes - multiphoton processes yield higher order correlation functions.

We shall now investigate how these correlation functions can be calculated theoretically by means of the Langevin equations below and above threshold. This means that we shall show how a model for the laser, (a model for the interaction of field + matter + reservoir) can be used to determine the correlation function $\langle \phi_F | \underline{A}^{-}(1)\underline{A}^{+}(2) | \phi_F \rangle$

or $\langle \phi_F | E_z^{(-)}\ (x,t)\ E_z^{+}\ (x',t') | \phi_F \rangle$ or,

for that matter, $\langle \phi_F | b^{+}(t)\ b(t') | \phi_F \rangle$, since these quantities are related through the expansion of the electric field operator

$$E_2(x,t) = (b^{+} - b)\ i\sqrt{2\pi\hbar\omega}\ \sqrt{\frac{2}{L}}\ \sin kx$$

(in the case of a single-mode electric field E_2 - just to remind the formalists, a multimode electric field is given by

$$E_2(x,t) = \sum_k (b_b^{+} - b_k)\ i\sqrt{2\pi\hbar\omega_k}\ \sqrt{\frac{2}{L}}\ \sin kx).$$

In order to calculate

$$\langle \phi_F | b^{+}(t)\ b(t^1) | \phi_F \rangle ,$$

we use the Langevin equation

$$\frac{db}{dt} = (-i\omega + \frac{g^2 D_o}{\gamma} - \kappa)b - Cb^+ b\, b + F_{tot},$$

again in two cases, namely (a) below threshold, and (b) G > 0, above threshold.

(a) Below threshold

$$\frac{db}{dt} = (-i\omega - |G|)b + F_{tot}$$

or $$b(t) = \int_o^t e^{(-i\omega-|G|)(t-\tau)} F_{tot}(\tau) dc + b(o) e^{(-i\omega-|G|)t} \tag{61}$$

and we take the time t large because we are interested in the stationary state of the electric field. In this case the homogeneous term in (61) can be neglected and the results for the correlation function is

$$< \phi_F b^\dagger(t)\, b(t')\phi_F > = \int_0^t d\tau \int_0^{t'} d\tau\; e^{(i\omega-|G|)(t-\tau)} e^{(-i\omega-|G|)(t'-\tau')}$$

$$\cdot < \phi_F | F^\dagger_{tot}(\tau) F_{tot}(\tau') | \phi_F >$$

To evaluate the expectation value in the integrand we remember that F_{tot} was compounded as follows

$$F_{tot} = F + (\frac{-ig}{\gamma}) \sum_\mu \Gamma_\mu$$

where F = noise source for field from, say, mirrors and

Γ_μ ≡ noise sources for matter field (atoms in heat bath, say)

and we note that the noise sources or reservoirs are statistically independent of another, i.e.

$$< F^\dagger \Gamma_\mu > = < F^\dagger > < \Gamma_\mu > = 0$$

and that means that cross products vanish. So we have

$$< \phi_F | F^\dagger_{tot}(t) F_{tot}(t') | \phi_F > = < \phi_F | F^\dagger(t) F(t') | \phi_F > + \frac{g^2}{\gamma} \sum_\mu < \phi\, | \Gamma^\dagger_\mu(t) \Gamma_\mu(t') >$$

$$= 2\kappa n_{th} \delta(t - t') + C\delta(t - t')$$

In the following paragraph we shall prove the δ-function property of $< \phi_F|F^\dagger(t)F(t')|\phi_F >$: Consider as on page 30 the model of a field b, $b^\dagger$, interacting with a heat bath. After eliminating the bath variables, we obtain the Heisenberg equation of motion (Haken 1970a, p40)

$$\frac{db^\dagger}{dt} = i\omega b^\dagger - \int_{t_o}^{t} b^\dagger(\tau) \sum_\omega |G_\omega|^2 e^{i\omega(t-\tau)} d\tau + i\sum_\omega G^*_\omega B^\dagger_\omega(t_o)e^{i\omega t} .$$

We assume that the G_ω's are of about equal amplitudes so that we have

$$\sum_\omega |G_\omega|^2 e^{i\omega(t-\tau)} = 2\kappa\delta(t-\tau)$$

with $\kappa = \pi|G\omega_o|^2$.

In this case we have

$$\dot{b}^\dagger = i\omega_o b^\dagger - \kappa b + \underbrace{i\sum_\omega G^*_\omega B^\dagger_\omega(t_o)e^{i\omega t}}_{F^\dagger(t)}$$

where the last term is evidently the fluctuating force. We determine the properties of the fluctuating force and evaluate

$$< \phi_F|F^\dagger(t')F(t'')|\phi_F > = \sum_\omega\sum_{\omega'} e^{i\omega t} e^{-i\omega' t'} G^*_\omega G_{\omega'} < \phi_F|B^\dagger_\omega(t_o)B_{\omega'}(t_o)|\phi_F>$$

where the expectation value in the "mixed state" is obtained by taking the trace

$$< \phi_F|B^\dagger_\omega (t_o) B_{\omega'} (t_o)|\phi_F >$$

$$= \frac{S\ [B^\dagger_\omega (t_o)B_{\omega'}(t_o)e^{-\Sigma\hbar\omega B^\dagger_\omega B_\omega|\lambda_1 T}}{S\ [\exp(- \Sigma\hbar\omega B^\dagger_\omega B_\omega|\kappa^\dagger)]}$$

$$= \begin{cases} 0 \text{ if } \omega \neq \omega' \\ \bar{n}_\omega (T)\delta_{\omega\omega'} \end{cases}$$

$$F^\dagger(t)F(t') = \sum_\omega |G_\omega|^2 e^{i\omega(t-t')}\bar{n}_\omega(T)$$

$\bar{n}_\omega(T)$ is the number of heat-bath bosons in the mode corresponding to ω. That is

$$< F^\dagger(t)F(t') > \; = \; 2\kappa\delta(t-t')\bar{n}_\omega(T).$$

Of course, this result normally appears under a time-integral which contains a factor $e^{i\omega_o t}$, where ω_o is the frequency of the light field. If this is taken into account, we obtain

$$< F^\dagger(t)F(t') > \; = \; 2\kappa\bar{n}_{\omega_o}(T)\delta(t - t')$$

and $< F(t)\, F^\dagger(t') > \; = \; 2\kappa(\bar{n}_{\omega_o}(T) + 1)\delta(t - t')$

The δ function here expresses the fact that the heat bath has a very short memory.

We now return to the calculation of the correlation function for the time ordering $(t > t')$:

$$< \phi_F|b^\dagger(t)b(t')|\phi_F > \; = \; \int_o^t d\tau \int_o^{t'} d\tau' \, e^{(i\omega-|G|)(t-\tau)} e^{(-i\omega-|G|)(t'-\tau')} \cdot C\,\delta(\tau-\tau')$$

$$= \frac{C}{2|G|} e^{i\omega(t-t')} e^{-|G|(t-t')}$$

$$= e^{i\omega(t-t')} e^{-|G|(t-t')} < \phi_F|b^\dagger(o)b(o)|\phi_F >.$$

From this follows, for the medium below threshold, that $< b^\dagger(t)\, b(t') >$ increases with the pumping rate ($|G| \rightarrow 0$) and $\gamma_{eff} = G$ decreases with increasing pumping, or the linewidth becames smaller

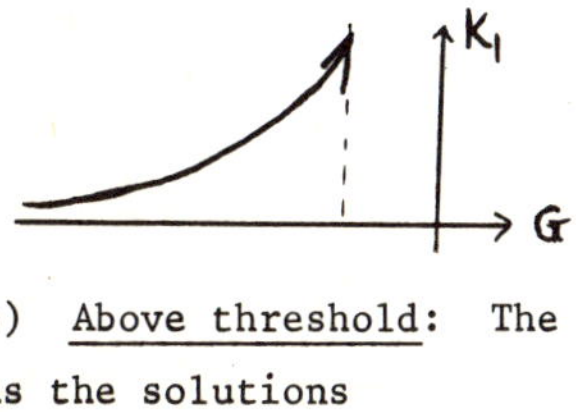

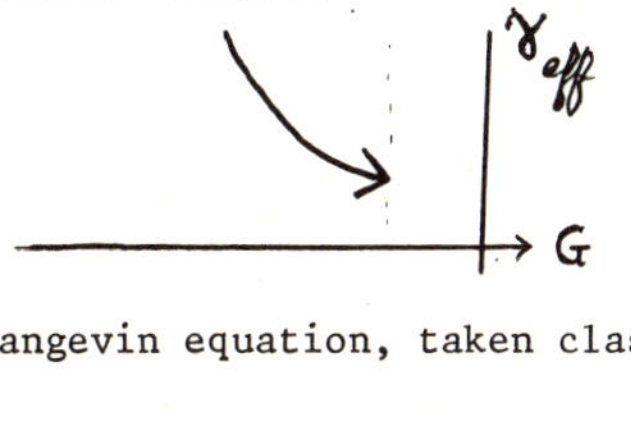

(b) Above threshold: The nonlinear Langevin equation, taken classically, has the solutions

$$b = (r_o + \rho)e^{i\phi} e^{-i\omega t}$$

where $\quad \dot{\phi} = \frac{1}{ir_o} \text{Im}\, \tilde{F}_{tot}, \qquad \phi = \frac{1}{ir_o} \int_o^t \text{Im}\, \tilde{F}_{tot}\, d\tau \perp \phi(o)$

$$r_o^2 = G/C\,.$$

In lowest order $\frac{<\rho^2>}{r_o^2} \ll 1$ the correlation function is given by

$$< b^{\dagger}(t)\, b(t) > \;=\; r_o^2\, e^{i\omega(t-t')} \left< e^{i\phi(t)}\, e^{-i\phi(t')} \right>$$

where $\phi(t) = \sum_{\mu} \phi_{\mu}(t)$ is a sum of independent stochastic variables (phase angles) $\phi_{\mu}(t)$, due to the different heat baths.

$$< e^{i\phi(t)}\, e^{-i\phi(t')} > \;=\; \prod_{\mu} < e^{i(\phi_{\mu}(t) - \phi_{\mu}(t'))} >$$

$$= \prod_{\mu} <(1 - i(\phi_{\mu}(t) - \phi_{\mu}(t')) - \tfrac{1}{2}(\phi_{\mu}(t) - \phi\mu(t'))$$

$$= \prod_{\mu} <(1 - \tfrac{1}{2}(\phi_{\mu}(t) - \phi_{\mu}(t'))^2) >$$

$$= e^{-\frac{1}{2}\sum_{\mu} < \phi_{\mu}(t) - \rho_{\mu}(t'))^2 >}$$

$$= e^{-\frac{1}{2}\, 2\gamma_{eff}(t-t')}$$

In the last step we have used the argument that the phase angle is a stochastic variable in a sense similar to the x-position of a particle performing a Brownian motion. If the density of particles n satisfies a diffusion equation $\frac{\delta n}{\delta t} = D\nabla^2 n$ the diffusion is adequately described in terms of the solution

$$n = N \frac{1}{\sqrt{4\pi D t}}\, e^{-x^2/4Dt}$$

and the average $< x^2 >$ is given by

$$< x^2 > \;=\; 2Dt$$

as is easily shown. Since $r_o\phi$ plays the role of x, we have

$$< r_o^2 (D\phi)^2 > \;=\; 2Dt \quad \text{or} \quad < \phi_{\mu}(t) - \phi_{\mu}(t'))^2 > \;=\; 2\gamma_{\mu}(t-t');\; \gamma_{\mu} = \frac{D}{r_o^2}$$

We finally obtain or $\sum_{\mu} < (\phi_{\mu}(t) - \phi_{\mu}(t'))^2 > \;=\; \gamma_{eff}(t-t')$

$$<|b^{\dagger}(t)\, b(t')|> \;=\; r_o^2\, e^{-\gamma_{eff}(t-t')}\, e^{i\omega(t-t')}$$

so that this function increases with increasing pumping, and also

$$\gamma_{eff} \to 0 \quad \text{as} \quad G \to \infty \quad (\gamma_{eff} \sim \frac{1}{g}).$$

Thus both below and above threshold κ_1 and γ_{eff} have the same monotonic behaviour, κ_1 increasing and γ_{eff} decreasing. In order to distinguish between the two states of the laser, (the thermal state and the coherent state) we must look for (a) a method to give information about $G = 0$, and (b) we must look for quantities which really behave differently below and above threshold. We need a higher order correlation function. The one that behaves differently is

$$\bar{\kappa}_2 = < b^\dagger(t) b^\dagger(t') b(t') b(t) >$$

(The quantum mechanical average of a triple product is not interesting - it vanishes). One looks for the deviation

$$\kappa_2 = < b^\dagger(t)\ b^\dagger(t')\ b(t')\ b(t) > - |< b^\dagger(t)\ b(t') >|^2.$$

Again we have a behaviour below and above threshold. When we are below threshold we get

$$\kappa_2 = 2|< b^\dagger(t) b(t') >|^2 - |< b^\dagger(t) b(t') >|^2$$

$$= |< b^\dagger(t) b(t') >|^2$$

$$= |\kappa_1(t_1 t')|^2$$

whereas, above threshold, we use the decomposition

$$b = (r_0 + \rho) e^{i\phi(t)}\ e^{-i\omega t}$$

The leading term will contain a r_0^2. All we have to calculate really is

$$< (r_0 + \rho)^4 > - (<(r_0 + \rho)^2 >)^2$$

$$\sim \psi r_0^2 < \rho(t)\rho(t') >.$$

We have derived the equation for ρ previously and we have found

$$\dot{\rho} = -2g\rho + \mathrm{Re}\,\tilde{F}_{tot}$$

The result is, as before

$$\kappa_2 - |\kappa_1|^2 = \text{const.}\ (n_{th} + n_{sp})\ e^{-2C\,\langle n\rangle\,|t-t'|}$$

where $\langle n\rangle$ is the average photon number. With rising photon number this function decreases.

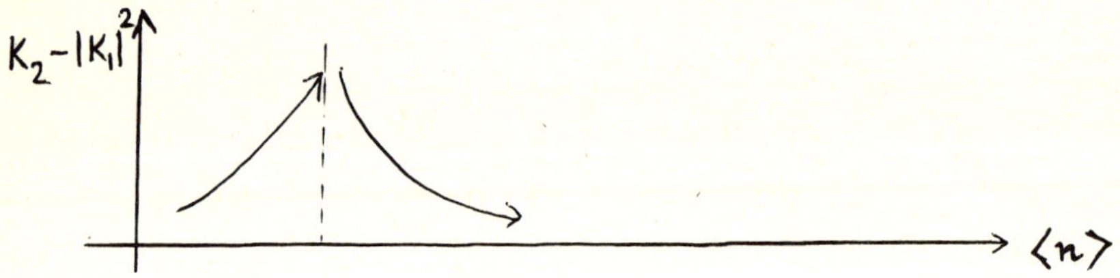

The second order correlation function K_2 behaves entirely different above and below threshold. Now the experimentalists can distinguish laser behaviour below and above threshold.

6. LAST LECTURE

Yesterday we looked at the statistical properties of laser light. But there was still a gap. Our method distinguished between pumping rates below and above threshold. For an improved theory we again consider, for the sake of simplicity, the one mode case, where we obtained, after elimination of the matter variables

$$\frac{db}{dt} = (-i\omega + G)b - Cb^{\dagger}b\,b + F_{tot}$$

or
$$\dot{q} = Gq - Cq^{\dagger}q\,q + F_{tot}\,e^{i\omega t}$$

where $b = q\,e^{-i\omega t}$. If we assume q to be a classical variable, we obtain the nonlinear stochastic equation

$$\dot{q} = Gq - C|q|^2\,q + \tilde{F}_{tot} \tag{61}$$

which is a generalisation of the equation for Brownian motion

$$\frac{dq}{dt} = -|G|q + F(t)$$

where q plays the role of particle velocity. As in the case of Brownian motion we can introduce a probability function $f(q,t)$ such that $f(q,t)\,dq$ is the probability of finding the particle at time t in the interval dq. The function $f(q,t)$ satisfies a Fokker Planck equation

$$\frac{df}{dt} = \frac{d}{dq}(|G|qf) + Q\frac{d^2f}{dq^2}$$

where $< F(t)F(t') > = Q\delta(t-t')$

In a similar sense the Fokker-Planck equivalent of the equation (61) is, in terms of the variables r,ϕ such that $q = re^{i\phi}$,

$$\frac{df}{\partial t} = -\frac{1}{r}\frac{\partial}{\partial r}[(Gr^2 - (r^4)f] + Q\{\frac{1}{r}\frac{\partial}{\partial r}(r\frac{\partial f}{\partial r}) + \frac{1}{r^2}\frac{\partial^2}{\partial\phi^2}\}, \qquad (62)$$

A simple explicit solution is the stationary solution ($\dot{f} = 0$) with ϕ-independent boundary conditions. In that case we may take $\frac{\partial}{\partial\phi} = 0$ and the resultant form has a first integral

$$(Gr^2 - (r^4)f = Qr\frac{\partial f}{\partial r}$$

or $$f(r) = Ne^{-V(r)/Q}$$

where $$V(r) = \frac{G}{2}r^2 + \frac{Cr^4}{4}$$

This is quite a nice result, which is, in this approximation, independent of the magnitude of the pumping rate G. Graphically we have the representations below:

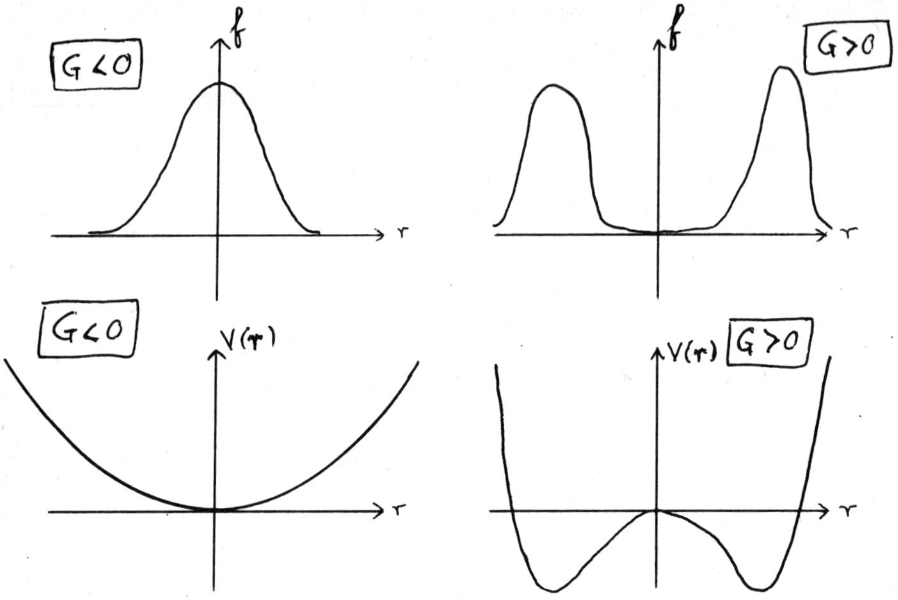

As bases we could also have used the variable q instead of $r = \mathrm{re}\, q$, since we have rotation symmetry $\frac{\partial}{\partial\phi} = 0$. The distribution function has a different behaviour below and above threshold and f is a continuous function of G.

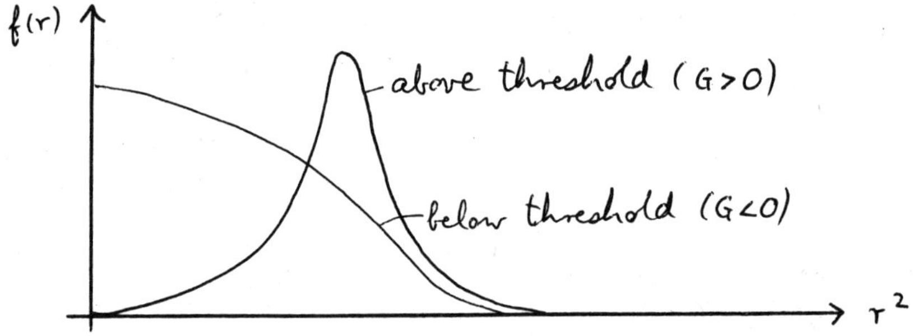

Can one measure these things? Yes. Consider again a correlation function at equal time (classical average)

$$< b^{\dagger} b\, b^{\dagger}\, b > \; = \int_0^{\infty} dr\; r^2 \int_0^{2\pi} d\phi\; r^4\, f$$

$$\equiv \; < n^2 >$$

or, consider the difference

$$< n^2 > - <n>^2 \;\equiv\; <n> (1 + <n>\, H_2)$$

to define a measurable quantity H_2. Arrechi (1968) has measured this quantity. A calculation of $H_2 = \frac{< n^2 > - < n >}{< n >^2} - 1$ can be performed and we btain below threshold:

$$< n^2 > - < n > = 2 < n >^2, \quad H_2 = 1$$

or $$< n^2 > - < n >^2 = < n >^2 + < n >$$

which is characteristic of Bose-Einstein statistics, whereas above threshold

$$< n^2 > - < n >^2 = < n >$$

which is characteristic of a Poisson distribution. How can we interpret this?

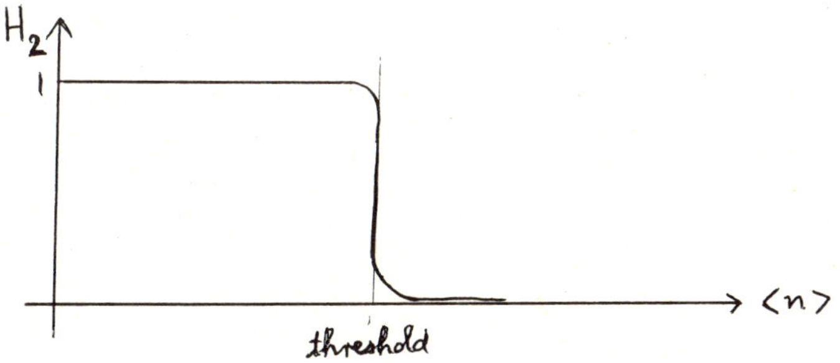

The new theory gives a smooth transition between the limiting cases $G < 0$ and $G < 0$. The Bose-Einstein distribution implies that below threshold the photons come in clusters, they come simultaneously and then for a while only a few. Should we plot the number of photons as a function of time, we get rather large fluctuations. Above threshold the fluctuations of photon number as function of time are much smaller. There is a certain mean time distance between the events, fluctuations are much smaller. The physical reason lies in the stabilization of the laser - the laser action can be explained by the depletion of atoms - while laser action takes place atoms are depleted and then they get filled up again by the pump mechanism. In the steady state you have a rather well defined depletion and whenever the photon number should become too big the atoms react correspondingly and emit less photons so that the old production rate is secured. On the other hand, if there are too few photons produced then the inversion will go up and produce more photons. That means laser light is well stabilised and we have a well stabilised amplitude.

REFERENCES

F T Arecchi (1970): In *Quantum Optics*, Proc. Intl. School "Enrico Fermi" XLII (ed. R J Glauber), Academic Press

A Einstein (1917): Phys. Zeitschr. 18: 121

A G Fox and T Li (1961): Bell System Tech. J. 40: 489

J P Gordon, H J Zeiger and C H Townes (1955): Phys. Rev. 99: 1 264

H Haken (1966): Z. Physik 190: 327
(1970a): Encycl. of Physics XXV/2c, Springer
(1970b): In *Quantum Optics*, 10th Scottish Univ. Summer School in Physics (eds. S M Kay and A Maitland), Academic Press
(1979): *Licht und Materie I*, BI-Wissenschaftsverslag, Mannheim

T H Maiman (1960): Nature 187: 494

H Risken (1965): Z. Physik 186: 85

A L Schawlow and C H Townes (1958): Phys. Rev. 112: 1940

QUANTUM STATISTICAL TREATMENT OF OPEN SYSTEMS , LASER DYNAMICS AND OPTICAL BISTABILITY

F . Casagrande and L.A. Lugiato
Istituto di Fisica dell'Università , Milano , Italy

Abstract

A typical problem of statistical mechanics is the dynamics of open systems, i.e. systems in interaction with a thermal reservoir. Namely, we illustrate a method to treat quantum open systems recently elaborated by one of us (L.A.L.) and apply it to problems in the field of quantum optics. When the reservoir is weakly perturbed by the system we recover the well known results,already obtained via treatments at second order in the reservoir-system coupling constant,concerning the radiative decay of a two-level atom à la Wigner and Weisskopf and the Brownian movement of a harmonic oscillator. On the other hand,our method can be applied also in the case of strong coupling between system and reservoir. This allows describing phenomena as the laser and optical bistability treating in detail the quantum statistical aspects.

1. Introduction

The study of the interaction between matter and radiation has always been one of the major topics in physics. At the beginning of the sixties, the applications of basic physical principles as well as of refined technologies allowed the first observations of laser operation. This event opened a new field, laser physics, which has focused a continuously growing interest and is nowadays the core of that branch of modern physics which is called quantum optics.

From a theoretical viewpoint, the rich phenomenology which is the object of quantum optics can be regarded as a stimulating field of application of quantum statistical mechanics. Or, more precisely, Synergetics [1] .

In fact, one deals with many-atom open systems far from thermal equilibrium whose dynamics is of a cooperative nature and which show phase-transition-like behaviors. Clearly, fluctuatios and correlations play a fundamental role. Hence a fully quantum statistical approach is the most suitable framework in order to describe these phenomena. However the most consolidated tools in the investigation of open systems are essentially weak coupling treatments. On the contrary a peculiarity of the phenomena in laser physics is the strong interaction between the atomic system and the electromagnetic field. This requires an extension of the previous methods to treat strong coupling situations. Recent progress in this sense has been obtained by a method to treat open systems developed by one of us [2]. In this paper we rewiew this approach and some recent applications in quantum optics. In this way, first we clearly show the underlying common quantum statistical basis and secondly we give a unified treatment of a variety of phenomena.

We start by illustrating in section 2 the quantum theory of open systems. We introduce the general formalism in a way that lends itself to the subsequent applications. The basic concepts and definitions of quantum statistical mechanics which are the background for our treatment are concisely reviewed in reference 3. In the following section we illustrate two simple applications, the decay of a two-level atom and the Brownian motion of a harmonic oscillator. This has not only illustrative purposes, since these examples are of fundamental relevance for the construction of the one-mode laser model which is described in detail in sec. 4.

By means of this model we treat two remarkable phenomena in quantum optics, i.e. the laser and optical bistability. In sec. 5 we recover the basic results of the semiclassical approximation, in which we neglect all fluctuations and correlations, both in the stationary and in the transient situation. Sec. 6 is the main part of this paper. We show that the usual assumption of weak interaction in the approach introduced in sec. 2 must be dropped if we want to describe these phenomena in a fully quantum context. Thus we follow the more powerful procedure of Ref.2, which applies in these cases as well. It follows a unified treatment of laser and optical bistability , in particular we give a detailed discussion of photon statistics in both phenomena.

2. Dynamics of open systems

Let us consider a quantum system S which is open, i.e. it interacts with a thermal reservoir R. S exchanges energy and possibly matter with R, which is a macroscopic system characterized by a very large number of degrees of freedom (ideally, infinites ones). The reservoir is initially in an equilibrium state at some temperature T. The interaction with the system S disturbs the reservoir R and makes it deviate from its equilibrium state. However, since the reservoir is very big, these disturbances are small and are rapidly eliminated by those dissipative mechanisms which sistematically lead the reservoir back to equilibrium. On the contrary, S is strongly affected by the interaction with R. This interaction ultimately leads S to reach the thermal equilibrium with the reservoir. Our aim is to study the dynamics of the open system S under the influence of R, i.e. the time evolution of S when it is driven by the reservoir.

To this end, let us consider the composite system Q= S + R. Its Hamiltonian has the structure

$$H = H_S + H_R + H_{SR} \tag{1}$$

where $H_S(H_R)$ is the Hamiltonian of S(R) and H_{SR} is the interaction Hamiltonian. The statistical operator W(t) of Q obeys the Von Neumann equation (VNE)

$$\frac{dW}{dt} = -i \left(\mathcal{L}_S + \mathcal{L}_R + \mathcal{L}_{SR} \right) W(t) \,, \tag{2}$$

$$\mathcal{L}_i W = \frac{1}{\hbar} \left[H_i \,, W \right] \quad (i = S, R, SR) \,. \tag{3}$$

We are interested only in the dynamics of the subsystem S. Now to calculate the mean values and the fluctuations of the observables of S it is not necessary to know the statistical operator W of the whole system S +R, but it is enough to know the "reduced" statistical operator of the subsystem S alone

$$\rho(t) = Tr_R \, W(t) \,, \tag{4}$$

where Tr_R means "partial trace" over the Hilbert space of R. Hence the problem arises of deriving the time evolution equation for the reduced statistical operator ρ from the VNE (2) for the full statistical operator W. This equation must be closed, i.e. it must be expressed only in terms of the variables of the subsystem S. Usually such a problem is solved by the so-called "projection operator technique" developed in general by Zwanzig [4] and adapted to open systems by Argyres and Kelley [5] (see also [6]).However, the projection technique is not suitable to treat the laser. Thus we shall follow a more powerful method [2].

2.1 Hierarchy of equations for the reduced statistical operator and for bath-system correlations.

For the sake of simplicity we shall take the reservoir R as a system of noninteracting harmonic oscillators or as a system of noninteracting two-level atoms. This is suitable to illustrate the method in the simplest possible situations. Furthermore, in the applications that we shall consider later the reservoir is just of this kind. When the reservoir is a set of noninteracting harmonic oscillators labeled by j=1,2,... n we call B_j and B_j^+ the annihilation and creation operators of the j-th oscillator. These operators obey the commutation rules

$$(5) \qquad \left[B_j , B_j^+\right] = \delta_{jj'} \quad , \quad \left[B_j , B_{j'}\right] = \left[B_j^+ , B_{j'}^+\right] = 0 .$$

On the other hand, when the reservoir is a set of noninteracting two-level atoms again labeled by j=1,2,.... n we call B_j and B_j^+ the lowering and raising operators of the j-th atom. In this case one has the anticommutation rules

$$(6) \qquad \left\{B_j , B_j^+\right\} = 1 \quad , \quad \left\{B_j , B_j\right\} = \left\{B_j^+ , B_j^+\right\} = 0 .$$

In both cases the Hamiltonian of the reservoir is given by

$$(7) \qquad H_R = \sum_j \hbar\omega_j \, B_j^+ B_j \quad ,$$

where ω_j is the frequency of the j-th oscillator or the transition frequency of the j-th two-level atom. It is not necessary to specify the Hamiltonian of the subsystem S.

Finally we assume the following structure for the interaction Hamiltonian:

$$(8) \qquad H_{SR} = i\hbar \sum_j g_j \left(A^+ B_j - A B_j^+ \right) ,$$

where g_j is a suitable coupling constant which we take real for simplicity and A is any operator in the Hilbert space of the system S. The structure (8) is admittedly not the most general, but it is enough to treat all the problems considered in this paper.

As initial condition for eq.(2) we assume that at time t=0 the subsystem and the reservoir are uncorrelated and furthermore that the reservoir is in equilibrium at inverse temperature β, so that

$$(9) \qquad W(0) = \rho(0) \cdot \left[\exp(-\beta H_R) \big/ Tr\left(\exp(-\beta H_R)\right)\right]$$

As a first step we take the partial trace of the VNE (2). We get

$$(10) \qquad \frac{d\rho}{dt} = -i \mathcal{L}_S \rho + \sum_j g_j \left\{ \left[A^+, W_j^{(-)}(t) \right] - \left[A, W_j^{(+)}(t) \right] \right\},$$

where eq. (4) has been used and

$$(11) \qquad W_j^{(+)}(t) = Tr_R \left(B_j^+ W(t) \right) \quad , \quad W_j^{(-)}(t) = Tr_R \left(B_j \, W(t) \right).$$

$W_j^{(+)}$, $W_j^{(-)}$ are operators in the Hilbert space of the subsystem S alone. Note that

$$(12) \qquad Tr \, W_j^{(+)}(t) = \langle B_j^+ \rangle (t) \quad , \quad Tr \, W_j^{(-)}(t) = \langle B_j \rangle (t) ,$$

where $\langle O \rangle$ indicates in general the mean value of the observable O. Furthermore, the quantities $W_j^{(\pm)}$ allow to calculate a suitable set of bath-system correlation functions. In fact, for any given observable O of S we have

$$(12') \qquad Tr\,(O_S W_j^{(+)}(t)) = \langle O_S B_j^{+} \rangle (t)\,, \quad etc.$$

For this reason, we call $W_j^{(\pm)}$ (as well as the following quantities $W_j^{(+-)}$ etc.) "bath-system correlations".
The relevant point is that eq. (10) is not closed in ρ, because there appear the quantities $W_j^{(\pm)}$. Therefore, we are led to derive from the VNE the time evolution equations for $W_j^{(\pm)}$. To this aim, we multiply eq.(2) times B_j^{+} or B_j and take the partial trace over the reservoir. After some calculations we find the following equation:

$$(13) \qquad \frac{dW_j^{(+)}}{dt} = -i\mathcal{L}_S W_j^{(+)}(t) + i\omega_j W_j^{(+)}(t) + \sum_{j'} g_{j'} \left(A^{+} W_{jj'}^{(+-)}(t) - W_{j'j}^{(-+)}(t) A^{+} - A W_{jj'}^{(++)}(t) + W_{j'j}^{(++)}(t) A \right)$$

Again we meet new quantities, namely $W^{(+-)}$ and $W^{(++)}$, where e.g. (cfr. (11),(12))

$$(14) \qquad \begin{aligned} W_{jj'}^{(+-)}(t) &= Tr_R\,(B_j^{+} B_{j'} W(t))\,, \quad etc. \\ Tr\,W_{jj'}^{(+-)}(t) &= \langle B_j^{+} B_{j'} \rangle (t)\,, \quad etc. \end{aligned}$$

The equation for $W_j^{(-)}$ is immediately obtained by operating hermitian conjiugation on eq.(13); in fact

$$(15) \qquad W_j^{(-)}(t) = [W_j^{(+)}(t)]^{+}.$$

At this point, we derive from the VNE the time evolution equations for the quantities $W_{jj'}^{(+-)}$ etc. These equations will contain in turn new quantities more, as e.g.

$$W_{jj'j''}^{(++-)}(t) = Tr\,(B_j^{+} B_{j'}^{+} B_{j''} W(t))\,,$$

where three operators B appear. In such a way, we obtain a hierarchy of linear equations for the reduced statistical operator ρ and for the bath-system correlations W with an arbitrary number of reservoir operators B. This hierarchy of equations is equivalent to the VNE (2). Clearly, one must suitably truncate this hierarchy, so that after elimination of the auxiliary quantities W one obtains a closed equation for ρ.

A general truncation prescription does not exist to our knowledge. One must choose the proper truncation procedure according to the individual problem that is studied. Now we shall illustrate the simplest truncation procedure. A more refined one will be shown in connection with the treatment of the laser and of optical bistability ,in sec.6.

2.2- The master equation for weak interaction between system and reservoir

Let us consider the two eqs.(10),(13). We can formally solve eq.(13) for $W_j^{(+)}(t)$ by treating the quantities as $W^{(+-)}$ as known quantities and taking into account that $W_j^{(+)}(0)= \rho(0) \langle B_j^+ \rangle_\beta =0$ due to eqs.(11),(9),(7) and to the fact that

$$\langle B_j^+ \rangle_\beta \stackrel{def}{=} Tr\left[B_j^+ e^{-\beta H_R} / Tr\, e^{-\beta H_R} \right]$$

Hence by substituting this expression for $W_j^{(+)}(t)$ into eq.(10) we obtain the following equation

$$\frac{d\rho}{dt} = -i \mathcal{L}_s \rho(t) + \int_0^t d\tau \sum_{jj'} g_j g_{j'} \cdot$$

(16)

$$\cdot \Big\{ \Big[A, e^{-i(\mathcal{L}_s - \omega_j)(t-\tau)} \Big(-A^+ W_{jj'}^{(+-)}(\tau) + W_{j'j}^{(-+)}(\tau) A^+ + A W_{jj'}^{(++)}(\tau)$$

$$- W_{j'j}^{(++)}(\tau) A \Big) \Big] + \Big[A^+, e^{-i(\mathcal{L}_s + \omega_j)(t-\tau)} \Big(W_{j'j}^{(+-)}(\tau) A$$

$$- A W_{jj'}^{(-+)}(\tau) - W_{jj'}^{(--)}(\tau) A^+ + A^+ W_{j'j}^{(--)}(\tau) \Big) \Big] \Big\} .$$

Eq.(16) is still exact but it is not closed in ρ owing to the presence of the quantities $W_{jj'}^{(+-)}$ etc. To get a closed equation, we must express these quantities in terms of ρ . This can be easily accomplished when the interaction between the subsystem and the reservoir is weak,so that the subsystem disturbs in a negligible way the reservoir. In this case one can extend to all times the factorization ansatz(9):

$$W(t) = \rho(t)\, e^{-\beta H_R} / Tr\, e^{-\beta H_R} \qquad (17)$$

Using the ansatz and the definition (14) of $W^{(+-)}$, one immediately has

$$W^{(+-)}_{jj'}(t) = \rho(t)\langle B_j^+ B_{j'}\rangle_\beta \,, \quad W^{(-+)}_{jj'}(t) = \rho(t)\langle B_j B_{j'}^+\rangle_\beta \,, \tag{18}$$

$$W^{(++)}_{jj'}(t) = \rho(t)\langle B_j^+ B_{j'}^+\rangle_\beta = 0 \,, \quad W^{(--)}_{jj'}(t) = \rho(t)\langle B_j B_{j'}\rangle_\beta = 0.$$

Substituting (18) into (16) we obtain the closed equation for $\rho(t)$

$$\frac{d\rho}{dt} = -i\mathcal{L}_s \rho(t) + \int_0^t d\tau \sum_{jj'} g_j g_{j'} \cdot$$

$$\Big\{\Big[A,\, e^{-i(\mathcal{L}_s-\omega_j)(t-\tau)}\Big(-\langle B_j^+ B_{j'}\rangle_\beta A^+\rho(\tau) + \langle B_{j'} B_j^+\rangle_\beta \cdot \tag{19}$$

$$\cdot \rho(\tau)A^+\Big)\Big] + \Big[A^+,\, e^{-i(\mathcal{L}_s+\omega_j)(t-\tau)}\Big(\langle B_{j'}^+ B_j\rangle_\beta\, \rho(\tau) A$$

$$- \langle B_j B_{j'}^+\rangle_\beta A\rho(\tau)\Big)\Big]\Big\}.$$

Eq. (19) has two contributions: the first term is the free evolution of the subsystem, while the second term comes from the interaction with the reservoir. The assumption of weak interaction is reflected by the fact that this term contains the coupling constant only at second order. When the interaction is strong, as in the laser, one finds contribution of all orders in the coupling constant [2].

Eq. (19) can be written in a more compact way by introducing time correlation functions for suitably defined quantities as it follows:

$$\sum_{jj'} g_j g_{j'} e^{i\omega_j t}\langle B_j^+ B_{j'}\rangle_\beta = \langle \mathcal{B}^+(t)\mathcal{B}(0)\rangle_\beta \,, \tag{20}$$

$$\mathcal{B} = \sum_j g_j B_j \quad , \quad \mathcal{B}^+ = \sum_j g_j B_j^+$$

By (20) and similar expressions, eq. (19) becomes

$$\frac{d\rho}{dt} = -i\mathcal{L}_s\rho(t) + \int_0^t d\tau \Big\{-\langle\mathcal{B}^+(t-\tau)\mathcal{B}(0)\rangle_\beta \cdot$$

$$\cdot \left[A, e^{-i\mathcal{L}_s(t-\tau)} A^{+} \rho(\tau) \right] + \langle \mathcal{B}(0) \mathcal{B}^{+}(t-\tau) \rangle_{\beta} \left[A, e^{-i\mathcal{L}_s(t-\tau)} \cdot \right.$$

$$(21) \qquad \left. \cdot \rho(\tau) A^{+} \right] + \langle \mathcal{B}^{+}(0) \mathcal{B}(t-\tau) \rangle_{\beta} \left[A^{+}, e^{-i\mathcal{L}_s(t-\tau)} \rho(\tau) A \right]$$

$$\left. - \langle \mathcal{B}(t-\tau) \mathcal{B}^{+}(0) \rangle_{\beta} \left[A^{+}, e^{-i\mathcal{L}_s(t-\tau)} A \rho(\tau) \right] \right\} .$$

Note that in eq. (21) the only quantities which refer to the reservoir are time correlation functions such as (20). In other words, the reser_ voir acts on the subsystem just via its time correlation functions. Eq. (21) is an integrodifferential equation in which the derivation of ρ at time t depends on the values of ρ on the whole interval from time 0 to time t. Hence one says that $\rho(t)$ has "memory" of its time evolu_ tionand eq.(21) is called "generalized master equation" or " nonmarkof_ fian master equation" [7,4] . However, in our case this memory ef_ fect or nonmarkoffian character is only apparent, because under our assumptions eq.(21) can be very well approximated by a differential equation as first shown by Van Hove [7].

To show this, first of all we pass to the interaction picture in order to eliminate the rapid variation of ρ in time due to the first (free evolution) term on the r.h.s. of (21). We recall that the relation between the statistical operator in the interaction picture $\rho^{(I)}$ and in the Schödinger picture ρ is

$$\rho^{(I)}(t) = e^{i\mathcal{L}_s t} \rho(t) = e^{\frac{i}{\hbar} H_s t} \rho(t) \, e^{-\frac{i}{\hbar} H_s t}$$

The time evolution of $\rho^{(I)}$ can be written in a compact way as it follows

$$(22) \qquad \frac{d\rho^{(I)}}{dt} = \int_0^t d\tau \, K(t-\tau) \, \rho^{(I)}(\tau) ,$$

where the operator K has the form

$$K(t)\cdot = \langle \mathcal{B}^+(t)\mathcal{B}(0)\rangle_\beta\, e^{-i\omega_0 t}\left[A^+\cdot, A\right] + \langle \mathcal{B}(0)\mathcal{B}^+(t)\rangle_\beta \cdot \tag{23}$$

$$\cdot e^{-i\omega_0 t}\left[A, \cdot A^+\right] + \langle \mathcal{B}^+(0)\mathcal{B}(t)\rangle_\beta\, e^{i\omega_0 t}\left[A^+, \cdot A\right] +$$

$$+ \langle \mathcal{B}(t)\mathcal{B}^+(0)\rangle_\beta\, e^{i\omega_0 t}\left[A\cdot, A^+\right].$$

In writing (23), we assumed that S is a harmonic oscillator or a two-level atom of frequency ω_0 (or a system of identical harmonic oscillators or two-level atoms) so that in the interaction picture the operators A and A^+ have the following time evolution:

$$A^{(I)}(t) = e^{-i\omega_0 t} A \quad , \quad (A^+)^{(I)}(t) = e^{i\omega_0 t} A^+ . \tag{24}$$

The variation of ρ in the interaction picture is slow, because it arises only from the weak interaction with the bath. Let us call τ_1 a time in_ terval which characterises the time variation of $\rho^{(I)}$. On the other hand the kernel K(t) in eq.(22) is the sum of four contributions, each of which contains a time correlation function. Since e.g.

$$\lim_{t\to\infty} \langle \mathcal{B}^+(t)\mathcal{B}(0)\rangle_\beta = \langle \mathcal{B}^+\rangle_\beta \langle \mathcal{B}\rangle_\beta = 0,$$

the kernel K vanishes for t tending to infinity. This is due to those dissipative mechanisms of the reservoir which damp the fluctuations systematically leading the reservoir back to equilibrium. This approach to equilibrium is very rapid since the very fast relaxation times are the peculiar feature of the reservoir R. If we call τ_0 a time interval which characterizes the approach to zero of the time correlations in play (and by (23) of K), we have the basic relation $\tau_0 \ll \tau_1$.
Now let us come back to eq. (22). We can write

$$\frac{d\rho^{(I)}}{dt} = \int_0^t d\sigma\, K(\sigma)\, \rho^{(I)}(t-\sigma)$$

$$\simeq \int_0^{\tau_0} d\sigma\, K(\sigma)\, \rho^{(I)}(t-\sigma) \quad , \qquad t > \tau_0 .$$

But $\rho^{(I)}(t-\sigma) \simeq \rho^{(I)}(t)$ for $0 \leqslant \sigma \leqslant \tau_0$. Hence

$$\frac{d\rho^{(I)}}{dt} \simeq \left[\int_0^{\tau_0} d\sigma \; K(\sigma)\right] \rho^{(I)}(t).$$

Finally, taking into account that $K(\sigma)$ is practically zero for $\sigma > \tau_0$, we get

$$\text{(25)} \qquad \frac{d\rho^{(I)}}{dt} = \mathcal{V} \rho^{(I)}(t) \qquad , \qquad \mathcal{V} = \int_0^{\infty} d\sigma \; K(\sigma).$$

Eq. (25) is a purely differential equation, which has no memory effect. This equation is called " Markoffian master equation" or "master equation" (ME) tout court. The Markoffian approximation is justified by the separation of two time scales: the time scale of the reservoir which is characterized by τ_0 and the time scale of the subsystem which is characterized by τ_1. By introducing the following symbols

$$\text{(26)} \qquad C^{(+-)}(\omega_0) = \int_0^{\infty} dt \, \langle \mathcal{B}^+(t)\, \mathcal{B}(0) \rangle_\beta \, e^{-i\omega_0 t} \, ,$$

$$C^{(-+)}(\omega_0) = \int_0^{\infty} dt \, \langle \mathcal{B}(0)\, \mathcal{B}^+(t) \rangle_\beta \, e^{-i\omega_0 t}$$

and coming back to the Schrödinger picture, the ME takes the final form

$$\text{(27)} \qquad \frac{d\rho}{dt} = -i\mathcal{L}_S \rho(t) + C^{(+-)}(\omega_0)\left[A^+\rho(t), A\right] +$$

$$+ C^{(-+)}(\omega_0)\left[A, \rho(t)A^+\right] + \left(C^{(+-)}(\omega_0)\right)^*\left[A^+, \rho(t)A\right]$$

$$+ \left(C^{(-+)}(\omega_0)\right)^*\left[A\rho(t), A^+\right]$$

Note that the ME (27) has a different structure from the VNE (2). In the latter one has only commutators with the Hamiltonians, describing a conservative dynamics. In the ME-apart from the free evolution term which is a conservative one-one finds terms with commutators of a more complicated structure. As we shall see from specific examples, these terms are not conservative but describe the dissipation of the energy of the subsystem into the reservoir.

As it follows from (27),(26), the coefficients of the ME are one-sided Fourier Transforms (i.e. Laplace Transforms) of time correlation functions of reservoir operators. Now, according to the well known theory of transport coefficients developed by Callen, Kubo, Green et al [8] the transport coefficients are just one-sided Fourier Transforms of time correlation functions. This fact establishes the connection between the theory of the ME and the theory of transports coefficients: the coefficients of the ME are transports coefficients of the reservoir.

Therefore we can apply some fundamental theorems of the theory of transport coefficients to show the remarkable links which exist between the coefficients of the ME. In fact, let us divide these coefficients into real and imaginary part

(28) $$C^{(+-)}(\omega_0) = \frac{1}{2}\gamma_\uparrow(\omega_0) + i\,\Delta^{(+-)}(\omega_0)\,,$$

$$C^{(-+)}(\omega_0) = \frac{1}{2}\gamma_\downarrow(\omega_0) + i\,\Delta^{(-+)}(\omega_0)\,.$$

First, the real and imaginary parts are linked by the so-called Kramers- Krönig relations.

(see p. 33)

Secondly, the fluctuation-dissipation theorem establishes the following link between the two real parts:

(29) $$\gamma_\uparrow(\omega_0) = \exp(-\beta\hbar\omega_0)\cdot\gamma_\downarrow(\omega_0)$$

Relation (29) is of fundamental importance, due to its universality. In fact, it does not depend on the peculiar features of the reservoir but only on its temperature.

3.- Two simple applications: the decay of a two-level atom and the Brownian motion of a harmonic oscillator.

We shall now consider two simple applications of the formalism developed in sec.2.2: i) the decay of a two-level atom and ii) the Brownian motion of a harmonic oscilator.

3.1-The decay of a two-level atom

Ingeneral, an atom or molecule has infinite energy levels. However, if the phenomenon that we want to describe involves the transition between two levels only, one can describe the atom as a two-level system, neglecting all the other levels [9]. We put for definiteness the zero of the energy of the atom half way between the two levels, so that they have energies $E_2 = \hbar\omega_0/2$ and $E_1 = -\hbar\omega_0/2$,where ω_0 is the Bohr transition frequency $\omega_0 = (E_2 - E_1)/\hbar$. The Hilbert space of the two-level atom is twodimensional; each ket $|\psi\rangle$ can be expanded as

(30) $$|\psi\rangle = a\,|1\rangle + b\,|2\rangle ,$$

where $|1\rangle$ ($|2\rangle$) is the lower (upper) state. One defines a raising (lowering) operator r^+(r^-) in the usual way:

(31) $$r^+|1\rangle = |2\rangle \quad , \quad r^+|2\rangle = 0 ,$$

(32) $$r^-|1\rangle = 0 \quad , \quad r^-|2\rangle = |1\rangle .$$

$r^\pm$ obey the anticommutation rule

(33) $$\{r^+, r^-\} = 1$$

It is easily seen that $r^+ = (r^-)^+$, $(r^+)^2 = (r^-)^2 = 0$. From eq. (33) it follows the physical interpretation of the operators r^+r^- and r^-r^+ as the number of particles in the upper and lower level, respectively. We introduce also the inversion operator

(34) $$r_3 = \frac{1}{2}\left(r^+r^- - r^-r^+\right) .$$

The three operators r^+, r^- and r_3 obey the angular commutation relations

(35) $$[r^+, r^-] = 2r_3 \quad , \quad [r_3, r^\pm] = \pm r^\pm .$$

This fact is obvious if we consider the matrix representation of these operators in the basis $|1>, |2>$:

(36) $$r^+ = \begin{pmatrix} 0 & 1 \\ 0 & 0 \end{pmatrix} , \quad r^- = \begin{pmatrix} 0 & 0 \\ 1 & 0 \end{pmatrix} , \quad r_3 = \begin{pmatrix} \frac{1}{2} & 0 \\ 0 & -\frac{1}{2} \end{pmatrix},$$

i.e. these operators correspond to Pauli matrices. Let us consider now the representation of the relevant observables of this system. The energy H_S is given by

(37) $$H_S = \begin{pmatrix} \frac{\hbar\omega_0}{2} & 0 \\ 0 & -\frac{\hbar\omega_0}{2} \end{pmatrix} = \hbar\omega_0 r_3 ;$$

the polarization $\mathcal{P}$ can be expressed as

(38) $$\vec{\mathcal{P}} = e\vec{x} = \begin{pmatrix} 0 & \vec{d} \\ \vec{d}^* & 0 \end{pmatrix} = \vec{d}r^+ + \vec{d}^* r^- , \quad \vec{d} = e <2|\vec{x}|1>,$$

where we have assumed that the two levels have opposite parity. Eq. (38) justifies calling r^+ and r^- polarization operators.

We now consider the interaction of our two-level atom with an electromagnetic field, which acts as a reservoir composed of noninteracting harmonic oscillators corresponding to the field modes. In the so-called dipole and rotating wave approximations the interaction Hamiltonian is given by

(39 $$H_{RS} = i\hbar \sum_j g_j (r^+ B_j - r^- B_j^+) ,$$

where B_j is the annihilation operator corresponding to the j-th mode of the e.m. field. Hamiltonian (39) is just of the kind (8) . Hence, as we have proven, the reduced statistical operator ρ of the two-level atom obeys the ME (27) which now describes the dynamics of the two-level atom under the influence of the e.m. field. By (37), (28) one rewrites the ME in the following form:

$$\frac{d\rho}{dt} = -i\left(\omega_0 + \Delta^{(+-)} - \Delta^{(-+)}\right)\left[r_3, \rho(t)\right] +$$

(40)

$$+\frac{\gamma_\downarrow}{2}\left(\left[r^-\rho(t), r^+\right] + \left[r^-, \rho(t) r^+\right]\right) + \frac{\gamma_\uparrow}{2}\left(\left[r^+\rho(t), r^-\right] + \left[r^+, \rho(t) r^-\right]\right).$$

From eq. (40) we see that the effect of the interaction with the reservoir is twofold. First, the atomic transition frequency ω_0 gets renormalized, i.e. one finds a frequency shift due to the imaginary parts of the coefficients of the ME. The second and more important effect of the reservoir arises from the terms containing the real parts $\gamma_\uparrow$, $\gamma_\downarrow$ which are called damping or dissipative terms. In fact, these terms determine the relaxation of the two-level atom towards the thermal equilibrium state. To see this point explicitly, we write the statistical operator ρ as it follows:

(41) $$\rho(t) = p_1(t)\,|1\rangle\langle 1| + p_2(t)\,|2\rangle\langle 2| + p_{12}(t)\,|1\rangle\langle 2| + p_{21}(t)\,|2\rangle\langle 1|,$$

where the conservation of probability requires that

(42) $$p_1(t) + p_2(t) = 1$$

and due to the hermiticity of $\rho(t)$

(43) $$p_{12}(t) = \left(p_{21}(t)\right)^* .$$

By taking the matrix elements of the ME, we obtain the following four equations for the matrix elements of ρ :

(44a) $$\dot{P}_2 = \gamma_\uparrow P_1(t) - \gamma_\downarrow P_2(t) ,$$

(44b) $$\dot{P}_1 = \gamma_\downarrow P_2(t) - \gamma_\uparrow P_1(t) ,$$

(44c) $$\dot{P}_{12} = \left[i\omega_0 - \frac{\gamma_\uparrow + \gamma_\downarrow}{2} \right] P_{12}(t) .$$

(44d) $$\dot{P}_{21} = \left[-i\omega_0 - \frac{\gamma_\uparrow + \gamma_\downarrow}{2} \right] P_{21}(t) ,$$

where $\omega = \omega_0 + \Delta^{(+-)} - \Delta^{(-+)}$.

Note that eqs. (44a,b) are independent of eqs. (44c,d) and that the two latter equations are ~~coplex conjigates~~ complex conjugates of each other. The two eqs. (44a,b) for the probabilities give the simplest example of rate eqnations and provide the physical interpretation of the two parameters $\gamma_\uparrow$ and $\gamma_\downarrow$:

$\gamma_\uparrow$ has the physical meaning of transition probability per unit time from the lower to the upper state. Similarly $\gamma_\downarrow$ is the transition probability per unit time from the upper to the lower state. In fact, eqs. (44a,b) have a very simple interpretation: e.g. eq.(44a) says that the variation per unit time of the probability of the upper state is equal to the probability of the lower state times the transition probability per unit time from the lower to the upper state, minus the probability of the upper level times the transition probability per unit time from theupper to the lower state. Eq.(44c) is immediately solved

(45a) $$P_{12}(t) = \exp\left\{ i\omega t - \frac{\gamma_\uparrow + \gamma_\downarrow}{2} t \right\} P_{12}(0) .$$

On the other hand Eqs.(44a,b) are also immediately solved recalling eq.(42):

(45b) $$P_2(t) = e^{-(\gamma_\uparrow + \gamma_\downarrow)t} P_2(0) + \frac{\gamma_\uparrow}{\gamma_\uparrow + \gamma_\downarrow} \left\{ 1 - e^{-(\gamma_\uparrow + \gamma_\downarrow)t} \right\} .$$

From Eq. (45a) it follows that

(46a) $$\lim_{t \to \infty} P_{12}(t) = \lim_{t \to \infty} P_{21}(t) = 0$$

with a relaxation rate

(46b) $$\gamma_\perp = \frac{\gamma_\uparrow + \gamma_\downarrow}{2} .$$

From (45b),(42) and the fluctuation-dissipation relation (29) one has

(47a) $$\lim_{t\to\infty} P_2(t) = \frac{\gamma_\uparrow}{\gamma_\uparrow + \gamma_\downarrow} = \frac{1}{e^{\beta\hbar\omega_0}+1} ,$$

$$\lim_{t\to\infty} P_1(t) = \frac{\gamma_\downarrow}{\gamma_\uparrow + \gamma_\downarrow} = \frac{\exp(\beta\hbar\omega_0)}{e^{\beta\hbar\omega_0}+1} ;$$

the relaxation rate is

(47b) $$\gamma_\parallel = \gamma_\uparrow + \gamma_\downarrow = 2\gamma_\perp .$$

The relevant point is that (46a)and(47a) turn out to be precisely the matrix elements of the canonical statistical operator of the two-level atom:

(48) $$\rho_\beta = \frac{\exp(-\beta H_s)}{\mathrm{Tr}\{\exp(-\beta H_s)\}} = \frac{\exp(-\beta\hbar\omega_0 r_3)}{\mathrm{Tr}\{\exp(-\beta\hbar\omega_0 r_3)\}} .$$

This shows that, as expected, the subsystem approaches a thermal equilibrium state in which it has the same temperature of the reservoir. In particular, let us consider the case that the reservoir has zero temperature, i.e. that the electromagnetic field is in the vacuum state. In this situation the atom simply decays exponentially to the lower state In fact, if we put $\beta = \infty$ in (47a), we obtain

(49) $$\lim_{t\to\infty} P_2(t) = 0 , \qquad \lim_{t\to\infty} P_1(t) = 1 .$$

This is essentially the well known Wigner-Weisskopf theory of the decay of the atom.

3.2. The Brownian motion of a harmonic oscillator

Let us consider a harmonic oscillator of frequency ω_0. The Hamiltonian can be written as

$$H_s = \hbar\omega_0 \left(A^+A + \frac{1}{2} \right) \tag{50}$$

where A^+ (A) is the raising(lowering) operator such that

$$\left[A , A^+ \right] = 1 \tag{51}$$

One has

$$\begin{aligned} H_s \, |n\rangle &= \hbar\omega_0 \left(n + \frac{1}{2} \right) |n\rangle \qquad (n = 0, 1, 2, \dots) \\ A^+ |n\rangle &= \sqrt{n+1} \; |n+1\rangle \, , \qquad A \, |n\rangle = \sqrt{n} \; |n-1\rangle \end{aligned} \tag{52}$$

Our harmonic oscillator interacts with a reservoir of noninteracting two-level atoms or harmonic oscillators. We assume as usual an interaction Hamiltonian of the form (8), namely

$$H_{RS} = i\hbar \sum_j g_j \left(A^+ B_j - A B_j^+ \right) . \tag{53}$$

Hence also in this case the reduced statistical operator ρ of the oscillator obeys the ME(27) that reads

$$\begin{aligned} \frac{d\rho}{dt} = &- i\omega \left[A^+A , \rho(t) \right] + \frac{\gamma_\uparrow}{2} \left(\left[A^+ \rho(t) , A \right] + \right. \\ &\left. + \left[A^+ , \rho(t) A \right] \right) + \frac{\gamma_\downarrow}{2} \left(\left[A \rho(t) , A^+ \right] + \left[A , \rho(t) A^+ \right] \right) , \end{aligned} \tag{54}$$

where ω is the renormalized frequency(cfr.Sec.3.1).
This equation can be suitably mapped into the Glauber diagonal representation [10], in which the quantum-mechanical ME (master equation - cf p.61) becomes a classical-looking partial differential equation. Let us briefly review the main properties of this representation [11]. First of all one considers the coherent states $|\alpha\rangle$, defined as the eigenstates of the annihilation operator A:

$$A\,|\alpha\rangle = \alpha\,|\alpha\rangle . \tag{55}$$

The spectrum of the nonhermitian operator A covers the whole complex plane. The mean values of A and A^+ in a coherent state are given simply by

$$\langle\alpha|A|\alpha\rangle = \alpha \quad , \quad \langle\alpha|A^+|\alpha\rangle = \alpha^* . \tag{56}$$

From the physical viewpoint, the most remarkable property of the coherent states is that the mean value of normal ordered products factorizes

$$\langle\alpha|(A^+)^m A^n|\alpha\rangle = \langle\alpha|A^+|\alpha\rangle^m \langle\alpha|A|\alpha\rangle^n = (\alpha^*)^m \alpha^n . \tag{57}$$

This implies that if we restrict ourselves to normal ordered quantities the coherent states are fluctuationless. This feature renders the coherent states the most"classical" states that one can find in quantum mechanics. We recall also that the coherent states least indetermination wave packets.

Now it has been proven [11] that there is a large class of statistical operators for the harmonic oscillator which can be represented in the following way:

$$\rho = \int d_2\alpha \;\; P(\alpha,\alpha^*)\;\; |\alpha\rangle\langle\alpha| , \tag{58}$$

where P is a c-number function and $d_2\alpha = d(Re\,\alpha)\cdot d(Im\,\alpha)$.

(58) is the Glauber diagonal representation of ρ, which maps the operator ρ into a classical function, the so-called P-function. This function is real but not always positive, hence it is called quasiprobability distribution. The mean value of the normal ordered products (57) is simply calculated from the P-function:

$$\langle (A^+)^m A^n \rangle = \int d_2\alpha \;(\alpha^*)^m \alpha^n \;\; P(\alpha,\alpha^*) . \tag{59}$$

When the operator ρ is mapped into the P- function according to eq. (58), the ME (54) turns out to be correspondingly mapped into the following partial differential equation [11] :

$$\frac{\partial P(\alpha,\alpha^*,t)}{\partial t} = -\frac{\partial}{\partial\alpha}\left(-i\omega+\frac{\gamma_\uparrow-\gamma_\downarrow}{2}\right)\alpha P(\alpha,\alpha^*,t) \tag{60}$$

$$-\frac{\partial}{\partial\alpha^*}\left(i\omega+\frac{\gamma_\uparrow-\gamma_\downarrow}{2}\right)\alpha^* P(\alpha,\alpha^*,t)+\gamma_\uparrow\frac{\partial^2}{\partial\alpha^*\partial\alpha}P(\alpha,\alpha^*,t).$$

Eq. (60) is a Fokker-Planck equation (FPE), as it was to be expected after the classical theory of Brownian motion. Let us consider the solution of the FPE (60) when the system is initially in a coherent state:

$$\rho(0)=|\alpha_0\rangle\langle\alpha_0| \quad , \quad P(\alpha,\alpha^*,0)=\delta(\alpha-\alpha_0)\,\delta(\alpha^*-\alpha_0^*), \tag{61}$$

i.e. the Green function of the FPE.

If we neglect the diffusion term (i.e. the second-order derivative term) the solution is

$$P(\alpha,\alpha^*,t)=\delta(\alpha-\tilde{\alpha}_0(t))\,\delta(\alpha^*-\tilde{\alpha}_0^*(t)), \tag{62}$$

$$\tilde{\alpha}_0(t)=\alpha_0\exp\left[-i\omega t+\frac{\gamma_\uparrow-\gamma_\downarrow}{2}t\right] , \tag{62'}$$

hence

$$\rho(t)=|\tilde{\alpha}_0(t)\rangle\langle\tilde{\alpha}_0(t)| \quad , \quad \langle A\rangle(t)=\tilde{\alpha}_0(t). \tag{63}$$

The P-function is still a δ-function, which means that the oscillator remains in a coherent state during the whole time evolution. The time evolution of the mean value -as ruled only by the drift terms(first-order derivative terms) of the FPE - is that of a classical damped harmonic oscillator. Note that no fluctuations arise during the time evolution. This is no longer true if we keep the diffusion term. In fact, in this case the solution is

$$P(\alpha,\alpha^*,t) = \left\{\pi n_{th}\left[1 - e^{-(\gamma_\downarrow - \gamma_\uparrow)t}\right]\right\}^{-1} \cdot \tag{64}$$

$$\cdot \exp\left\{-\frac{(\alpha - \tilde{\alpha}_0(t))(\alpha^* - \tilde{\alpha}_0^*(t))}{n_{th}\left[1 - \exp(-(\gamma_\downarrow - \gamma_\uparrow)t)\right]}\right\},$$

where, using the fluctuation-dissipation relation (29),

$$n_{th} = \langle A^+ A\rangle_\beta = \frac{\gamma_\uparrow}{\gamma_\downarrow - \gamma_\uparrow} = \frac{1}{e^{\beta\hbar\omega} - 1} . \tag{65}$$

One sees that for $t>0$ the P-function is no longer a δ-function, i.e. the oscillator is no longer in a coherent state. The P-function broadens in time. Initially thre are no fluctuations, but the diffusion term creates fluctuations which increase in time. On the other hand, the drift terms rule the mean motion, which is still given by (62'). As a last point, let us consider the solution (64) for $t \longrightarrow \infty$. One finds

$$\lim_{t\to\infty} P(\alpha,\alpha^*,t) = \frac{\exp\left[-|\alpha|^2/n_{th}\right]}{\pi n_{th}} , \tag{66}$$

which can be shown to be the P-function $P_\beta(\alpha,\alpha^*)$ that corresponds to the canonical operator [11]. Again one finds as expected that the system approaches the thermal equilibrium state in the long time limit.

4. The one-mode laser model

At this point we have all the elements to treat the laser systems [12,13,11,9]. Let us start by illustrating the one-mode laser model of the Stuttgart school [12,14].

In a laser system we have a resonant cavity with mirrors of reflectivity coefficient R and transmittivity T=1-R. This cavity contains N atoms homogeneously distributed in a pencil-shaped sample of volume $V=Ld^2$, where L is the length and d^2 the section. The atoms are assumed to be two-level atoms with the same transition frequency ω (homogeneously broadened atomic system). We consider only the resonant electromagnetic field mode of the cavity, and let A, A^+ be the annihilation and creation operators of this mode which obey the usual boson commutation relations

(67) $$[A, A^+] = 1.$$

Let W(t) be the statistical operator of the system atoms+field. The time evolution equation for W consists of three different groups of terms, which describe the dynamics of atoms and of the field and the interaction between the atoms and the radiation mode ,respectively. Let us write concisely

(68) $$\frac{dW}{dt} = \left(\frac{dW}{dt}\right)_A + \left(\frac{dW}{dt}\right)_F + \left(\frac{dW}{dt}\right)_{AF} ,$$

where $(dW/dt)_A$ describes the time evolution of the atoms, $(dW/dt)_F$ the time evolution of the field mode and $(dW/dt)_{AF}$ the atom-field interaction.

Let us consider the three groups in (68) separately. First of all, we have N two-level atoms labeled by i=1,2,....N. According to sec.3, the i-th atom is associated with the raising and lowering operators r_i^+, r_i^- and with the inversion operator r_{3i}. These operators obey the angular momentum commutation relations

(69) $$[r_i^+, r_j^-] = 2r_{3i}\delta_{ij} , \quad [r_{3i}, r_j^{\pm}] = \pm r_i^{\pm}\delta_{ij} .$$

As long as we do not consider the interaction with the cavity mode the atoms evolve independently of one another. Their time evolution arises from the free time evolution, the decay which can be purely radiative or due to collisions, and from the pump action that we exert on the atoms. Since these decay and pumping mechanisms can be described as due to the interaction with suitable reservoirs, we can immediately describe the dynamics of the atomic system:

$$(70)\quad \left(\frac{dW}{dt}\right)_A = -i\mathcal{L}_A W + \Lambda_A W ,$$

$$(71a)\quad \mathcal{L}_A W = \frac{1}{\hbar}\left[H_A , W\right] = \omega \sum_{i=1}^{N} \left[r_{3i} , W\right] ,$$

$$(71b)\quad \Lambda_A W = \sum_{i=1}^{N} \Big\{ \frac{\gamma_\uparrow}{2}\left(\left[r_i^+ , W r_i^-\right] + \left[r_i^+ W , r_i^-\right]\right) + \frac{\gamma_\downarrow}{2}\left(\left[r_i^- , W r_i^+\right] + \left[r_i^- W , r_i^+\right]\right) + \frac{\eta}{2}\left(\left[r_{3i} , W r_{3i}\right] + \left[r_{3i} W , r_{3i}\right]\right) \Big\} .$$

Thre are two differences with respect to eq.(40) for the decay of the single atom . The first difference is the presence of the terms containing η in (71b), which are the dephasing terms arising from the elastic collisions. Furthermore, in (71b) $\gamma_\uparrow$ and $\gamma_\downarrow$ must be considered as independent parameters because $\gamma_\downarrow$ is determined by the decay whereas $\gamma_\uparrow$ is controlled by the pump. In particular, laser operation requires $\gamma_\uparrow > \gamma_\downarrow$.

If one neglects the interaction with the cavity mode, the time evolution of the atoms is described by the ME (70). It is easy to solve it and verify that the atoms approach an asymptotic situation in which the polarization is zero while the inversion of the single atom has the value

$$(72a)\quad \sigma = \frac{\gamma_\uparrow - \gamma_\downarrow}{\gamma_\uparrow + \gamma_\downarrow}$$

In particular, to get a positive inversion one must have $\gamma_\uparrow > \gamma_\downarrow$. The polarization tends to zero with a rate

$$\text{(72b)} \qquad \gamma_\perp = \frac{\gamma_\uparrow + \gamma_\downarrow + \eta}{2}$$

whereas the inversion tends to its asymptotic value σ with a rate

$$\text{(72c)} \qquad \gamma_\parallel = \gamma_\uparrow + \gamma_\downarrow .$$

For a purely radiative decay $\eta=0$ so that $\gamma_\parallel = 2\gamma_\perp$ as in eq. (47b). For $\eta \neq 0$ one has instead $\gamma_\parallel < 2\gamma_\perp$.
Let us now consider the radiation mode. It is simply a quantum harmonic oscillator of frequency ω. It is damped because the photons escape from the cavity with a rate

$$\text{(73)} \qquad K = cT/\mathcal{L}$$

due to the interaction with the exterior of the cavity, which acts as a reservoir at zero temperature. Hence the dynamics of the radiation mode is described by the following equation (cfr.eq.(54)):

$$\text{(74)} \qquad \left(\frac{dW}{dt}\right)_F = -i\,\mathcal{L}_F W + \Lambda_F W ,$$

$$\text{(75a)} \qquad \mathcal{L}_F W = \frac{1}{\hbar}\left[H_F , W\right] = \omega\left[A^+A, W\right],$$

$$\text{(75b)} \qquad \Lambda_F W = K\left(\left[A, WA^+\right] + \left[AW, A^+\right]\right).$$

Note that with respect to eq.(54) we write k instead of $\gamma_\downarrow$ and we have $\gamma_\uparrow = 0$ because the reservoir has zero temperature. If we neglect the interactions with the atoms the time evolution of the field mode is described by the ME(74). By solving this ME (e.g. in the Glauber representation) one finds that the mode approaches the vacuum state with the rate k, as it must be.

Finally we consider the atom-field interaction. In the dipole and rotating wave approximation, the interaction between the atom and the cavity mode is given by the Hamiltonian

$$(76)\qquad H_{AF} = i\hbar\bar{g}\sum_{i=1}^{N}\left(e^{-i\vec{K}\cdot\vec{x}_i}A^+ r_i^- - e^{i\vec{K}\cdot\vec{x}_i}A r_i^+\right)$$

where $\bar{g}$ is the coupling constant

$$(77)\qquad \bar{g} = \left(\frac{2\pi\omega}{\hbar V}\right)^{1/2}\mu$$

and μ is the modulus of the dipole moment of the two-level atoms. $\vec{K}$ is the wave vector of the radiation mode and $\vec{x}_i$ the position of the i-th atom. Therefore one has the interaction term for the dynamics of W(t)

$$(78)\qquad \left(\frac{dW}{dt}\right)_{AF} = -i\mathcal{L}_{AF}W = -\frac{i}{\hbar}\left[H_{AF}, W\right].$$

In conclusion from (68),(70),(74),(78) we obtain the ME that fully describes the dynamics of the coupled system atoms+cavity mode

$$(79)\qquad \frac{dW}{dt} = \left\{-i\left(\mathcal{L}_A + \mathcal{L}_F + \mathcal{L}_{AF}\right) + \Lambda_A + \Lambda_F\right\}W(t),$$

where the terms on the r.h.s. are explicitly given in (71a,b),(75a,b) and(78). This ME is the one-mode laser model formulated by Weidlich and Haake [14]. This model has been recently generalized by Bonifacio and Lugiato [15] in order to take into account the possible presence of an external coherent field which is injected into the cavity. This generalization is necessary to treat the so-called optical bistability. In fact, let us assume that a coherent monochromatic field enters into the cavity and that this external field is perfectly resonant with the atomic system and the cavity. It can be shown that the effect of the external field can be taken into account by simply changing the term Λ_F which describes the damping of the cavity mode. Precisely one must put

$$\Lambda_F W = \kappa \left\{ \left[\left(A - \alpha_0 e^{-i\omega t} \right), W \left(A - \alpha_0 e^{-i\omega t} \right)^+ \right] + \right.$$

(80)

$$\left. + \left[\left(A - \alpha_0 e^{-i\omega t} \right) W, \left(A - \alpha_0 e^{-i\omega t} \right)^+ \right] \right\}.$$

If we put $\alpha_0 = 0$ in (80) we recover (75b). Furthermore, if we neglect the interaction with the atoms so that the dynamics of the cavity mode is described by the ME (74) and we solve it by using (80) instead of (75b), we find that the mode does no longer approach the vacuum state $|0\rangle\langle 0|$ but the coherent state $|\alpha_0\rangle\langle\alpha_0|$ of the external laser field. This explains the generalization of eq.(75b) into(80).

As a last point in the construction of the one-mode laser model, we pass to the interaction picture so that we get rid of the free time evolution terms. In this picture the ME reads

(81) $$\frac{dW^{(I)}}{dt} = \left(\Lambda_A + \Lambda_F - i \mathcal{L}_{AF} \right) W^{(I)}(t)$$

where now

$$\Lambda_F W^{(I)} = \kappa \left\{ \left[\left(A - \alpha_0 \right), W^{(I)} \left(A - \alpha_0 \right)^+ \right] + \right.$$

(82)

$$\left. + \left[\left(A - \alpha_0 \right) W^{(I)}, \left(A - \alpha_0 \right)^+ \right] \right\}.$$

In the following, we drop the index (I) everywhere.

We shall treat the model (81) first at the semiclassical level and secondly at a fully quantum statistical level. At both levels we are interested in studying the behaviour of the macroscopic observables of our system, which are essentially the field, the total inversion of the atomic system and its macroscopic polarization. The field is associated to the operators A and A^+, the total inversion is associated to the collective inversion operator

$$(83) \qquad R_3 = \sum_{i=1}^{N} r_{3i}$$

and finally the macroscopic polarization is associated to the collective dipole operators

$$(84) \qquad R^{\pm} = \sum_{i=1}^{N} r_i^{\pm} \, e^{\pm i \vec{k} \cdot \vec{x}_i} .$$

By means of the definitions (83),(84) we can rewrite

$$(85) \qquad H_A = \hbar \omega R_3 ,$$

$$(86) \qquad H_{AF} = i \hbar \bar{g} \left(A^+ R^- - A R^+ \right) .$$

A very remarkable fact is that these macroscopic operators obey the angular momentum commutation relations

$$(87) \qquad \left[R^+, R^- \right] = 2 R_3 , \quad \left[R_3 , R^{\pm} \right] = \pm R^{\pm} .$$

5.- Semiclassical treatment of laser and optical bistability

From the ME(81) one can derive the time evolution equations for the mean values of the macroscopic quantities $R^{\pm}$, R_3, A, A^+. One obtains a system which is not closed because it contains mean values of products of these quantities. However, in the semiclassical approximation in which one neglects all the fluctuations and correlations, all the mean values of products factorize into the products of mean values. Thus, by introducing the c-number variables

$$\tilde{A} = \langle A \rangle = Tr(AW) \quad , \quad \tilde{A}^+ = \langle A^+ \rangle \quad , \tag{88}$$

$$\tilde{R}^{\pm} = \langle R^{\pm} \rangle \quad , \quad \tilde{R}_3 = \langle R_3 \rangle$$

one gets the following closed system of equations(dot=derivative with respect to time)

$$\dot{\tilde{R}}^- = 2\bar{g}\tilde{A}\tilde{R}_3 - \gamma_\perp \tilde{R}^- \quad , \tag{89a}$$

$$\dot{\tilde{R}}_3 = -\bar{g}\left(\tilde{A}\tilde{R}^+ + \tilde{A}^+\tilde{R}^-\right) - \gamma_\parallel \left(R_3 - \frac{\sigma N}{2}\right) \quad , \tag{89b}$$

$$\dot{\tilde{A}} = \bar{g}\tilde{R}^- - \kappa\left(\tilde{A} - \alpha_0\right) \tag{89c}$$

The equations for $\tilde{R}^+$ and $\tilde{A}^+$ are complex conjugates of eqs. (89a) and (89c), respectively.

In each equation we recognize a coherent interaction term and an in_ coherent damping term. On the basis of eqs. (89a -c), we shall treat two different phenomena: the laser and so-called optical bistability (OB).

5.1- The laser

First we analyze eqs. (89) in the stationary situation, which is also the asymptotic situation that the system approaches for $t \longrightarrow \infty$ .From eqs. (89a,b) one obtains the following expressions of polarization and inversion as functions of the field:

$$\tilde{R}^- = \frac{\sigma N}{2}\sqrt{\frac{\gamma_\parallel}{\gamma_\perp}}\frac{x}{1+|x|^2} \quad , \quad \tilde{R}_3 = \frac{\sigma N}{2}\frac{1}{1+|x|^2} \quad , \tag{90}$$

where x is the normalized field amplitude

$$x = 2\bar{g}\tilde{A} \Big/ \sqrt{\gamma_\perp \gamma_\parallel} \quad . \tag{91}$$

If we substitute (90) into the field equation (89c), we obtain the equation that determines the stationary values of the field variable x:

(92) $$\sigma \frac{\bar{g}^2 N}{\gamma_\perp} \frac{x}{1+|x|^2} - \kappa(x-y) = 0 ,$$

where y is the normalized amplitude for the incident field

(93) $$y = \frac{2\bar{g}\alpha_0}{\sqrt{\gamma_\perp \gamma_\parallel}} .$$

Let us analyze eq. (92) separately in the cases of the laser and of OB. In the case of the laser $\alpha_0=0$, which implies $y=0$. Hence the stationary equation can be written as it follows:

(94) $$x\left(1 - \frac{\sigma}{\sigma_T} \frac{1}{1+|x|^2}\right) = 0 ,$$

where

(95) $$\sigma_T = \kappa\gamma_\perp / \bar{g}^2 N$$

is the threshold inversion per atom as we show below. Clearly eq.(94) has two solutions. The first is the trivial solution x=0, which means $\tilde{A}=0$, i.e. no radiation emission. Substituting the solution X=0 into eq.(90) we obtain $\tilde{R}=0$ and $\tilde{R}_3 = \sigma N/2$. The other solution is $|x|^2 = (\sigma/\sigma_T) - 1$. This solution exists provided $\sigma/\sigma_T \geq 1$, i.e.

(96) $$\sigma \geq \sigma_T .$$

Note that the phase is left completely arbitrary. For this solution there is radiation emission, i.e. the laser is operating. Now a stationary solution is physically meaningful only if it is stable. The stability can be checked by means of a standard linear stability analysis. Let us consider the trivial solution ($\tilde{A}=0$, $\tilde{R}=0$, $\tilde{R}_3=\sigma N/2$).

One introduces the deviations of the macroscopic quantities from their stationary values

$$(97)\quad \Delta R^- = \tilde{R}^- \,, \qquad \Delta R_3 = \tilde{R}_3 - \frac{\sigma N}{2} \,, \qquad \Delta A = \tilde{A} \,.$$

Substituting (97) into (89) and keeping only the terms which are linear in the deviations, we get the linearized system

$$(98a)\quad \Delta\dot{R}^- = \sigma N \bar{g}\, \Delta A - \gamma_\perp \Delta R^- \,,$$

$$(98b)\quad \Delta\dot{R}_3 = -\gamma_\parallel \Delta R_3$$

$$(98c)\quad \Delta\dot{A} = \bar{g}\, \Delta R^- - \kappa \Delta A \,.$$

The system (98) admits solution of exponential form

$$(99)\quad \begin{Bmatrix} \Delta R^-(t) \\ \Delta R_3(t) \\ \Delta A(t) \end{Bmatrix} = e^{\lambda t} \begin{Bmatrix} \Delta R^-(0) \\ \Delta R_3(0) \\ \Delta A(0) \end{Bmatrix} \,.$$

If we substitute ansatz (99) into the system (98), we obtain a linear algebraic homogeneous system, which has nontrivial solutions if the determinant of the matrix $\mathbb{M}$ of the coefficients vanishes, where

$$(100)\quad \mathbb{M} = \begin{vmatrix} \lambda + \gamma_\perp & 0 & -\sigma N \bar{g} \\ 0 & \lambda + \gamma_\parallel & 0 \\ -\bar{g} & 0 & \lambda + \kappa \end{vmatrix}$$

From the condition $\det \mathbb{M} = 0$ it follows an algebraic equation for λ – the characteristic equation– which determines the possible values of the complex constant λ. According to (99), the stationary solution is stable if and only if the real part of λ is negative for all roots of the characteristic equation. Now the solutions of the characteristic equations are

(101a) $$\lambda_1 = -\gamma_{\parallel}$$

(101b) $$\lambda_{\substack{2\\3}} = \mp \left[(\kappa+\gamma_\perp)^2 + 4\kappa\gamma_\perp \left(\frac{\sigma}{\sigma_T} - 1\right) \right]^{1/2} - (\kappa+\gamma_\perp).$$

λ_1 and λ_2 are always negative, whereas $\lambda_3 > 0$ if $\sigma > \sigma_T$. Therefore the trivial solution is stable only if $\sigma < \sigma_T$. Quite similarly, one can analyse the stability of the nontrivial solution. One finds that it can become unstable only for very high pumping, and in this case, according to the values of the parameters, one finds the so-called Lorentz instability [16] or a self-pulsing behaviour [17]. In conclusion, if we increase the pump parameter σ, we have first the trivial solution which is stable, until in correspondence to the threshold $\sigma = \sigma_T$ it becomes unstable and the laser begins to operate. Here one finds a discontinuity in the derivative of $|x|^2$ vs. σ, so that the behaviour of the laser in the threshold region closely resembles a 2nd-order phase transition [18].

5.2- Optical bistability

In thiscase the optical cavity is filled with purely absorbing material, i.e. we do not pump the atoms ($\gamma_\uparrow = 0$). Hence the pump parameter σ (eq.(72)) must be put equal to -1. The problem of OB is the following: a coherent monochromatic field α_0 enters into the cavity which is filled with absorbing resonant atoms. Part of this light is transmitted by the cavity, and we want to find the behaviour of the transmitted field as a function of the incident field. The stationary equation (92) links the transmitted field A and the incident field α_0, since $x \propto A$ and $y \propto \alpha_0$. For definiteness, we take α_0 real and positive, so that also y is real and positive. It follows that also the solutions x of eq.(92) are real so that we drop the modulus symbol. Introducing the parameter

(104) $$C = \frac{\bar{g}^2 N}{2\kappa \gamma_\perp} ,$$

we have the state equation [15]

(105) $$y = x + \frac{2Cx}{1+x^2}$$

Eq.(105) expresses the incident field as a function of the transmitted field. Actually we want just the inverse function. Let us first analyse the function y(x) defined by the state equation. We have a linear term and a nonlinear term which arises from atomic cooperation and in fact is proportional to the number of atoms. In the case of an empty cavity this term vanishes, so that eq.(105) reduces to y=x, i.e. transmitted field=incident field as it is well known. On the other hand for a cavity filled with absorbing atoms just the nonlinearity of eq.(105) introduces all the interesting features. Let us consider this function for large and for small x. In the first case, eq.(105) is approximated by y=x, i.e. the empty cavity solution. The atoms are completely saturated so that the medium is completely transparent. In these conditions, each atom interacts with the incident field as if the other atoms would not exist: this is the noncooperative situation, and in fact in this case one can prove that there are no correlations between atom and atom. For small x, eq.(105) is approximated by y=(2C+1)x so that we obtain a linear relation. But now the linearity is not due to the lack of atomic cooperation but to the fact that we are considering a system driven by a weak external field so that the response is linear. In this situation for $C \gg 1$ the atomic cooperation is dominant, and in fact one can prove that one has relevant atom-atom correlations.

The form of the curve y(x) between the two linear asymptotic behaviours is qualitatively different in the two cases $C<4$, $C>4$, as shown in fig. 1. For $C<4$ y is a monotonic function of x so that there is no bistability. However, also in this case there is a very interesting phenomenon. In fact, if one plots x vs.y as in fig. 2, one finds a portion of the curve where $dx/dy > 1$, so that a slow modulation of the incident field gets amplified. In these conditions, the system behaves as an optical transistor [19]. For C=4 the curve has an inflection point with horizontal tangent and finally for $C > 4$ the function y(x) has a maximum and a minimum. Therefore in the latter case there is a suitable range of values of y,i.e. of the incident field, in correspondence of which we find three different values of x,i.e. of the transmitted field. The points which lie on the part with negative slope are unstable. In fact, these curves are analogous to the Van der Waals curves for the liquid-vapor phase transition, and this part is analogous to the portion of the Van der Waals curve with negative compressibility. Here a decrease of the incident field would imply an increase of the trnsmitted field, which is impossible. Hence in the case $C > 4$ we have a bistable situation, with a solution x_1 in which the atomic cooperation is important and a solution x_3 in which atomic cooperation is negligible. Therefore we shall call x_1 "cooperative stationary state" and x_3 "one-atom stationary state". If we exchange the axes to have a plot of transmitted light as a function of incident light (Fig.3) we obtain immediately a hysteresis cycle. In fact, if we start from low values of the incident field, we see that the transmission is very low. Nearly all the light is reflected. Increasing the incident field the transmitted field increases very slowly until at a certain point the incident light increases abruptly and nearly all the incident light is trasmitted. Coming back, the transmitted field decreases continuously also when we cross the previous upper threshold, until we reach another lower threshold, where the transmitted light suddenly jumps to the low transmission branch.

The presence of hysteresis effects is typical of first-order phase transitions in equilibrium systems. In fact, OB is prototype of first order phase transitions in quantum optics, exactly as the laser is the prototype of 2nd-order phase transitions. This analogy between OB and first-order phase transitions will be further developed when we shall treat the photon statistics of the transmitted light.

5.3- Transient behaviour: the adiabatic elimination

Until now, we have treated the semiclassical equations (89a-c) at steady state. The treatment of the transient, i.e. of the approach to the steady state, is much more complicated. In fact, the solution of the nonlinear system (89) can be found only numerically. However, it must be mentioned that the semiclassical equations can be simplified in some limit situations which are commonly found in quantum optics, namely when the damping rates which appear in these equations are such that either $k \ll \gamma_\perp, \gamma_\parallel$ or conversely $k \gg \gamma_\perp, \gamma_\parallel$. The first situation is typical of the laser, and in general of all situations in which the quality of the cavity is good. In fact, k is proportional to the transmission coefficient T, so that if the mirrors have a good reflectivity k can be made much smaller than $\gamma_\perp$ and $\gamma_\parallel$. Therefore we shall call this case "good quality cavity case". The opposite situation is typical of superfluorescence [20] in which the cavity has no mirror at all. We shall call this "bad cavity case".

In the latter case ($k \gg \gamma_\perp, \gamma_\parallel$) the atomic variables vary much more slowly than the field variables. By integrating the field equation (89c) we obtain the expression of the field as a function of the atomic variables

$$(106) \quad \tilde{A}(t) = e^{-kt}\tilde{A}(0) + \int_0^t d\tau\, e^{-k(t-\tau)}\left[\bar{g}\tilde{R}^-(\tau) + k\alpha_0\right].$$

Now let us consider times $t \gg k^{-1}$, where k^{-1} characterizes the field time scale, so that the first term can be dropped, and note that in the integral the term $\tilde{R}^-(\tau)$ varies on a different time scale, namely $\gamma_\perp^{-1}, \gamma_\parallel^{-1}$. This is just the same type of situation that we have discussed in detail in illustrating the markoff approximation (see the passage from eq.(22) to eq.(25). Hence performing the markoff approximation on eq.(106) we obtain the result

$$(107) \quad \tilde{A}(t) = \frac{\bar{g}}{k}\tilde{R}^-(t) + \alpha_0 .$$

Now we can replace this expression into eqs.(89a,b), obtaining the reduced system of two differential equations for the atomic variables only

$$\dot{\tilde{R}}^- = 2\bar{g}\alpha_0 \tilde{R}_3 + \frac{2\bar{g}^2}{K} \tilde{R}^- \tilde{R}_3 - \gamma_\perp \tilde{R}^- ,$$

(108) $$\dot{\tilde{R}}_3 = -\bar{g}\alpha_0 (\tilde{R}^+ + \tilde{R}^-) - \frac{2\bar{g}^2}{K} \tilde{R}^+ \tilde{R}^- - \gamma_{\parallel} \left(\tilde{R}_3 - \frac{\sigma N}{2}\right).$$

From eq. (107) we see that the field variables follow adiabatically, i.e. without retardation, the motion of the atomic variables. Hence the approximation (107) is called adiabatic elimination of the field variable. As we have seen it coincides with the markoff approximation. Note that there is a quicker procedure to perform the adiabatic elimination. In fact, if we consider the field equation, i.e. the equation for the "fast" variable, and put $\dot{\hat{A}}(t)=0$ we obtain eq.(107) directly. Similarly in the good cavity case $K \ll \gamma_\perp, \gamma_\parallel$ we can adiabatically eliminate the atomic variables. Thus we put $\dot{\tilde{R}}^- = \dot{\tilde{R}}_3 = 0$ in the atomic equations (89a,b) and substitute the stationary expressions of $\tilde{R}^-$ and R_3 into the field eq.(89c), obtaining the differential equation for the field variable only

(109) $$\dot{x} = -K\left(x - y - \sigma \frac{\bar{g}^2 N}{K\gamma_\perp} \frac{x}{1+x^2}\right).$$

6.- Quantum statistical treatment of laser and optical bistability

In this section we treat the one-mode laser model on a fully quantum statistical level, i.e. considering fluctuations and correlations. The ME (81) cannot be solved exactly even at steady state. However, the situation is considerably simplified in the two limit cases of good cavity and bad cavity. Let us consider the good cavity case (see [21a-c] for the laser; [22] for OB)

(110) $$K \ll \gamma_\perp, \gamma_\parallel .$$

In this situation, we can consider the atomic system as a reservoir for the radiation field. In fact, the most important feature is to have fast relaxation rates with respect to the system. Therefore we can apply the method illustrated in sec.2 to derive from the one-mode laser model (81) a time evolution equation for the reduced statistical operator of the field only

$$\rho = \mathrm{Tr}_A W , \tag{111}$$

where Tr_A means partial trace over the atomic Hilbert space. We introduce the operators

$$\tilde{r}_j^{\pm} = r_j^{\pm} \exp\left[\pm i \vec{k}\cdot\vec{x}_j\right] \tag{112}$$

in order to write the interaction Hamiltonian (76) in the simpler form

$$H_{AF} = i\hbar \bar{g} \sum_{j=1}^{N} \left(A^{+}\tilde{r}_j^{-} - A\tilde{r}_j^{+}\right). \tag{113}$$

These operators satisfy the same commutation relations(69) as the old operators $r_j^{\pm}$, and the expression of the operator Λ_A in terms of $\tilde{r}_j^{\pm}$ is the same as before as one sees by inspection of eq.(71b). We recall that the method described in sec. 2 consists of deriving from the starting equation a hierarchy of equations for the reduced statistical operator ρ of the subsystem and for some atom-field correlations, the simplest of which are $W_i^{(\pm)}(t) = \mathrm{Tr}_A(\tilde{r}_i^{\pm} W(t))$. In view of the applications of sec.3, we truncated the hierarchy using a procedure which was based on the assumption that the reservoir was not appreciably disturbed by the interaction with the subsystem. However, in the case of the laser the reservoir, i.e. the atomic system, is notably disturbed by the strong interaction with the field. Therefore this truncation procedure does not work for the laser. Hence we adopt another truncation method, which is the following. In the interaction between the field and the atoms, we consider only those processes in which the field interacts with one atom at a time and neglect those higher-order processes in which the field interacts with two or more atoms simultaneously. This assumption amounts to consider, in our hierarchy, only the auxiliary quantities which involve a single atom. More precisely, we consider only the operators ρ, $W_i^{(\pm)}$, $W_i^{(+-)} = Tr_A(\tilde{r}_i^+ \tilde{r}_i^- W)$, $W_i^{(-+)} = Tr_A(\tilde{r}_i^- \tilde{r}_i^+ W)$. On the other hand we neglect quantities such as $W_{ij}^{(+-)} = Tr_A(\tilde{r}_i^+ \tilde{r}_j^- W)$, $i \neq j$, which contain the variables of two different atoms. Keeping into account the relations $W_i^{(-)} = [W_i^{(+)}]^+$ and $W_i^{(-+)} = \rho - W_i^{(+-)}$, we derive the following closed system of equations for ρ, $W_i^{(+)}$ and $W_i^{(+-)}$:

$$\frac{d\rho}{dt} = \Lambda_F \rho + \bar{g} \sum_{i=1}^{N} \left([A^+, W_i^{(-)}] - [A, W_i^{(+)}] \right), \tag{114a}$$

$$\frac{dW_i^{(+)}}{dt} = (\Lambda_F - \gamma_\perp) W_i^{(+)} + \bar{g} \left\{ A^+ W_i^{(+-)} - (\rho - W_i^{(+-)}) A^+ \right\}, \tag{114b}$$

$$\frac{dW_i^{(+-)}}{dt} = (\Lambda_F - \gamma_\parallel) W_i^{(+-)} + \gamma_\parallel \frac{1+\sigma}{2} \rho - \bar{g} \left(A W_i^{(+)} - W_i^{(-)} A^+ \right). \tag{114c}$$

The physical meaning of the three operator equations (114) is transparent because there is a straightforward connection with the three semiclassical equations(89), expressed by the relations

$$\langle A \rangle = Tr(A\rho), \quad \langle \tilde{r}_i^\pm \rangle = Tr\, W_i^{(\pm)}, \quad \langle \tilde{r}_i^+ \tilde{r}_i^- \rangle = Tr\, W_i^{(+-)}. \tag{115}$$

to obtain a closed equation for ρ, by virtue of (110) we perform the adiabatic approximation(which as we have seen in the previous section, is equivalentto the Markoff approximation) by setting $dW_i^{(+)}/dt = dW_i^{(+-)}/dt = 0$. Furthermore, since Λ_F is proportional to k we neglect $\Lambda_F W^{(+)}$ with respect to $\gamma_\perp W^{(+)}$ in eq. (114b) and $\Lambda_F W^{(+-)}$

with respect to $\gamma_{\parallel} W^{(+-)}$ and $(\gamma_{\parallel}/2)(1+\sigma)\rho$ in eq.(114c). Hence the system(114) takes the form

$$\text{(116a)} \quad \frac{d\rho}{dt} = \Lambda_F \rho + \bar{g} N \left(\left[A^+, W^{(-)} \right] - \left[A, W^{(+)} \right] \right) ,$$

$$\text{(116b)} \quad 0 = -\gamma_{\perp} W^{(+)} + \bar{g} \left\{ A^+ W^{(+-)} - \left(\rho - W^{(+-)} \right) A^+ \right\} ,$$

$$\text{(116c)} \quad 0 = -\gamma_{\parallel} W^{(+-)} + \gamma_{\parallel} \frac{1+\sigma}{2} \rho - \bar{g} \left(A W^{(+)} + W^{(-)} A^+ \right) ,$$

where we have taken into account that the atoms are identical, so that we have dropped the atomic index i everywhere and replaced $\sum_{i=1}^{N}$ by N.

If we substitute for $W^{(+)}$ in eqs.(116a,c) the expression obtained from eq.(116b), we are left with the following system for ρ and $W^{(+-)}$:

$$\text{(117a)} \quad \frac{d\rho}{dt} = \Lambda_F \rho + \frac{\bar{g}^2 N}{\gamma_{\perp}} \left\{ \left[A^+, W^{(+-)} A \right] + \left[A^+ W^{(+-)}, A \right] + \left[A, \left(\rho - W^{(+-)} \right) A^+ \right] + \left[A \left(\rho - W^{(+-)} \right), A^+ \right] \right\} ,$$

$$\text{(117b)} \quad 0 = -\gamma_{\parallel} W^{(+-)} + \gamma_{\parallel} \frac{1+\sigma}{2} \rho + \frac{\bar{g}^2}{\gamma_{\parallel}} \left\{ 2 A \left(\rho - W^{(+-)} \right) A^+ - A A^+ W^{(+-)} - W^{(+-)} A A^+ \right\} .$$

To eliminate $W^{(+-)}$, it is suitable to translate the operator equations (117) into partial differential equations by using the Glauber diagonal representation. Therefore we associate to the reduced statistical operator ρ the Glauber P-distribution(58), and likewise we associate to $W^{(+-)}$ a function $\tilde{W}$:

$$\text{(118)} \quad W^{(+-)} = \int d_2\alpha \, |\alpha\rangle\langle\alpha| \, \tilde{W}(\alpha, \alpha^*) .$$

With the change of variables

$$\alpha = r e^{i\varphi} \tag{119}$$

eqs.(117) are mapped into the following c-number equations:

$$\frac{\partial P(r,\varphi,t)}{\partial t} = \frac{\kappa}{r}\left\{\frac{\partial}{\partial r} r(r-\alpha_0\cos\varphi) + \frac{\partial}{\partial\varphi}\alpha_0\sin\varphi\right\}P(r,\varphi,t) \tag{120a}$$

$$+\frac{\bar{g}^2 N}{\gamma_\perp r}\left\{\frac{\partial}{\partial r} r^2 P(r,\varphi,t) - 2\left(\frac{\partial}{\partial r} r^2 - \frac{1}{4}\frac{\partial}{\partial r} r \frac{\partial}{\partial r} - \frac{1}{4r}\frac{\partial^2}{\partial\varphi^2}\right)\tilde{W}\right\},$$

$$0 = -\gamma_\parallel \tilde{W}(r,\varphi,t) + \gamma_\parallel \frac{1+\sigma}{2} P(r,\varphi,t) \tag{120b}$$

$$+\frac{\bar{g}^2}{\gamma_\perp}\left\{2r^2 P(r,\varphi,t) - 4r^2\tilde{W}(r,\varphi,t) + r\frac{\partial}{\partial r}\tilde{W}(r,\varphi,t)\right\}.$$

Fromeq.(120b) we obtain

$$\tilde{W}(r,\varphi,t) = \frac{1}{2}\left\{1 + \frac{r^2}{N_s} - \frac{1}{4N_s} r \frac{\partial}{\partial r}\right\}^{-1}\left(1+\sigma+\frac{r^2}{N_s}\right)P(r,\varphi,t) \tag{121}$$

where

$$N_s = \frac{\gamma_\perp\gamma_\parallel}{4\bar{g}^2} \tag{122}$$

is the saturation photon number ($N_s \gg 1$).
Finally by substituting (121) into(120a) we obtain the required equa_ tion for the P-function of the laser mode:

$$\frac{\partial P(r,\varphi,t)}{\partial t} = \frac{\kappa}{r}\left\{\frac{\partial}{\partial r} r\left(r-\alpha_0\cos\varphi + \frac{\bar{g}^2 N}{\kappa\gamma_\perp} r\right) + \frac{\partial}{\partial\varphi}\alpha_0\sin\varphi\right. \tag{123}$$

$$\left. - \frac{\bar{g}^2 N}{\kappa\gamma_\perp}\cdot\left(\frac{\partial}{\partial r} r^2 - \frac{1}{4}\frac{\partial}{\partial r} r\frac{\partial}{\partial r} - \frac{1}{4r}\frac{\partial^2}{\partial\varphi^2}\right)\left[1 + \frac{r^2}{N_s} - \frac{1}{4N_s} r \frac{\partial}{\partial r}\right]^{-1}\right.\cdot$$

$$\cdot\left(1+\sigma+\frac{r^2}{N_s}\right)\Big\}\,P(r,\varphi,t)\,.$$

Note that eq.(123) gives a <u>strong</u> <u>coupling</u> treatment of the laser. In fact, since N_s is inversely proportional to $\bar{g}^2$, it contains the coupling constant at all orders, contrary to what occurs in the master equation(27), which describes the system-reservoir interaction at second order in the coupling constant. A weak coupling treatment is unable to describe the laser above threshold and OB.

Equation (123) is very complicated owing to the inverse operator. In fact, if we expand this operator into a geometric series as it follows:

$$\left[1+\frac{r^2}{N_s}-\frac{1}{4N_s}\,r\frac{\partial}{\partial r}\right]^{-1}=\sum_{n=0}^{\infty}\left(\frac{1}{4N_s}\,\frac{r}{1+\frac{r^2}{N_s}}\,\frac{\partial}{\partial r}\right)^{n}\cdot\,\frac{1}{1+r^2/N_s}\,, \qquad (124)$$

we realize that eq.(123) contains derivatives of all orders in the variable r. The exact stationary solution of eq.(124) can be however exactly calculated, as it is shown in [21b]. However this equation can be simplified. In fact, since N_s is a very large number, we can neglect all the terms of the expansion(124) except the first two. Introducing the scaled variable

$$x = r/\sqrt{N_s}\,, \qquad (125)$$

eq.(123) is very well approximated by the Fokker-Planck equation

$$\frac{\partial P(x,\varphi,t)}{\partial t}=\frac{\kappa}{x}\Big\{\frac{\partial}{\partial x}\,x\left(x-y\cos\varphi-\sigma\,\frac{\bar{g}^2 N}{\kappa\gamma_\perp}\,\frac{x}{1+x^2}\right)+\frac{\partial}{\partial\varphi}\,y\sin\varphi$$
$$+\,q\frac{\partial}{\partial x}\,x\,\frac{\partial}{\partial x}\,\frac{1+\sigma+x^2}{(1+x^2)^2}+\frac{q}{x}\,\frac{\partial^2}{\partial\varphi^2}\,\frac{1+\sigma+x^2}{1+x^2}\Big\}\,P(r,\varphi,t) \qquad (126)$$

where

$$q = \frac{\bar{g}^2 N}{4 K \gamma_\perp N_s} \tag{127}$$

is the diffusion constant($q \ll 1$)

6.1- Photon statistics of the laser

In the case of the laser, one has y=0(no external field) and by recalling (95) theFPE(126) takes the form

$$\frac{\partial P(x,\varphi,t)}{\partial t} = \frac{K}{x}\frac{\partial}{\partial x} x\left\{\left(1-\frac{\sigma}{\sigma_T}\frac{1}{1+x^2}\right)x + q\frac{\partial}{\partial x}\frac{1+\sigma+x^2}{(1+x^2)^2}\right\}P(x,\varphi,t) + Kq\frac{1+\sigma+x^2}{x^2(1+x^2)}\frac{\partial^2}{\partial\varphi^2}P(x,\varphi,t). \tag{128}$$

The drift term is immediately linked to the semiclassical stationary equation(94). On the other hand we note that both the amplitude and the phase diffusion coefficients tend to zero for $x\to\infty$ i.e. for high intensity. This feature has two basic physical consequences for the laser: 1) the extreme smallness of amplitude fluctuations, which implies that the photon statistics in the laser correspond pratically to a Poisson distribution and 2) the very slow diffusion of the phase,which implies that the laser light is nearly monochromatic. These two properties give the "coherence" of the laser light [10].
In the threshold region, i.e. for $\sigma \simeq \sigma_T$, one has that $x^2 \ll 1$, so that one neglect x^2 with respect to unity in the diffusion terms and perform the socalled "cubic" approximation in the drift term

$$x - \frac{\sigma}{\sigma_T}\frac{x}{1+x^2} \simeq x - \frac{\sigma}{\sigma_T}x\left(1-x^2\right),$$

recovering the well-known Risken equation [23], which has been the first FPE considered for the laser

$$(129)\quad \frac{\partial P(x,\varphi,t)}{\partial t} = \frac{\kappa}{x}\frac{\partial}{\partial x} x \left\{\left(1-\frac{\sigma}{\sigma_T}+\frac{\sigma}{\sigma_T}x^2\right)x + q(1+\sigma)\frac{\partial}{\partial x}\right\} \cdot P(x,\varphi,t) + \frac{\kappa q}{x^2}\frac{\partial^2}{\partial \varphi^2} P(x,\varphi,t).$$

To obtain the photon statistics of the laser at steady state, we calculate the stationary solution of the FPE. In this case $\partial P/\partial t = \partial P/\partial \varphi = 0$. The result can be written in the form

$$(130)\quad P_{st}(x) = \mathcal{N} \exp\left\{-V(x)/q\right\},$$

where $\mathcal{N}$ is a normalization constant and

$$(131)\quad V(x) = \int dx \, \frac{(1+x^2)^2}{1+\sigma+x^2}\left(1-\frac{\sigma}{\sigma_T}\frac{1}{1+x^2}\right)x - q \ln \frac{(1+x^2)^2}{1+\sigma+x^2}$$

$$\overset{q\ll 1}{\simeq} \int dx \, \frac{(1+x^2)^2}{1+\sigma+x^2}\left(1-\frac{\sigma}{\sigma_T}\frac{1}{1+x^2}\right)x$$

$$= \frac{x^4}{4} + \frac{x^2}{2}\left(1-\sigma-\frac{\sigma}{\sigma_T}\right) + \frac{\sigma^2}{2}\left(1+\frac{1}{\sigma_T}\right)\ln\frac{1+\sigma+x^2}{1+\sigma}.$$

The function V(x) plays the role of a generalized free energy for this system far from thermal equilibrium. The maxima of the distribution function coincide with the minima of the free energy. In turn, the minima of V(x) coincide with the semiclassical stationary solutions as one sees from (131).

Both P_{st} and V are functions of x^2 for a given value of the pump parameter σ. In the case $\sigma<\sigma_T$ i.e. laser below threshold, the free energy has one minimum at the origin, and correspondingly the P-function has the exponential behaviour which is typical of blackbody distribution (state 1 in fig.4). Above threshold, i.e. $\sigma > \sigma_T$, the free energy has one minimum in correspondence to the semiclassical value of the intensity. Accordingly the P-function has a Gaussian-like shape (state 2 in fig. 4). Well above threshold, the peak is so narrow that it pratically coincides with a δ - function. Hence such a P-function corresponds to a coherent state, i.e. to a Poisson photon distribution [10] . On the contrary, the P-function below threshold corresponds to a Bose-Einstein photon distribution.

6.2- Photon statistics of transmitted light in optical bistability

In this case $\sigma = -1$, and using definition (104) the FPE (126) becomes

$$\frac{\partial P(x,\varphi,t)}{\partial t} = \frac{K}{x}\frac{\partial}{\partial x}x\left\{x - y\cos\varphi + \frac{2Cx}{1+x^2} + q\frac{\partial}{\partial x}\frac{x^2}{(1+x^2)^2}\right\}\cdot \tag{132}$$

$$\cdot P(x,\varphi,t) + \frac{K}{x}\frac{\partial}{\partial\varphi}\left\{y\sin\varphi + q\frac{\partial}{\partial\varphi}\frac{x}{1+x^2}\right\}P(x,\varphi,t).$$

The amplitude drift term is clearly linked to the state equation for OB (105). Contrary to what occurs in the case of the laser, in the case of OB the stationary P-function depends on the phase. In fact, the incident field has a well defined phase that we have chosen equal to zero for definiteness. The phase of the transmitted field is equal to the phase of the incident field, with small fluctuations around this value. We cannot find the stationary solution of the FPE (132) exactly. However, the amplitude distribution can be calculated with excellent approximation as it follows. One can show that the phase fluctuations have a negligible influence on the amplitude distribution. Thus we drop the phase diffusion term. This amounts to assume that the phase of the transmitted field is strictly locked to the phase of the incident field, which is zero. Accordingly we put $\varphi = 0$ in the remaining terms of eq.(132), obtaining the following FPE at steady state for the amplitude distribution

$$(133)\quad 0 = \frac{\kappa}{x}\frac{d}{dx}\, x\left\{ x - y + \frac{2Cx}{1+x^2} + q\,\frac{x^2}{(1+x^2)^2} \right\} P_{st}(x).$$

The solution is

$$(134)\qquad P_{st}(x) = \mathcal{N}\exp\left\{ -V_y(x)/q \right\},$$

where

$$(135)\qquad \begin{aligned} V_y(x) &= \int dx \left(\frac{1+x^2}{x}\right)^2 \left(x - y + \frac{2Cx}{1+x^2}\right) - 2q \ln\frac{1+x^2}{x} \\ &\overset{q\ll 1}{\simeq} \int dx \left(\frac{1+x^2}{x}\right)^2 \left(x - y + \frac{2Cx}{1+x^2}\right) \\ &= \frac{x^4}{4} + x^2(C+1) + (2C+1)\ln x - 2xy - \frac{1}{3}yx^3 + \frac{y}{x}. \end{aligned}$$

Again the maxima of the stationary distribution P_{st} coincide with the minima of V_y and in turn the maxima and the minima of V_y coincide with the stationary semiclassical solutions. Fig. 5 shows the shapes of the free energy and of the distribution function for different values of the incident field y. For the values of y for which one has only one semiclassical solution, the potential has a single minimum and correspondingly the probability distribution has a single peak. On the other hand for values of y for which one has three stationary solutions, the potential has two minima corresponding to the stable solutions and one maximum corresponding to the unstable solution. Accordingly the distribution has two peaks. Crossing the lower bistability threshold , the distribution develops a second peak until at the upper bistability threshold, the first peak disappears and we get again a one-peaked distribution. In Fig. 6 the semiclassical stationary solution is compared with the mean value of the field. The mean value coincides with the semiclassical solution everywhere except where it jumps from the low to the high transmission branch of the semiclassical solution(transition region). This behaviour clearly resembles the first-order phase transitions in equilibrium systems. Figure 6 gives us essentially a generalized Maxwell rule. However this rule does not coincide with that of equilibrium thermodynamics, which prescribes to cut the semiclassical curve in such a way that one obtains two regions of equal area. This difference arises from the fact that the amplitude diffusion term of the FPE depends on the field amplitude. If this diffusion term were a constant, we would obtain just the usual Maxwell rule [22].

To end the discussion of the photon statistics of the transmitted light, let us consider the fluctuations of the transmitted field. Fig.7 shows that these fluctuations are very small everywhere except in the narrow transition region, where they are quite large. This feature arises from the fact that in the transition region the probability distribution has two peaks of comparable area, and the competition between the two peaks creates the anomalous fluctuations.

References

[1] H.Haken(1977)-"Synergetics",Springer-Verlag,Berlin.
G.Nicolis and I.Prigogine(1977)-"Self-organization in nonequilibrium systems". Wiley,NewYork.

[2] L.A.Lugiato(1975)-Physica 81A,565.

[3] P.Caldirola and L.A.Lugiato(1977)-Memoria negli Atti della Accademia Nazionale dei Lincei, Vol.XIV, Sez. II, 227 .

[4] R.Zwanzig(1960)-Lectures in theoretical Physics (Boulder) 3,106 .

[5] P.N.Argyres and P.L.Kelley(1964)-Phys.Rev. A 134,98 .

[6] F.Haake(1973)-Springer Tracts in Modern Physics n.66, Springer-Verlag, Berlin.
G.S.Agarwal(1974)-Springer Tracts in Modern Physics n.70, Springer-Verlag, Berlin.

[7] L.Van Hove(1955)-Physica 21,517 ;(1957)-Physica 23,441 .
I.Prigogine and P.Resibois(1971)-Physica 27,629 .

[8] R.Kubo(1957)-J.Phys.Japen 12,570 ;(1965)-in Statistical Mechanics of

Equilibrium and nonequilibrium , S.Meixner ed., North-Holland,Amsterdam and references quoted therein.
[9] L.Allen and J.H.Eberly(1975)-Optical resonance and two-level atoms, Wiley, New York.
[10] R.J.Glauber(1963)-Phys.Rev.130, 2529 ; 131,2766 .
[11] W.H.Louisell(1973)- Quantum statistical properties of radiation , Wiley,New York.
[12] H.Haken(1970)- Handbuch der Physik Vol.XXV/2c , Springer-Verlag,Berlin
[13] M.Sargent III, M.O.Scully and W.E.Lamb Jr.(1974)- Laser Physics, Addison-Wesley,Reading,Mass.
[14] W. Weidlich and F. Haake(1965)- Z.Phys. 185,30.
[15] R.Bonifacio and L.A.Lugiato(1976)- Opt.Comm. 19,172;(1978)- Phys.Rev. A18,1129.
[16] H.Haken(1975)- Phys.Lett. 53 A,77.
[17] H.Risken and K. Nummedal(1968)- Jour.Appl.Phys. 39,4662.
R.Graham and H.Haken (1968)- Z.Phys. 213,420.
[18] R.Graham and H.Haken(1970)- Z.Phys. 237,31.
V.Degiorgio and M.O.Scully(1970)- Phys.Rev. A2,1170.
[19] H.M.Gibbs,S.L.Mc Call and T.N.C. Venkatesan(1976)- Phys.Rev.Lett. 36,1135.
[20] R.Bonifacio,P.Schendimann and F.Haake(1971)- Phys.Rev. A4,302.
R.Bonifacio and L.A.Lugiato(1975)- Phys.Rev. A11,1507;A12,587.
[21] a) L.A.Lugiato(1976)- Physica 82 A,1.
b)F.Casagrande and L.A.Lugiato(1976)- Phys.Rev. A14,778;(1977)- Phys.Rev. A15,429; (1978)- Nuovo Cimento 48B,287.
c) F.Casagrande and R.Cordoni(1978)- Phys.Rev. AI8,1628.
[22] R.Bonifacio, M.Gronchi and L.A.Lugiato(1978)- Phys.Rev. A 18,2266.
[23] H.Risken(1965)-Z.Phys.186,85.

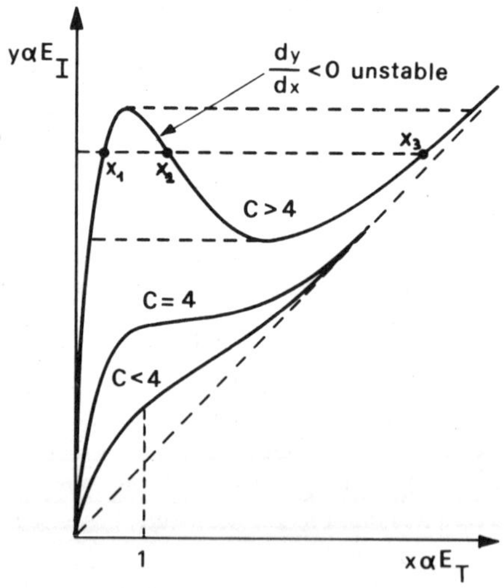

Fig.1) Plot of the function y(x) as given by the state equation (105) for different values of the parameter C . The curve for C=4 is the critical"isotherm".

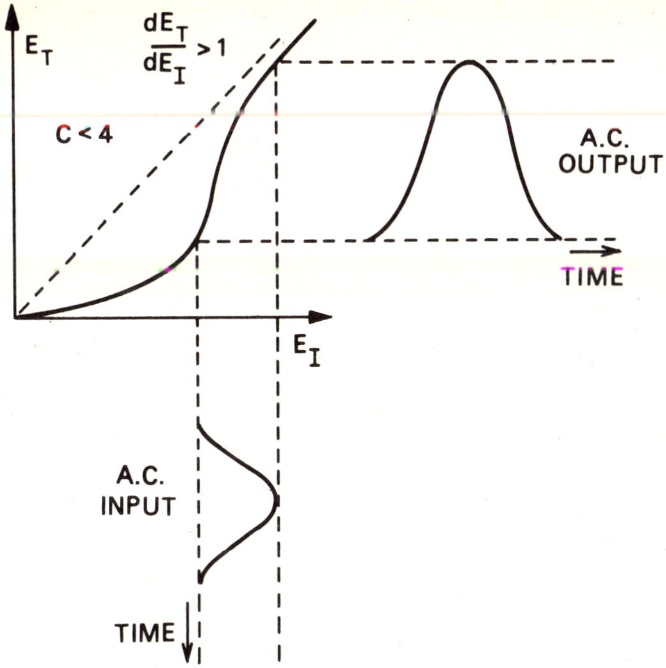

Fig.2) Explaining how the system works as an optical transistor: an input signal comes out amplified if $dE_T / dE_I \ll 1$.

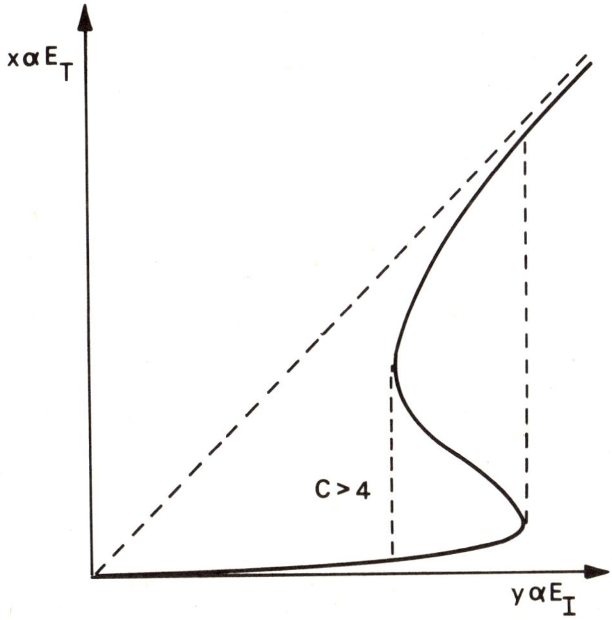

Fig.3) Hysteresis cycle of transmitted vs. incident light .

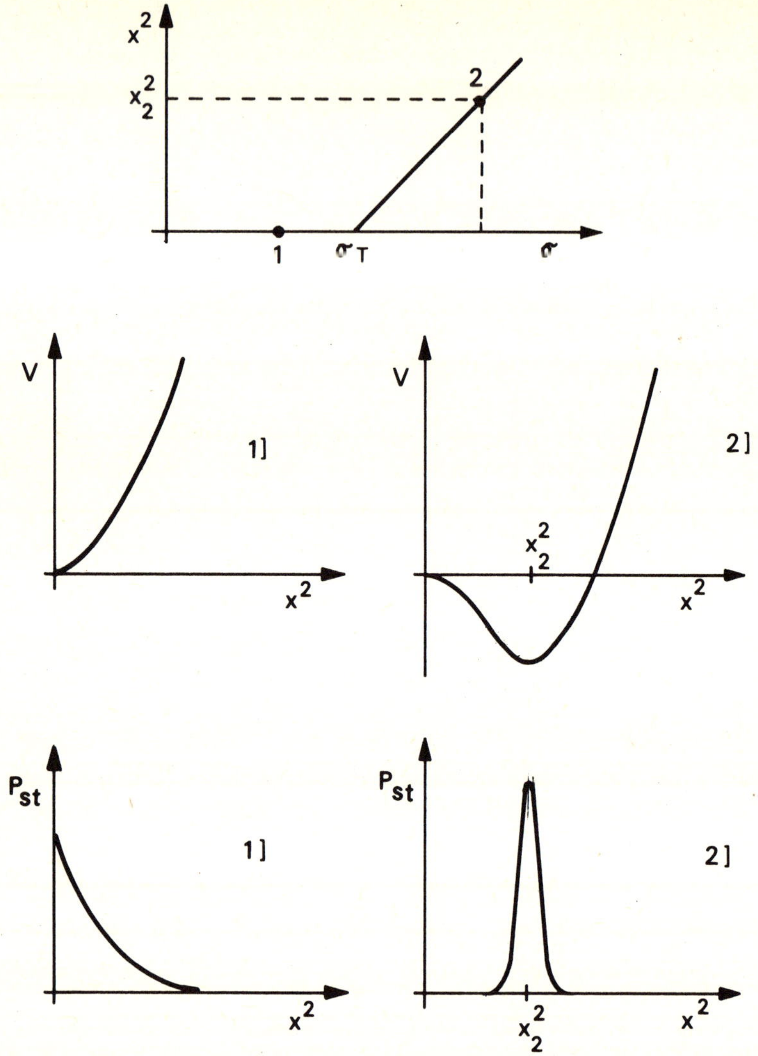

Fig.4) The generalized free energy V and the stationary distribution function P_{st} vs. normalized laser photon number x^2 (eqs. (130), (131)) are shown in correspondence of the points 1,2 in the (x^2, σ) plane, σ being the pump parameter. Note the blackbody distribution when the laser is below threshold and the Gaussian-like one above threshold.

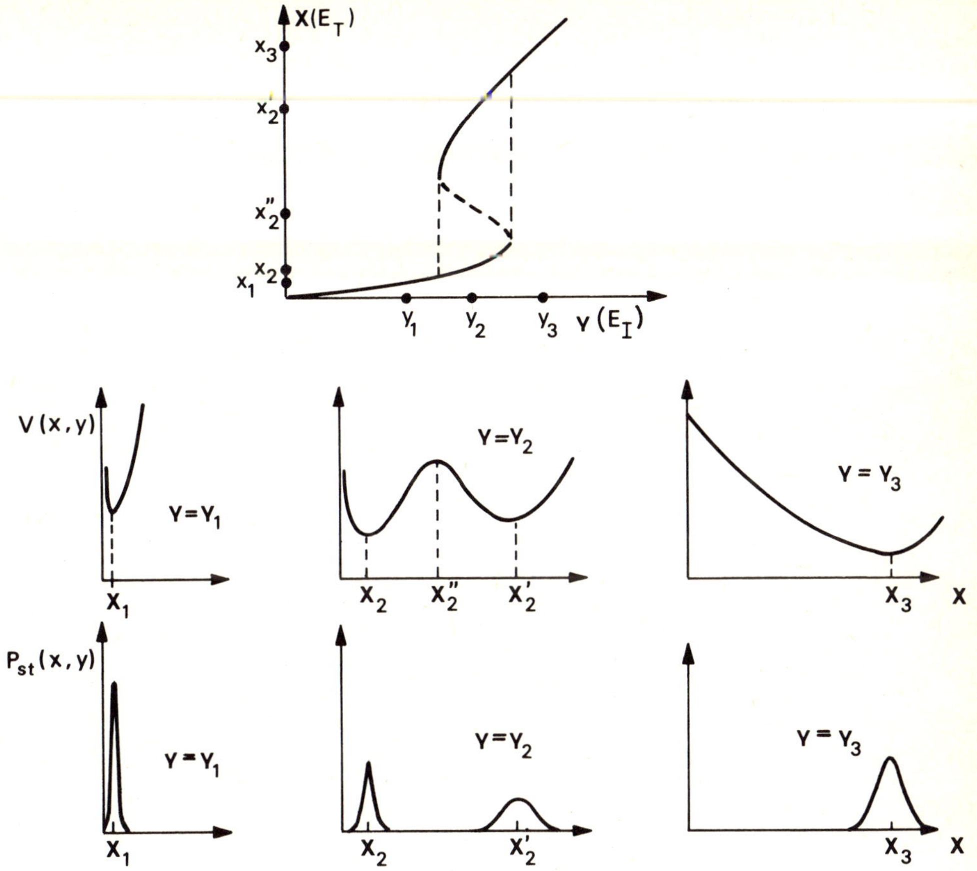

Fig.5) The generalized free energy and the stationary distribution P_{st} vs. transmitted field x are shown for different values of the incident field y in optical bistability (eqs. (134) , (135)). In the case $y=y_1$ one has only the lower-branch semiclassical solution; for $y=y_2$ one has three stationary solutions (bistable region) ; for $y=y_3$ one has only the upper-branch solution .

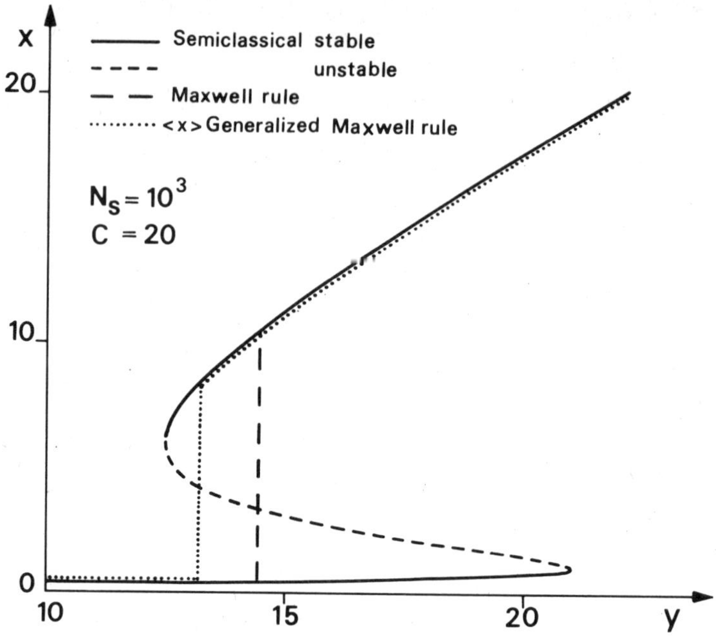

Fig.6) Comparison of the mean value ⟨x⟩ of the transmitted field with the semiclassical stationary solution.

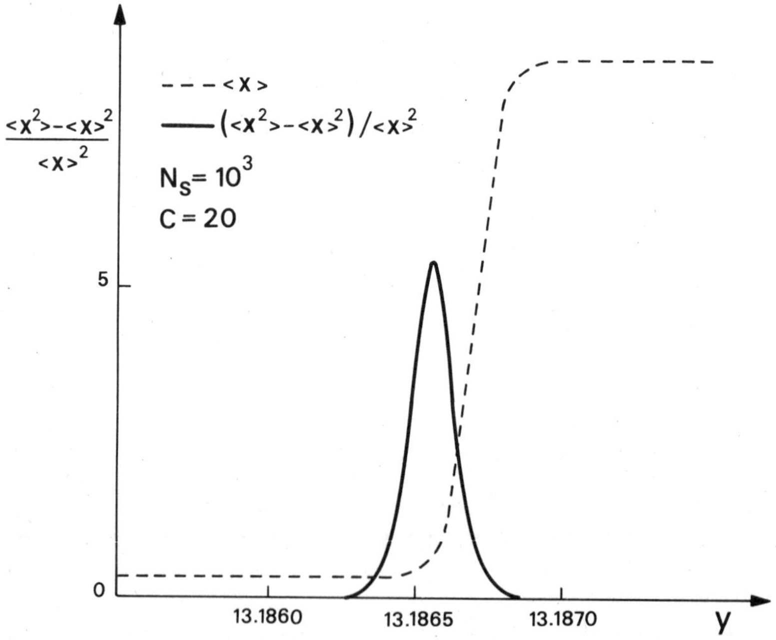

Fig.7) Relative fluctuations of the transmitted field amplitude vs. the incident field.

NONLINEAR OPTICAL PHENOMENA AND FLUCTUATIONS

Axel Schenzle
University of ESSEN
ESSEN, West Germany

INDEX

D. Fluctuations

a. Langevin Formalism

b. Fokker Planck Equations

E. Fluctuations in Nonlinear Optics

1. Parametric Oscillators
   a) Quantum Fluctuations below Threshold
   b) Nonlinear Classical Model

2. Subharmonic Generation
   a) Additive Fluctuations
   b) Additive and Multiplicative Fluctuations

3. Absorptive Optical Bistability
   a) Stationary Properties
   b) Dynamic Properties

4. Dispersive Optical Bistability
   a) The Special Case of Equal Detuning $\Delta=\delta$
   b) Approximate Solution in the Limit of Weak Fluctuations

Appendix A

1. Theorem on Optical Coherent Transients
2. Oscillatory Free Induction Decay
3. Optical Coherent Transients by Phase Switching

Appendix B

Steady States without Detailed Balance

A. INTRODUCTION

With the discovery of the laser principle, a whole new field of physics has been opened to experimental as well as theoretical research, while well established disciplines in physics gained a totally new dimension. The laser had a tremendous impact on many fields, as e.g. what may vaguely be characterized as spectroscopy, because the laser supplies the long wanted versatile source of intense and tunable coherent radiation. With its unique properties the laser has become a generally used tool in experimental as well as applied research.

But already from the beginning, the laser itself - the principle of creating highly coherent light fields through an ensemble of atoms which is driven far away from thermal equilibrium by a continuous flux of energy through the system - has attracted the interest of theorists. It has been - and still is - an intriguing question to understand in detail how order is generated from disorder in many-body systems far from thermal equilibrium. The transition of the laser from a thermal light source through a critical regime to the quasi classical coherent state bears a strong analogy to equilibrium phase transitions of the second order, with the breaking of a continuous symmetry and the phenomenon of critical slowing down.

Among the processes which have been made possible by the development of the laser, are the various nonlinear optical phenomena which can be observed inspite of the weak nonlinear response of matter, due to the enormous field intensities available. Most of these devices operate only above a certain critical threshold by emitting partially coherent light. Below threshold the emitted field has the properties of thermal light. This transition is as well analogous to second order phase transitions as it spontaneously breaks the rotational phase invariance.

In the last years simple optical devices have been found which can operate in different states depending on the history of preparation.

The existence of metastable states and the observation of hysteresis cycles makes these systems attractive for applications, but they are also attractive from a theoretical point of view because of their resemblance to first order phase transitions.

In the threshold regime where these processes become "soft" and susceptible to external perturbations, microscopic fluctuations which also probe the system are enhanced up to a macroscopic scale and become experimentally observable. For a general description which should allow to follow the details of these "phase changes" through the threshold regime, a statistical nonequilibrium theory has to be developed.

In this article we want to give an introduction into the basic aspects of nonlinear optical phenomena from first principles. We start with a brief discussion of the interaction of light and matter in order to formulate the language used throughout the paper. The principle of the nonlinear field-field interaction which comes about through the nonlinear response of gaseous or solid materials is discussed on a microscopic level. Here we have the aim to give only a qualitative introduction of how these processes can be understood, but leave aside all quantitative aspects.

Before discussing the different nonlinear devices with respect to their statistical properties, in the last chapter we have inserted a paragraph where we summarize some basic tools from the theory of stochastic processes.

B. INTERACTION OF FIELD AND MATTER

a) Hamiltonian

In classical electrodynamics the evolution of electromagnetic fields is described by the set of coupled Maxwell equations for the electric and magnetic field strength. As long as the charge- and current densities, which are the sources of the field, can be considered as externally prescribed functions of space and time, the entire theory of electromagnetic fields is linear and the principle of super position holds. In terms of the scalar potential $\varphi(\underset{\sim}{x},t)$ and the vector potential $\underset{\sim}{A}(\underset{\sim}{x},t)$, the fundamental field equations can be condensed into the variational principle

$$\delta\int L(A,\varphi)\,dt = 0 \qquad (1)$$

where the Lagrangian L assumes the form

$$L = \frac{1}{8\Pi}\int \left(\frac{1}{c^2}\frac{\partial A_i}{\partial t}\frac{\partial A_i}{\partial t} - \frac{\partial A_i}{\partial x_i}\frac{\partial A_i}{\partial x_i} \right) dV$$

$$+ \frac{1}{c}\int j_i^T(\underset{\sim}{x},t)\,A_i\,dV + \frac{1}{8\Pi}\int \frac{\partial\varphi}{\partial x_i}\frac{\partial\varphi}{\partial x_i}\,dV - \int \rho(\underset{\sim}{x},t)\,\varphi\,dV \qquad (2)$$

With the special choice of the Coulomb gauge we are left with the vector potential $\underset{\sim}{A}$ as the only dynamical field variable, while the scalar potential φ follows without retardation the changes of the charge distribution in a quasi static way.

Introducing the canonical momentum associated with the field $\underset{\sim}{A}$ through the relation

$$\Pi_i = \frac{\delta L}{\delta \dot{A}_i} = \frac{1}{4\Pi c^2}\frac{\partial A_i}{\partial t} \qquad (3)$$

we find for the free field Hamiltonian H the following expression:

$$H = 2\Pi c^2 \int \Pi_i \Pi_i \, dV + \frac{1}{8\Pi} \int \varepsilon_{ijk} \varepsilon_{ilm} \frac{\partial A_k}{\partial x_j} \frac{\partial A_m}{\partial x_l} dV + \int \varphi\rho dV$$

$$- \frac{1}{8\Pi} \int \frac{\partial\varphi}{\partial x_i} \frac{\partial\varphi}{\partial x_i} dV - \frac{1}{c} \int j_i^T A_i \, dV \qquad (4)$$

The corresponding Hamilton equations of motion

$$\frac{\partial \Pi_i}{\partial t} = - \frac{\delta H}{\delta A_i} \quad \text{and} \quad \frac{\partial A_i}{\partial t} = \frac{\delta H}{\delta \Pi_i} \qquad (5)$$

reproduce the wave equation for the vector potential $\underset{\sim}{A}$. This equation is linear and contains the sources through the inhomogeneity $\underset{\sim}{j}^T(\underset{\sim}{x},t)$. If, however, the response of the medium is taken into account, the transverse current density $\underset{\sim}{j}^T$ does not only describe the given distribution of the external sources, but describes as well the distortion of the effective charge densities in matter as a reaction on the polarizing external fields. These polarization currents in the electronic wave functions themselves follow dynamical equations of motion in which the electromagnetic field is the source of external perturbation.

The dynamic time evolution of the electronic degrees of freedom is governed by a Hamiltonian H. In second quantization, this Hamiltonian assumes the following general form:

$$H = \int dV \, \psi^+(\underset{\sim}{x}) \left(\frac{p^2}{2m} + V(\underset{\sim}{x}) \right) \psi(\underset{\sim}{x})$$

$$+ \frac{1}{2} \int dV dV' \, \psi^+(\underset{\sim}{x}) \, \psi^+(\underset{\sim}{x}') \frac{e^2}{|\underset{\sim}{x}-\underset{\sim}{x}'|} \psi(x') \, \psi(x) \qquad (6)$$

where $V(\underset{\sim}{x})$ is the potential of the external forces like e.g. the electron nucleus potential and ψ^+, ψ are the fermion field operators. This is the many particle formulation of the isolated atomic system.

In the presence of the electromagnetic field $\underset{\sim}{A}$ we have to identify the momentum P with the canonical momentum

$$P_j = \frac{\hbar}{i} \frac{\partial}{\partial x_j} - \frac{e}{c} A_j \tag{7}$$

and obtain for the general problem of the field-matter interaction the following Hamiltonian:

$$H = \int dV\, \psi^+(\underset{\sim}{x}) \frac{1}{2m} \left(\frac{\hbar}{i} \frac{\partial}{\partial x_j} - \frac{e}{c} A_j\right) \left(\frac{\hbar}{i} \frac{\partial}{\partial x_j} - \frac{e}{c} A_j\right) \psi^+(\underset{\sim}{x})$$

$$+ \int dV\, \psi^+(\underset{\sim}{x})\, V(\underset{\sim}{x})\, \psi(\underset{\sim}{x}) + \int dV dV'\, \psi^+(\underset{\sim}{x}) \psi^+(\underset{\sim}{x}') \frac{e^2}{|\underset{\sim}{x}-\underset{\sim}{x}'|} \psi(\underset{\sim}{x}') \psi(\underset{\sim}{x})$$

$$+ 2\Pi c^2 \int \Pi_i \Pi_i\, dV + \frac{1}{8\Pi} \int dV\, \varepsilon_{ijk}\, \varepsilon_{ilm} \frac{\partial A_k}{\partial x_j} \frac{\partial A_m}{\partial x_l} \tag{8}$$

For more details see e.g.[(1), (2), (3), (4)]. This is the most general microscopic formulation for this problem which we will use in the subsequent paragraphs as a common starting point for the description of various nonlinear optical phenomena. In order to formulate a specific problem we will have to specialize the Hamiltonian (8). The physical information about the special problem in mind can be introduced into the theory by choosing a convenient set of basis functions for the representations of the fields $\underset{\sim}{A}(\underset{\sim}{x})$ and $\psi(\underset{\sim}{x})$, $\psi^+(\underset{\sim}{x})$

$$\psi^+(\underset{\sim}{x}) = \sum_k \varphi_k^*(\underset{\sim}{x})\, a_k^+ \tag{9}$$

where the operators a_k^+, a_k obey the common Fermi commutation relations, e.g.

$$a_k a_{k'}^+ + a_{k'}^+ a_k = \delta_{k,k'}$$

while the electromagnetic field can be expressed in normal modes:

$$\underset{\sim}{A}(\underset{\sim}{x}) = \sum_{\underset{\sim}{q}} \underset{\sim}{u}_{\underset{\sim}{q}} b_{\underset{\sim}{q}} + \underset{\sim}{u}^*_{\underset{\sim}{q}} b^+_{\underset{\sim}{q}} \tag{10}$$

For the expansion of the electromagnetic field - if not a very specific geometry is taken into account - a plain wave expansion is the most appropriate choice for the complex vector function $\underset{\sim}{u}_{\underset{\sim}{q}}$:

$$\underset{\sim}{A}(\underset{\sim}{x}) = \sum_{\underset{\sim}{q},j} \sqrt{\frac{2\Pi\hbar c^2}{V\omega_q}} \left(\underset{\sim}{e}_{\underset{\sim}{q},j}\, e^{i\underset{\sim}{q}\underset{\sim}{x}}\, b_{\underset{\sim}{q},j} + \underset{\sim}{e}_{\underset{\sim}{q},j}\, e^{-i\underset{\sim}{q}\underset{\sim}{x}}\, b^+_{\underset{\sim}{q},j}\right) \tag{11}$$

where $\underset{\sim}{e}_{q,j}$ is the normalized basis of polarization vectors subject to the transversality condition $\underset{\sim}{e}_{\underset{\sim}{q},j} \cdot \underset{\sim}{q} = 0$.So far, the electromagnetic field has been considered as a classical variable. For the quantization of the field, this seems to be the most convenient point in the general formulation to do so, because we have no longer to deal explicitely with the fields themselves but only with a discrete set of mode amplitudes $b_{q,j}$, $b^+_{q,j}$. The quantized description is achieved by imposing the condition of local noncommutability of the field and the canonical momentum eq. 3. Expressed in the mode amplitudes b_{qj}, b^+_{qj} we obtain the well known Bose commutation relations:

$$[b_{qj}, b^+_{q'j'}] = \delta qq' \ \delta jj', \qquad [b_{qj}, b_{q'j'}] = 0 \tag{12}$$

and the free field Hamiltonian assumes the following simple form:

$$\sum_{\underset{\sim}{q}j} \hbar\omega_q\, b^+_{\underset{\sim}{q}j}\, b_{\underset{\sim}{q}j} \qquad \text{with} \qquad \omega_q = c|q| \tag{13}$$

For the expansion of the matter fields $\psi^+(\underset{\sim}{x})$, $\psi(\underset{\sim}{x})$ we may use different representations depending on the physical nature of the problem in question. To give just an idea of what we mean by an appropriate choice, we want to indicate a few physical examples.

i. Free Atoms

$\varphi \equiv \varphi_{\{n\}}(\underset{\sim}{x})$ is a single electron orbital in an effective core potential characterized by the quantum numbers $\{n\}$, which diagonalizes (approximately) the electronic Hamiltonian

$$H = \sum_{\{n\},l} E^l_{\{n\}} a^{+l}_{\{n\}} a^l_{\{n\}} \tag{14}$$

The index l characterizes the individual atoms.

ii. Free Molecules

$\varphi \equiv \varphi_{\{n\},\nu}(\underset{\sim}{x},Q)$ may be identified with the molecular orbitals in Born-Oppenheimer approximation, describing the electronic and vibrational degrees of freedom.

iii. Crystalline Solids

For the delocalized states in solids an appropriate choice for the expansion is $\varphi \equiv \varphi_{k,\mu}(\underset{\sim}{x}) = \exp i\underset{\sim}{k}\underset{\sim}{x}\, u_{\underset{\sim}{k},\mu}(\underset{\sim}{x})$ the Bloch States in single electron approximation in the effective periodic potential of the core electrons

$$H = \sum_{\underset{\sim}{k},\mu} E^\mu_k a^+_{k,\mu} a_{k,\mu} \tag{15}$$

For narrow band insulators like e.g. molecular crystals, a convenient choice for the basis set are the localized Wannier functions

$\varphi \equiv \varphi_{\underset{\sim}{l},\mu}(\underset{\sim}{x})$ where $\underset{\sim}{l}$ is a localization index.

$$H = \sum_{\underset{\sim}{l},\mu} E_{\mu}^{\underset{\sim}{l}} \, a^{+}_{\underset{\sim}{l},\mu} \, a_{\underset{\sim}{l},\mu} + \sum_{\underset{\sim}{l},\underset{\sim}{l}',\mu} D_{\mu}(\underset{\sim}{l}-\underset{\sim}{l}') \, a^{+}_{\underset{\sim}{l}\mu} \, a_{\underset{\sim}{l}'\mu} \tag{16}$$

iv. Solids with Lattice Vibrations

If we expand the electron field operators in Bloch states and quantize the harmonic motion of the lattice we obtain the following electron-phonon Hamiltonian (Fröhlich-Hamiltonian):

$$H = \sum_{\underset{\sim}{k}\mu} E^{\mu}_{\underset{\sim}{k}} \, a^{+}_{\underset{\sim}{k}\mu} \, a_{\underset{\sim}{k}\mu} + \sum_{\underset{\sim}{q}} \hbar\Omega_{\underset{\sim}{q}} \, c^{+}_{\underset{\sim}{q}} \, c_{\underset{\sim}{q}}$$

$$+ \sum_{\underset{\sim}{k},q,\mu} \hbar \, G_{q} \, a^{+}_{\underset{\sim}{k}+\underset{\sim}{q},\mu} \, a_{\underset{\sim}{k},\mu} \, (c_{\underset{\sim}{q}} + c^{+}_{-\underset{\sim}{q}}) \tag{17}$$

where the Bose operators describe the quantized motion of the lattice.

This list of possible basis functions φ_n is certainly not complete, but here we only wanted to give a brief indication of the idea and have to leave the details to the special literature [(3),(4),(5),(6)].

So far we have derived the Hamiltonian of the uncoupled free motion of the electromagnetic field and the atoms and have entirely disregarded the coupling term

$$H_{int} = \int \psi^{+}(\underset{\sim}{x}) \left(\frac{-e}{mc}\right) A_j(\underset{\sim}{x}) \frac{\hbar}{i} \frac{\partial}{\partial x_j} \psi(\underset{\sim}{x}) \, dV \tag{18}$$

Introducing the field expansions in one of the representations, we obtain the following interaction Hamiltonian:

In case (i), for free atoms eq. (16)

$$H_{int} = \sum_{nn',q,jl} \hbar\, g^{nn'}_{qjl}\; a^{+}_{n}\; a_{n'}\; b_{q,j} + hc \qquad (19)$$

with

$$g^{nn'}_{qjl} = i \sqrt{\frac{2\Pi\hbar}{V\omega q}}\, \frac{e}{m} \int \varphi^{*}_{n}\, (\mathbf{x}-\mathbf{x}_{l})\; (e_{qj})_{s}\; e^{-i\mathbf{q}\mathbf{x}}\, \frac{\partial}{\partial x_{s}}\, \varphi_{n'}\, (\mathbf{x}-\mathbf{x}_{l})\, dV \qquad (20)$$

where n,n' stands for all the quantum numbers characterizing the atomic orbitals and $\mathbf{x}_l$ localizes the atom number l.

In case (vi) for Bloch electrons eq. (15)

$$H_{int} = \sum_{\mathbf{k},\mathbf{q},\mu,\mu'} \hbar\, g^{\mu\mu'}_{\mathbf{k},\mathbf{q}}\; a^{+}_{\mathbf{k}+\mathbf{q},\mu}\; a_{\mathbf{k},\mu'}\; (b_{\mathbf{q}} + b^{+}_{-\mathbf{q}}) \qquad (21)$$

with

$$g^{\mu\mu'}_{\mathbf{k},\mathbf{q}} = i \sqrt{\frac{2\Pi\hbar}{V\omega q}}\, \frac{e}{m} \int u^{\mu *}_{k+q}(\mathbf{x})\; e^{i\mathbf{k}\mathbf{x}}\; (e_{qj})_{l}\; \frac{\partial}{\partial x_{l}}\, u^{\mu'}_{k}\, (\mathbf{x})\; e^{-i\mathbf{k}\mathbf{x}} dV \qquad (22)$$

The free atom case (i) which is the most elementary example will be used in the next chapter to introduce some basic concepts and to derive elementary examples in order to explain some fundamental properties of the field-matter interaction. It is also a realistic formulation for gaseous samples at low pressure, where collective electronic interactions can safely be neglected and only collisions have to be taken into account. An example is the entire field of atomic spectroscopy using gaseous samples and its Fourier transformed analogue, the "Optical Coherent Transients"[7].

The molecular basis (ii) allows to describe transitions in electronic states of molecular systems that are accompanied by changes in the vibrational motion of the molecules, as it is the case e.g. in processes like Raman scattering in molecular samples.

For the formulation of nonlinear processes in dense solid media the Bloch picture serves as a starting point. Parametric three-wave mixing and the processes of frequency up- and down conversion in crystals lacking inversion symmetry will be described in this picture. When optical and acoustical phonons are taken into account explicitly, a formulation of spontaneous and stimulated Raman- and Brioullin scattering will emerge from the Bloch picture with a Fröhlich type electron-phonon interaction.

b) Maxwell-Bloch Equations for Multilevel Atoms

Most of the basic properties of the interaction of electromagnetic field and matter can be understood by studying a single multilevel atom interacting with a small number of coherent travelling modes of the electromagnetic field. The dynamic time evolution of this problem is governed by the Hamiltonian eq. 14, 19

$$H = \sum_q \hbar\omega q \, b_q^+ b_q + \sum_{n,l} E_n^l a_n^{l+} a_n^l$$

$$+ \sum_{n,n',q,l} \hbar \, g_{ql}^{nn'} a_n^{+l} a_{n'}^l b_q + hc \qquad (23)$$

where we dropped the polarization index j in order not to overload the notation. The product operators $a_n^{+l} a_{n'}^l = P_{nn'}^l$ in the single electron Hilbert space have the properties of projection operators and are transformed into one another under the dynamic evolution of the Hamiltonian eq. 23. For $n \neq n'$, $P_{nn'}$ describes an electronic transition form level n' to level n, while the expectation value of P_{nn} is a measure of the occupation probability of level n. The Heisenberg equations of motion for the projectors $P_{nn'}$ assume the following form:

$$\dot{P}^{l}_{nn'} = \frac{i}{\hbar} [H, P^{l}_{nn'}]$$

$$= \frac{i}{\hbar} [E^{l}_{n} - E^{l}_{n'}] P^{l}_{nn'}$$

$$+ i \sum_{m,q} (g^{mn}_{q} P^{l}_{m,n'} - g^{n'm}_{q} P^{l}_{n,m}) (b^{+}_{q} + b_{q}) \qquad (24)$$

where n and n' run over the values 1,2...N; N is the number of atomic levels considered. A closed system of evolution equation will therefore consist at the most of (N·N-1) equations for the atomic degrees of freedom.

For the electromagnetic field, we obtain the equation

$$\dot{b}^{+}_{q} = i\omega q\, b^{+}_{q} + i \sum_{n,n',l} g^{nn'}_{q,l} P^{l}_{nn'} \qquad (25)$$

We notice that due to the projection operator properties of $P^{l}_{nn'}$ the equations (24) became nonlinear equations of motion, while eq. (25) - which is nothing else than the Maxwell equation in quantized form - is still linear as it must.

We can now draw some elementary conclusions about the field-matter interaction by investigating some special cases.

i. Coherent States

Let us replace for a moment the operator $P^{l}_{nn'}$ in eq. (25) by its expectation value $P^{l}_{nn'}$ and assume that not all nondiagonal elements of this matrix vanish. We further assume that we can neglect in eq. (24) the coupling to the field which allows us to solve this equation immediately

$$<P^{1}_{n,n'}>_t = <P^{1}_{n,n'}>_{t=0} \exp \frac{i}{\hbar} (E_n - E_{n'})t$$

With this assumption, we simulate a classical source for the electromagnetic field and suppress the quantum fluctuations of the atomic ensemble entirely. Under this assumption, eq. (25) is integrated immediately into

$$b^{+}_{q} (t) = b^{+}_{q} (0) e^{i\omega t} + i \sum_{n,n',1} g^{nn'}_{q,1} <P^{1}_{nn'}>_0$$

$$\int_0^t e^{i\omega (t-t')} e^{i/\hbar (E_n - E_{n'})t'} dt' \qquad (26)$$

and we obtain the formal result

$$b^{+}_{q} (t) = b^{+}_{q} (0) e^{i\omega q t} + E^{*}_{q} (t) \qquad (27)$$

where $E^{*}_{q}(t)$ is a classical function of time which breaks the phase invariance in course of time

$$<b^{+}_{q} (t)> = E^{*}_{q} (t) \qquad (28)$$

if the initial state was invariant under phase transformations like e.g. the vacuum state. The corresponding wave function $\psi(t)$ which is created in this process from the vacuum state, is a quantum-mechanical coherent state $|E^{*}(t)>$. If the external source is coupled strongly to the field close to resonance, the field b_q, b^{+}_{q} behaves in many aspects like a classical electromagnetic field with the energy

$$\hbar\omega_q <b^{+}_{q} b_q> = \hbar\omega \mid E_q(t) \mid^2 \qquad (29)$$

The expectation value in reverse order

$$\hbar \omega_q \langle b_q b_q^+ \rangle = \hbar \omega \left(|E_q(t)|^2 + 1 \right) \tag{30}$$

is almost identical with (29) if $|E_q|^2 >> 1$.

ii. Semiclassical Maxwell-Bloch Equations

In case of strong coherent electromagnetic fields as indicated in the previous paragraph, we may replace the field amplitudes b_q and b_q^+ by their expectations values and consider them as classical fields.

In this way we suppress the quantum fluctuations of the fields:

$$\dot{P}^l_{n,n'} = \frac{i}{\hbar} (E^l_n - E^l_{n'}) P^l_{n,n'} + i \sum_{m,q} (g_q^{mn} P^l_{m,n'} - g_q^{n'm} P^l_{n,m}) (E_q + E_q^*) \tag{31}$$

$$\dot{E}^*_q = i\omega_q E^*_q + i \sum_{n,n',l} g^{n,n'}_{q,l} P^l_{n,n'} \tag{32}$$

We are now in a position where we can take the expectation values of the transition operators $P_{n,n'}$ with respect to an arbitrary initial state characterized by the density operator ρ:

$$P^l_{n,n'} \to \langle P^l_{n,n'} \rangle = \mathrm{tr}\rho \quad P^l_{n,n'}$$

and obtain a coupled set of c-number equations. In order not to overload the notation, we will drop the angular brackets <> for further discussion, keeping in mind that $P^l_{n,n'}$ is no longer an operator.

The sum over the field modes q extends only over the small number of coherent fields which come into play, while the bulk of the q-sum has been averaged to zero. This procedure is not quantum mechanically consistent, as we will see later, and a more careful elimination of this background of incoherent field modes is necessary in order not to violate quantum mechanical commutation relations.

A correct elimination of these modes which are not macroscopically occupied introduces dissipative terms in the matter equation eq. (31) accounting for spontaneous emission. These terms can be lumped together with other sources of dissipation, like collision induced transition and random phase shifts into effective damping constants.

Dissipative terms have also to be included in the field equations (32) when the imperfect reflectitivity, i.e. the finite quality of the optical cavity which has been used as a quantization volume, is taken into account. We then obtain the system of generalized Maxwell-Bloch equations for a multilevel ensemble:

$$\dot{P}^{l}_{n,n'} = \left(i\Omega^{l}_{n,n'} - \frac{1}{T_{n,n'}}\right) P^{l}_{n,n'} + i\sum_{mq} \left(g^{mn}_{q,l}\, P^{l}_{m,n'} - g^{n'm}_{q,l}\, P^{l}_{n,m}\right)\left(E_q + E^*_q\right)$$

and

$$\dot{P}^{l}_{nn} = -\Gamma_n P^{l}_{nn} + \sum_{n' \neq n} \Gamma_{nn'} P^{l}_{n'n'} + i\left(\sum_{m,q} g^{mn}_{q,l}\, P^{l}_{mn} - hc\right) \qquad (33)$$

with $\Gamma_n = \sum_{n'} \Gamma_{n',n}$ the classical condition of detailed balance and

$T_{nn'} = T_{n'n}$. The field equations read

$$\dot{E}^*_q = (i\omega_q - \varkappa_q)\, E^*_q + i \sum_{n,n',l} g^{nn'}_{q,l}\, P_{n,n'} \qquad (34)$$

The most widely discussed special case is the two-level system in dipol approximation

$$g_{q,l}^{nn'} = (g_{q,l}^{n'n})^*, \qquad g_{q,l}^{nn'} = g\, e^{iqx_l}$$

interacting with a single mode of the field $E_q = E$

$$\dot{P}^l_{12} = (-i\Omega^l - \frac{1}{T_2})P^l_{12} + 2ige^{-iqx_l}\, W^l E \tag{35}$$

$$\dot{W}^l = -\frac{1}{T_1}(W^l - W_o) + i\, g(e^{iqx_l}\, P^l_{12}\, E^* - e^{-iqx_l}\, P^l_{21}\, E) \tag{36}$$

$$\dot{E}^* = (i\omega - \varkappa)\, E + i \sum_l g\, P^l_{21}\, e^{-iqx_l} \tag{37}$$

where $W^l = P^l_{22} - P^l_{11}$. We have introduced the abbreviations $T_2 = T_{12}$, $\frac{1}{T_1} = \Gamma_1 + \Gamma_{12}$, $W_o = (\Gamma_1 - \Gamma_{12})/\Gamma_1 + \Gamma_{12}$. The rapidly oscillating terms have been neglected (rotating wave approximation) because they only lead to minute level shifts.

These equations are the starting point for a semiclassical model of the single mode laser, for optical bistability and other resonant coherent optical phenomena, where in lowest approximation the fluctuation phenomena have been neglected.

For optically thin samples in an external coherent field, E and E* can safely be identified with the external field amplitude and can be regarded as a given function of time. The coherent response of the atomic sample is then governed by eqs. (35) and (36) with

$$E = \tilde{E}(t)\, \exp - i\omega t$$

where E is still a function of time varying slowly on a time scale ω^{-1}. These equations are e.g. the basis for the description of "Optical Coherent Transients" of two-level systems. The generalization to multilevel atoms is straightforward.

The index l which runs over the individual atoms could be dropped, if not in general Ω^l the transition frequency would depend on l explicitly even for identical atoms, due to a nonidentical environment or thermal motion which leads to individual Doppler shifts. If we know the physical origin and the statistics of the frequency shifts, we can introduce the following averaged variables:

$$P_q(\Delta) = D^{-1}(\Delta) \sum_l P_{12}^l \; \delta\,(\Omega^l - \omega - \Delta)\; \exp i \underset{\sim}{q} \underset{\sim}{x}_l \tag{38}$$

with

$$D(\Delta) = \sum_l \; \delta(\Omega^l - \omega - \Delta)$$

which characterizesthe shape of the inhomogeneous line, and obtain the familiar Bloch equations in the following form:

$$\dot{P}_q = (-i(\omega+\Delta) - \frac{1}{T_2})\, P_q + 2ig \quad W\, \tilde{E}\, e^{-i\omega t} \tag{39}$$

$$\dot{W} = -\frac{1}{T_1}\,(W - W_o) + ig\,(P_q \tilde{E}{}^* \, e^{i\omega t} - P_q^* \, \tilde{E}\, e^{-i\omega t}) \tag{40}$$

W is the difference of the populations in the upper and lower state "averaged" in the same way as eq. (38).

The oscillating atomic polarization P_q driven solely by the given external field amplitude E is again the source of an emission field which follows from the Maxwell equation (25)

$$\dot{b}_q^+ = i\omega_q\, b_q^+ - \varkappa_q\, b_q^+ + i g_q^{12} \int D(\Delta)\, P_q^*\,(\Delta,t)\, d\Delta \tag{41}$$

As P_q is a classical source, we conclude from our previous considerations eq. (28) that the reradiated field in this approximation is coherent as well.

The eq. (39) and (40) in this approximation are linear-differential equations which depend nonlinearly on the external field E. The basic physical information about this nonlinear response to an external field can be found by setting the field amplitudes E and E* equal to a constant. The eigenvalues z_i that govern the dynamic evolution are determined by the following cubic equation[8]

$$0 = f(z) = \left(z+\frac{1}{T_1}\right)\left\{\left(z+\frac{1}{T_2}\right)^2 + \Delta^2\right\} + 4g^2\,|E|^2\left(z+\frac{1}{T_2}\right) \tag{42}$$

The three roots of the cubic equation in general assume a rather horrible form and only in special cases reduce to simple tractable expressions like e.g. $T_1=T_2=T$

$$z_1 = -{}^1\!/_T\,,\qquad z_{2,3} = -{}^1\!/_T \pm i\,(\Delta^2+4g^2|E|^2)^{1/2} \tag{43}$$

As the Bloch equation (40) is inhomogeneous, we find a nontrivial steady state solution which is assumed asymptotically in the limit $t \to \infty$[7]

$$P_q^*\,(t\to\infty) = \left(i\Delta-\frac{1}{T_2}\right)gE^*W_o \Big/ \left(\Delta^2+\left(\frac{1}{T_2}\right)^2 + 4g^2\,\frac{T_1}{T_2}\,|E|^2\right) \tag{44}$$

If we can approximate the field envelope E(t) reasonably by a piecewise constant function, the scope of the above results is already sufficient to discuss all the basic optical coherent transient experiments like: optical free induction decay, optical nutation or photon echoes etc. For the comparison between theory and experiment where the reemitted field is observed, one has to insert the solution of eq. (39) and eq. (40) into eq. (41) where the collective polarization

$$\int D(\Delta)\ P_q^*(\Delta)\ d\Delta$$

acts as a source for the coherent emission field. The time dependence of the source is influenced essentially by this averaging procedure if the range of integration is large compared to the time constants $\mathrm{Re}\, z_i$.

Some rather unexpected and new results have been obtained just recently[9],[10] which have been verified experimentally in all essential details[11]. A brief summary of some recently discussed coherent transient effects which facilitate an insight into the variety of effects associated with these elementary equations, is given in Appendix A.

Through this paragraph it has become clear that for a great variety of experimental situations like atomic gases, molecular samples or solids, the basic formulations are essentially identical and only the details of the representations are different from case to case. The results which have been explicitely formulated here, using the language of an isolated atomic sample, can be reformulated in the same way e.g. for band to band transitions in solids and, from a mathematical point of view, similar results emerge.

C. NONLINEAR OPTICAL PHENOMENA

In the previous chapter we have outlined the basic mathematical framework to describe the interaction of electromagnetic radiation and matter. To get an introductory insight into this question, we have described the nonlinear dynamical response of the atomic degrees of freedom to a given time dependent external field. In this chapter now, we want to focus our attention primarily on the dynamical evolution of the electromagnetic fields while the atomic polarizations only mediate the nonlinear interaction between the fields via the nonlinearity of the generalized Maxwell-Bloch equations 25, 24.

We recall that these nonlinear interaction terms have been of the general form (eq. 24)

$$\sum_{q,l,n,m} (g^{mn}_{q,l}\, P^{l}_{mn'} - g^{n'm}_{q,l}\, P^{l}_{n,m})\,(b^{+}_{q} + b_{q}) \tag{45}$$

Without going into any specific details we can in principle distinguish between two different fundamental nonlinear processes:

i. Coupling through Diagonal Elements:

$$P_{ij} \sim (P_{ii} - P_{jj})\, b^{+}_{q} \tag{46}$$

The nonlinearity is based on dynamic changes in the level population induced by the field.

ii. Coupling through Nondiagonal Elements:

For systems with more than two levels, a nonlinearity can be mediated e.g. by a quasiresonant nonradiating transition

$$P_{il} \sim P_{ij}\, b_q^+ \qquad i \neq j$$

while for a two level system (j=l) a nonlinearity of this type is in general strongly nonresonant and can only be obtained in dipole approximation, $g^{n,n} \neq 0$, when the states involved have no parity.

We will see that these different nonlinearities will give rise to different physical effects. While in a perturbation expansion in the coupling constant g the coupling through the offdiagonal elements leads in lowest order to a cubic nonlinearity $O(g^3)$, the nonlinear diagonal response only comes into play in higher order $O(g^4)$.

a) The Idea of the Effective Hamiltonian

The coupled system of equations which we derived in the previous chapter (eq. 24 and 25) describe, under the approximation of negligible damping and fluctuations, the interaction of a discrete set of field modes with an ensemble of atoms in a quantum mechanical formulation. This is a very general description but much too involved to draw any immediate physical conclusions. If we realize, however, by comparing the different time scales of the dynamical evolution of the ensemble that we can differentiate between fast and slow variables, it is rather obvious that for the special physical processes of interest a simplified formulation can be derived from first principles.

In contrast to the resonant absorption described above, e.g. eq. 39 and eq. 40, in most nonlinear optical phenomena the fields and the atomic transitions are far from resonance and no appreciable population redistribution is brought about the irradiation even of intense fields, and only minute fractions of light intensities are absorbed.

In this limit we gain a small expansion parameter $\varepsilon = g/_{\Omega_{ij}-\omega}$, the ratio of the dipol coupling constant versus a typical detuning frequency. On the reference frame of external field oscillations, the atomic polarisation is a fast variable which is controlled by the slow motion of the electromagnetic field amplitudes and follows its evolution adiabatically. Under this assumption we can eliminate the electronic degrees of freedom in the equations of motion, and will be left with a small number of equations for the coherent fields alone, which, by the elimination procedure, suffer a nonlinear field-field interaction. These equations can then be interpreted as the Heisenberg equations of motion for the field operators under the dynamics of an effective nonlinear field-Hamiltonian. It is the aim of this chapter to derive the effective Hamiltonians for the fundamental nonlinear processes in Quantum Optics.

As the expansion in the coupling parameter ε is at the same time an expansion in the field strength $\langle b_q \rangle$, we will therefore obtain an effective Hamiltonian which emerges as a power series in the field amplitudes b_q or E, and can formally be written as follows:

$$H = \left(\frac{1}{4\Pi}\delta_{ij} + \chi^{(1)}_{ij}\right) E_i E_j + \chi^{(2)}_{ijk} E_i E_j E_k + \chi^{(3)}_{ijkl} E_i E_j E_k E_l + \ldots \tag{48}$$

We may identify in a classical way this Hamiltonian with the field energy $H = {}^1/_{4\Pi}\, ED$, where we use the standard definitions

$$D = E + 4\Pi P \qquad \text{and} \qquad P = \chi E$$

and find that this expansion is equivalent with the definition of a nonlinear polarization P or a field dependent susceptibility

$$P_i = \chi^{(1)}_{ij} E_j + \chi^{(2)}_{ijk} E_j E_k + \chi^{(3)}_{ijkl} E_j E_k E_l \tag{49}$$

In this way the nonlinear atomic response obviously leads to an explicit field-field interaction:

$\chi_{ij}^{(1)}$ - gives rise to the familiar linear dispersion relation.

$\chi_{ijk}^{(2)}$ - leads to a cubic nonlinearity which is responsible for the three-wave mixing processes in parametric devices, as well as second harmonic and subharmonic generation. As this term does not leave the Hamiltonian invariant under spatial inversion transformations, $\chi^{(2)}$ is non-vanishing only for crystalline samples lacking inversion symmetry.

$\chi_{ijkl}^{(3)}$ - is responsible for four-wave mixing processes which have recently found considerable interest in connection with the phenomenon of optical phase conjugation[12],[13] and different multiphoton spectroscopy techniques. A more familiar four-wave process is stimulated Raman- and Brioullin scattering, and the parametric coupling of Stokes and antistokes radiation.

There has been a great effort to calculate these higher order susceptibilities from first principles, to get a feeling of the effects which lead to large nonlinearities.

Explicit but enormously complicated expressions for $\chi^{(n)}$ can be found in many original publications and in the standard book on nonlinear optics by Bloembergen[14].

In the following chapters we only want to give an idea of the microscopic background of these nonlinearities and how they arise in a straightforward way by deriving the field-field interaction from first principles, but leaving aside all technical details.

b) Parametric Oscillators

As a first example of nonlinear field interactions, we want to derive the effective Hamiltonian for parametric processes which give rise to parametric up- and down conversion of the incident field frequency. Three-wave mixing processes, as explained previously, can be observed only in solid materials, due to the requirement of lacking inversion symmetry. We will therefore start from the field-matter Hamiltonian for an insulator in Bloch-function representation (eq. 15) and will keep in mind that the equilibrium electron wave function will describe an occupied valence band and an empty conduction band.

It is obvious from the Hamiltonian eq. (15) that the scattering processes that give rise to an interaction between three fields must be of the following structure:

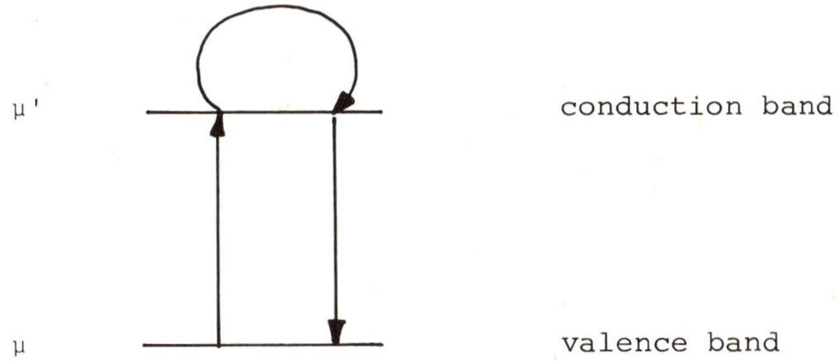

Now we separate the interaction Hamiltonian into two contributions, one which creates interband transitions, and one which describes intraband scattering processes. We therefore write formally

$$H = H_o + \lambda_1 H_{\mu\mu'} + \lambda_2 H_{\mu\mu}$$

where the last part lives from the assumption of missing inversion symmetry of the crystal lattice and is assumed to be the smallest term under consideration.

with the help of the canonical transformation

$$\tilde{H} = e^{\lambda s} H e^{-\lambda s} \tag{50}$$

we can eliminate the interband interaction characterized by the coupling constant λ_1 to first order, and obtain, by performing the straightforward expansion in λ_1:

$$\tilde{H} = H_o + \frac{1}{2} \lambda_1^2 \lambda_2 [s,[s, H_{\mu\mu}]] + O(\lambda_1^3) \tag{51}$$

where s is defined through the relation

$$H_{\mu\mu'} = [H_o, s] \tag{52}$$

Evaluating eq. (52) we obtain for the antihermitian operator s

$$s = \sum_{k,q} \hbar g^{cv}_{k,q} (E^c_{k+q} - E^v_k - \hbar\omega_q)^{-1} a^{+c}_{k+q} a^v_k b_q - hc \tag{53}$$

We insert eq. (53) into eq. (51) and take the expectation value of the result with respect to the equilibrium electron distribution ψ

$$a^+_{k,v} \psi = 0 \qquad a_{k,c} \psi = 0$$

The term which gives rise to the three-wave interaction reads:

$$\tilde{H} = H_o - \sum_{kqq'} \left(\frac{\hbar^3 g^{cc}_q g^{cv}_{q'} g^{*cv}_{q'+q} \qquad b_q b_{q'} b^+_{q'+q}}{(E^c_{k+q'} - E^v_k - \hbar\omega_{q'}) \; (E^c_{k+q'-q} - E^v_k - \hbar\omega_{q'+q})} + hc \right) \tag{54}$$

For the interaction of three specific modes q_1, q_2, q_3 we obtain the following effective Hamiltonian:

$$H = \sum_{i=1}^{3} \hbar\omega_{qi}\, b^+_{qi}\, b_{qi} + \hbar\,\{g b^+_{q_3}\, b_{q_2}\, b_{q_1} + hc\} \tag{55}$$

In a uniform medium translational invariance requires quasi-momentum conservation $q_3=q_1+q_2$

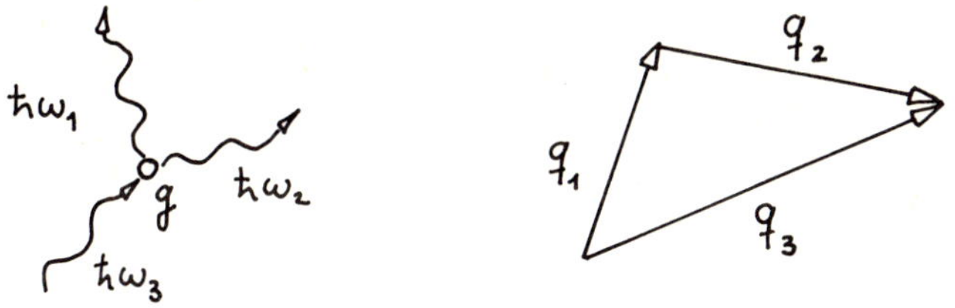

This Hamiltonian describes a spontaneous breakup of a photon of frequency ω_3 into two photons of frequency ω_1 and ω_2. For this interaction to be efficient, the near resonance condition $\omega_3 \approx \omega_2+\omega_1$ is required, otherwise the effective interaction g exp $i(\omega_3-\omega_1-\omega_2)t$ is averaged to zero.

In a dispersive medium, this condition is in conflict with the conservation of quasi-momentum k. Media with large nonlinearities, however, usually are birefringened and momentum as well as energy conservation can be satiesfied approximately for a special ratio of the frequencies ω_1 and ω_2 by choosing appropriate noncolinear directions of propagation.

It is then obvious that the ratio by which the incoming field frequency is split into the subharmonic frequencies, is strongly dependent on the crystal parameters and the geometry. By rotating the crystal axes with respect to the direction of propagation of the external field or by changing e.g. the temperature of the crystal, the ratio of ω_1 and ω_2 can be changed continuously. This aspect is the basis for the great practical importance of this effect, as it allows to create coherent light fields that can be tuned continuously but with spectral properties similar to those of the driving laser field.

The dynamical evolution of the fields under the unitary transformation of this effective Hamiltonian eq. (55) can most easily be visualized by writing down the corresponding Heisenberg equations of motion:

$$\dot{b}^+_{q_i} = \frac{i}{\hbar} [H, b^+_{q_i}] \tag{56}$$

$$\dot{b}^+_{q_1} = i\omega_{q_1} b^+_{q_1} + i g b^+_{q_3} b_{q_2}$$

$$\dot{b}^+_{q_2} = i\omega_{q_2} b^+_{q_2} + i g b^+_{q_3} b_{q_1} \tag{57}$$

$$\dot{b}^+_{q_3} = i\omega_{q_3} b^+_{q_3} + i g^* b^+_{q_1} b^+_{q_2}$$

These nonlinear operator equations can only be solved for a number of limiting cases where a linearization around some steady state solution is possible. For the initiation of this spontaneous breakup quantum fluctuations are inevitable. The onset of parametric oscillation e.g. can be studied by disregarding for a moment the dynamics of the driving laser field b_{q_3}. We replace it by a classical amplitude E exp $-i\omega_3 t$. This assumption is appropriate as long as the amplitudes of the fields b_{q_1}, b_{q_2} remain small. In the "rotating frame" where the explicit time dependence of the resulting linear problem disappears, the eigenvalues can be written in the form

$$\lambda_{1/2} = \frac{1}{2} \Delta \pm i \left(gg^* |E|^2 - \left(\frac{\Delta}{2}\right)^2\right)^{1/2} \tag{58}$$

where $\Delta = \omega_3 - \omega_1 - \omega_2$.
For an external field intense enough, i.e. $|E|^2 > \Delta^2/4gg^*$, we find no longer periodic solutions but exponentially divergent amplitudes, indicating a threshold beyond which a new type of steady state solution will be observed and where the small signal approximation will break down. These equations which we will generalize somewhat at the end of the next chapter, together with the equations of de-

generate processes discussed there, lay the basis of the quantum mechanical as well as classical formulation of parametric interactions.

In this course we have chosen not to discuss the classical problems like wave propagation in nonlinear media etc., but want to focus our attention on the statistical properties of these nonlinear models. We therefore leave the discussion of optical parametric processes here and come back to this problem only after the introduction of fluctuations in the chapter D.

c) Second Harmonic and Sub-Harmonic Generation

The nonlinear interaction of only two fields is contained in the previous results, when we consider the degenerate case where the fields b_{q_1} and b_{q_2} become identical, i.e. photons of the frequency ω_3 decay spontaneously into two photons of half this frequency $\omega_1=\omega_2=\omega_3/2$, generating thereby the so called subharmonic field. The time inverted process is certainly possible as well where from a field of frequency ω_1 the second harmonic frequency of twice this frequency is generated. An important difference between these processes lies in the fact that the generation of sub-harmonic fields is obtained only above a certain critical threshold intensity, while second-harmonic fields are generated without threshold and can be understood in classical terms. This difference is not only important for practical purposes but is also of interest from a theoretical point of view. While sub-harmonic generation can be described in terms of an instable amplifier which is triggered by vacuum fluctuations, second-harmonic generation is analogous to an oscillator driven by a resonant external force. These properties can easily be derived from the Heisenberg equations of motion for the degenerate parametric oscillator.

The Hamiltonian eq. (55) reduces to

$$H = \sum_{i=1}^{3} \hbar\omega_{q_i} b^+_{q_i} b_{q_i} + \hbar (g\, b^+_{q_3} b^2_{q_1} + g^* b^{+2}_{q_1} b_{q_3}) \qquad (59)$$

and we obtain the following set of coupled equations of motion:

$$\dot{b}^+_{q_3} = i\omega_{q_3} b^+_{q_3} + i\, g^* b^2_{q_1} \tag{60}$$

$$\dot{b}^+_{q_1} = i\omega_{q_1} b^+_{q_1} + 2ig\, b^+_{q_3} b_{q_1} \tag{61}$$

Eq. (61) describes the parametric generation of the sub-harmonic field.

At this point, we would like to be a little bit more precise about the experimental setup used in these experiments, in order to complete the equations of motion. We assume that the nonlinear medium is contained in an optical cavity with dielectric mirrors which allow - depending on the frequency of the fields - some percentage of the light field intensity to leak out of the cavity. The external fields, provided by strong coherent laser sources, are coupled into the system. We will simulate this coupling to the external fields by additive source terms, circumventing in this way the delicate boundary value problem.

The general parametric process will then be described by the following somewhat more realistic set of equations:

$$\begin{aligned}
\dot{b}^+_{q_1} &= (i\omega_{q_1} - \varkappa_1) b^+_{q_1} + i\, g\, b^+_{q_3} b_{q_2} + P_1\, e^{i\omega_{q_1} t} \\
\dot{b}^+_{q_2} &= (i\omega_{q_2} - \varkappa_2) b^+_{q_2} + i\, g\, b^+_{q_3} b_{q_1} \\
\dot{b}^+_{q_3} &= (i\omega_{q_3} - \varkappa_3) b^+_{q_3} + i\, g^* b^+_{q_1} b^+_{q_2} + P_3\, e^{i\omega_{q_3} t}
\end{aligned} \tag{62}$$

By merely adding dissipative terms, as we will discuss in the next chapter, we loose the quantum mechanical consistency of the equations because we have to consider fluctuations as well. If we interprete, however, eq. (62) in classical terms, we have not to worry about consistency and obtain a realistic picture as long as the fields are not amplified from vacuum fluctuations.

Depending on which of the previously discussed process we want to consider, we will assume for the source terms:

1. Parametric amplifier: $P_1, P_3 \neq 0, P_3 >> P_1$
2. Parametric Oscillator: $P_1 = 0$ $P_3 \neq 0$
3. Second-harmonic generation: $P_1 \neq 0$ $P_3 = 0$
4. Sub-harmonic generation: $P_1 = 0$ $P_3 \neq 0$
5. Sub-harmonic bistability: $P_1 \neq 0$ $P_3 \neq 0$

d) Raman Scattering

So far, we have focussed our attention only on the electronic degrees of freedom, disregarding entirely e.g. the vibrational motion of the nuclei in molecules or equivalently the phonons in solid materials. Electronic transitions, however, can be accompanied by transitions in the vibrational manifold of states. If in a scattering process - which can be visualized as a nonresonant absorption and mediate reemission process - vibrational quanta or phonons are created or absorbed, the emitted light field will be shifted in frequency according to the overall energy conservation.

This scattering process is known under the name of Raman process, when molecular vibrations or optical phonons are involved, and is called Brioullin-scattering when acoustical phonons are emitted or absorbed. See pp. 287–301

We are not going into great technical details here but only want to make the basic ideas transparent, by deriving the effective interactions that are responsible for the Raman and Brioullin scattering from first principles. We will do this separately for the vibrational processes in molecules as well as for the electron-phonon interaction in condensed materials.

i. Vibrational Raman Effect in Single Molecules

An elementary access to the Raman effect, as it is observed by scattering light from gaseous molecular samples, can be obtained by simply considering a three level system:

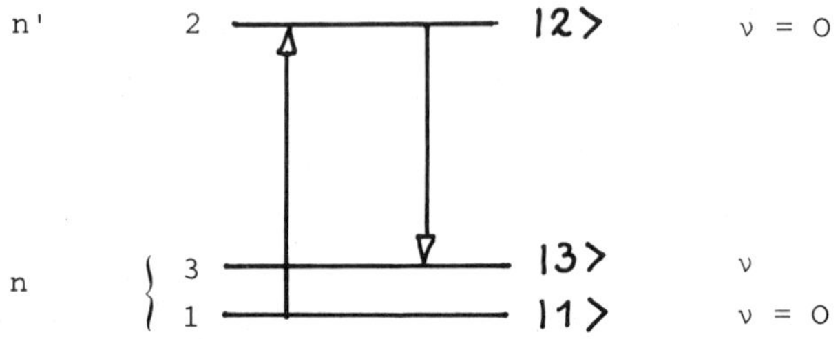

made up by the electronic ground state n, ν=0, an electronic excited state n', ν=0, and a state of molecular vibration in the electronic groundstate n, ν=0. We will assume that the symmetry of the states allows optical dipol transition between state 1 and 2 and state 2 and 3, while no allowed dipol transitions occur between state 1 and 3.

Using the molecular wave function basis, a first principle description will start from the following Hamiltonian:

$$H = \sum_{n,\nu} E_{n,\nu} a^{+}_{n,\nu} a_{n,\nu} + \hbar \sum_{\substack{n,n'\\ \nu,\nu'}} g^{\nu,\nu'}_{n,n'q} a^{+}_{n,\nu} a_{n',\nu'} b_q + hc \qquad (63)$$

with

$$g^{\nu,\nu'}_{n,n'q} = i \sqrt{\frac{2\Pi\hbar}{V\omega_q}} \frac{e}{m} \int \varphi^{*}_{n,\nu} \frac{\partial}{\partial x_l} \varphi_{n',\nu'} (e_q)_l \, e^{-i\underset{\sim}{q}\underset{\sim}{x}_l} \, dV \qquad (64)$$

For the simplified problem we have in mind here, this general Hamiltonian can be reduced essentially by considering only three states $|1>$, $|2>$, $|3>$ explicitly, which interact with only two modes of the field, b_{q_1}, b_{q_2}. We then obtain the following Hamiltonian

$$H = H_o + \lambda H_1$$

$$= \sum_{n=1}^{3} E_n |n><n| + \hbar \sum_{i=1}^{2} \omega_{q_i} b^+_{q_i} b_{q_i}$$

$$+ \hbar g_1 |2><1| b_{q_1} + \hbar g_2 |2><3| b_{q_2} + hc \tag{65}$$

where the identification of the coefficients is obvious.

By a canonical transformation quite analogous to our previous procedure eq. (50), we will eliminate the interaction to first order and will obtain an effective Hamiltonian which is of second order in the coupling constant g_q. For that purpose we define a unitary transformation

$$\tilde{H} = e^{\lambda s} H e^{-\lambda s} \tag{66}$$

where λ is assumed to be of the order of g_i and find to lowest order in λ

$$H = H_o + \frac{1}{2} \lambda^2 [s, H_1] \tag{67}$$

where s is defined through

$$H_1 = [H_o, s]$$

It is obvious that the antihermitian operator s can be written in the form

$$S = \frac{\hbar}{E_2 - E_1 - \hbar\omega q_1} (g_1 \ |2\rangle\langle 1| \ b_{q_1} - g_1^* \ |1\rangle\langle 2| \ b_{q_1}^+)$$

$$+ \frac{\hbar}{E_2 - E_3 - \hbar\omega_{q_2}} (g_3 \ |2\rangle\langle 3| \ b_{q_2} - g_3^* \ |3\rangle\langle 2| \ b_{q_2}^+) \tag{68}$$

and we find for the transformed Hamiltonian eq. (67)

$$\tilde{H} = \sum_{i=1}^{2} \hbar\omega_{q_i} \ b_{q_i}^+ \ b_{q_i} + E_1 \ |1\rangle\langle 1| + E_3 \ |3\rangle\langle 3|$$

$$+ \hbar\tilde{g} \ |3\rangle\langle 1| \ b_{q_2}^+ \ b_{q_1} + \hbar\tilde{g}^* \ b_{q_1}^+ \ b_{q_2} \ |1\rangle\langle 3| \tag{69}$$

where $\tilde{g}$ is to be identified with

$$\tilde{g} = g_1 \ g_3^* \left(\frac{\hbar}{E_2 - E_1 - \hbar\omega_{q_1}} + \frac{\hbar}{E_2 - E_3 - \hbar\omega_{q_2}} \right)$$

The effective interaction obtained in this way describes the scattering of a photon from state q_1 into state q_2 by the "emission" of a vibrational quantum:

Formally, this scattering process is identical with the previously described parametric three-wave mixing processes, where one field operator has been replaced by the polarization operator $|3\rangle\langle 1|$ which connects the vibrationally excited state $|3\rangle$ with the ground state $|1\rangle$. As this single-molecule problem is not translationally invariant, momentum conservation does not play a role, and no phase matching condition is required. For an ensemble of independent atoms in an optical cavity e.g., this can be stated somewhat differently. In each localized molecule the fields q_1 and q_2 will create an oscillatory offdiagonal element $P_{3,1} = |3\rangle\langle 1|$ with a phase retardation of exp i $(\underset{\sim}{q}_1-\underset{\sim}{q}_2)\ \underset{\sim}{x}_l$ where $\underset{\sim}{x}_l$ is the localization of the lth molecule. In this way a standing "polarization wave" is created which picks up the difference in photon momenta $(q_2-q_1)\hbar$.

In the derivation of the effective scattering Hamiltonian, we have stopped at this intermediate step to point out the formal analogy to the parametric processes. Physically, however, there is a drastic difference due to the different origin of the coupled fields. One important difference is the fact that while the electronic transitions are driven far from resonance, the vibrational transition is driven on resonance or at least close to resonance, and the damping of this mode comes into play determining the linewidth of the scattered field.

If we consider the damping of the polarization $|3\rangle\langle 1|$ explicitly, we can eliminate the atomic degrees of freedom and obtain equations of motion for the fields alone. As we have to include the damping Γ of this resonant process, we can not expect that this interaction will be of Hamiltonian form.

We eliminate the polarization $|3\rangle\langle 1|$ adiabatically by assuming that this mode is strongly damped:

$$|3\rangle\langle 1| = (E_3 - E_1 - \hbar\omega_1 + \hbar\omega_2 + i\Gamma)^{-1}\ \tilde{g}\ b^+_{q_1}\ b_{q_2}\ (n_1 - n_3) \qquad (70)$$

We insert this algebraic relation into the Heisenberg equations of motion for the two modes b_{q_1}, b_{q_2}, and obtain

$$\dot{b}^{+}_{q_1} = i\omega_{q_1} b^{+}_{q_1} - i\Lambda\, b^{+}_{q_1} b^{+}_{q_2} b_{q_2}$$

$$\dot{b}^{+}_{q_2} = i\omega_{q_2} b^{+}_{q_2} - i\Lambda^{*}\, b^{+}_{q_2} b^{+}_{q_1} b_{q_1} \qquad (71)$$

where the effective coupling constant Λ assumes the form

$$\Lambda = \tilde{g}\tilde{g}^{*}\ (\Delta - i\Gamma)/(\Delta^2 + \Gamma^2)\ (n_1 - n_3)$$

with $\hbar\Delta = E_2 - E_1 - \hbar\omega_1 + \hbar\omega_2$ and n_i=thermal population of level i.

The real part of Λ leads to a renormalization of the frequencies and is not of importance. The imaginary part gives rise to Raman Scattering.

From the explicite appearance of the equation (71) it is obvious that they cannot be interpreted as the equation of motion of an effective Hamiltonian. This is not surprising, however, because we have broken the time invariance of the process by introducing dissipation. This is also easily understood in physical terms when we rewrite the equation (71) for the case of resonance and $n_1=1$, $n_3=0$:

$$\dot{b}^{+}_{q_1} = i\omega_{q_1} b^{+}_{q_1} - \frac{\tilde{g}\tilde{g}^{*}}{\Gamma} b^{+}_{q_1} b^{+}_{q_2} b_{q_2}$$

$$\dot{b}^{+}_{q_2} = i\omega_{q_2} b^{+}_{q_2} + \frac{\tilde{g}\tilde{g}^{*}}{\Gamma} b^{+}_{q_2} b^{+}_{q_1} b_{q_1} \qquad (72)$$

The Stokes photons are created with a rate proportional to the laser intensity, while the laser field is depleted by the effective interaction. $b^{+}_{q_1} b_{q_1} + b^{+}_{q_2} b_{q_2}$ is a constant of motion.

ii. Raman Effect in Solids

In solid materials, light scattering can be associated with the emission or absorption of optical phonons. This process can be understood as the sequence of a nonresonant band-to-band absorption, emission or absorption of a phonon in the optical branch of the phonon dispersion relation, and recombination back to the valence band. If the phonon is created in the acoustical branch, the effect is called Brillouin Scattering. These processes are formally very similar; from an experimental point of view, however, there exists a tremendous difference between these two effects due to the different dispersion relations of optical and acoustical phonons. While momentum conservation in these scattering process is easily satisfied for the dispersionless optical branch, the acoustic dispersion relation $\omega_{Ph} = v\, q$ with the sound velocity v does not allow to satisfy momentum and energy conservation simultaneously for phonon frequencies that exceed a critical value of

$$\omega_{Ph} > \omega_{Ph}^{c} = 2\omega \frac{v}{c} \tag{73}$$

where ω is the frequency of the incident field.

We will briefly derive the corresponding effective Hamiltonian, starting from the Bloch picture of the electrons and include the electron photon (eq. (21)) and electron phonon (eq. (17)) interaction. As the idea of the transformation will be very similar to the one of our previous derivations, we will therefore indicate here only the main steps.

The Hamiltonian can be written in the form

$$H = H_o + \lambda_1 H_1 + \lambda_2 H_2 \tag{74}$$

where H_1 and H_2 describes the electron photon and the electron-phonon interaction respectively. We assume that the coupling constant λ_2 which characterizes the strength of the electron phonon interaction is small compared to λ_1. We eliminate therefore H_1 through a unitary transformation to first order in λ_1. This is achieved by identifying

$$\tilde{H} = e^{\lambda_1 S}\, H\, e^{-\lambda_1 S} \tag{75}$$

where the operator S satisfies the relation

$$H_1 = [H_o, S] \tag{76}$$

If we collect only the terms which give rise to Raman Scattering and disregard all the contributions which e.g. only renormalize the photon dispersion relation, we obtain

$$\tilde{H} = H_o + \frac{1}{2} \lambda_1^2 \quad \lambda_2 \quad [S, [S, H_2]] \tag{77}$$

This is the only nontrivial interaction term which survives averaging over the equilibrium electron states.
Schematically this interaction process can be visualized in the following picture:

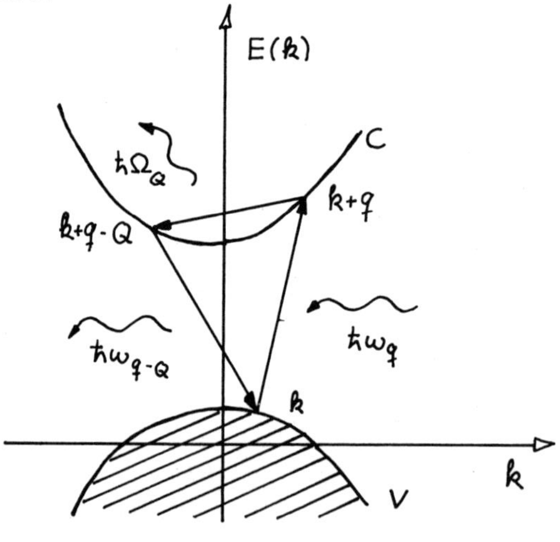

We evaluate the commutators in eq. (77) and average the result over the occupied valence band states. We find:

$$H_{int} = \frac{\hbar}{2} \sum_{q,Q} \Lambda_{q,Q} \, b_q^+ \, b_{q-Q} \, C_Q + \Lambda_{q,Q}^* \, b_{q-Q}^+ \, b_q \, C_Q^+ \tag{78}$$

where we used the abbreviation

$$\Lambda_{Q,q} = \sum_k \frac{g_q \, g^*_{q-Q} \, G_Q}{(E^c_{k+q} - E^v_k - \hbar\omega_q)(E^c_{k+q-Q} - E^v_k - \hbar\omega_{q-Q})} \tag{79}$$

The formal analogy to the optical parametric process is evident in eq. (78) where only one of the lower frequency photons has been replaced by a phonon.

An effective interaction between the electromagnetic fields alone can be derived also in this case when the dissipation of the phonon field is considered explicitly.

The Heisenberg equation for the phonon field c^+_Q can be written in the form:

$$\dot{c}^+_Q = i\,\Omega_Q\, c^+_Q + \frac{i}{2} \sum_q \Lambda_{Qq}\, b^+_q\, b_{q-Q} \tag{80}$$

This process is assumed to be quasi resonant

$$\omega_q - \omega_{q-Q} \approx \Omega_Q$$

An effective dynamics of the fields can therefore only be derived when the damping of the phonons is taken into account through an additional term $-\Gamma c^+_Q$ in eq. (80). In adiabatic approximation, we find for the phonon fields:

$$\dot{c}^+_Q = \frac{1}{2} \sum_q \Lambda_{Qq} \,(\omega_q - \delta_{q-Q} - \Omega_Q - i\Gamma)^{-1}\, b^+_q\, b_{q-Q}$$

Under this approximation we can eliminate the phonons explicitly from the equations of motion for the Stokes and the laser field, and are again left with equations of the form eq. (71) with an effective interaction

$$\sim |g_q\, g^*_{q-Q}|^2\, (E^c_{k+q} - E^v_{k+q} - \hbar\omega_q)^{-2}\, (E_{k+q-Q} - E_k - \hbar\omega_{q-Q})^{-2}$$

$$* \, (\omega_q - \omega_{q-Q} - \Omega_Q - i\Gamma)^{-1}$$

These equations again can *not* be rewritten in terms at an effective Hamiltonian. From a macroscopic point of view, the cubic response can be associated with a nonlinear susceptibility $\chi^{(3)}$.

$$P_i \sim \chi^{(3)}_{ijkl}\, E_j\, E_k\, E_l \tag{81}$$

e) Parametric Interaction of Stokes and Antistokes Radiation

A quartic interaction of the form obtained for the Raman effect, can in principle couple more than just two fields. That only the external laser field has been coupled to the stokes shifted radiation eq. (72) and eq. (80), was an assumption made explicitly to simplify the equations of motion. This assumption was also motivated by the requirement of energy conservation for the overall scattering process. If we go back to the Hamiltonian eq. (65), it is easily verified that we can include another overall resonant process when we allow for a third field (b_{q_3}, $b^+_{q_3}$) in the following way:

$$H_{int} = \hbar g_1\, |2\rangle\langle 1|\, (b_{q_1} + b_{q_3}) + hc$$

$$\hbar g_3\, |2\rangle\langle 3|\, (b_{q_2} + b_{q_1}) + hc \tag{82}$$

The three fields are still assumed to be far off resonance with the dipole transitions in order not to cause any measurable population transfer and absorption in the three level structure. If we go through the same unitary transformation procedure as in eq. (66), the effective Hamiltonian eq. (69) will contain two additional terms which read

$$\hbar\, \hat{g}\, b^{+}_{q_1}\, b_{q_3}\, |3\rangle\langle 1| + \hbar\, \hat{g}^{*}\, b^{+}_{q_3}\, b_{q_1}\, |1\rangle\langle 3| \qquad (83)$$

with the coupling constant $\hat{g}$ defined analogous to eq. (69):

$$\hat{g} = g_1\, g_3\, \hbar \left(\frac{1}{E_2 - E_1 - \hbar\omega_{q_3}} + \frac{1}{E_2 - E_3 - \hbar\omega_{q_1}} \right)$$

For the overall scattering process including the field b_{q_3} to be resonant, ω_{q_3} has to be upshifted from the laser frequency by a vibrational frequency quantum. The interaction eq. (83) then is responsible for the creation of antistokes radiation. By the same line of arguments as used previously, we can derive an explicit field-field interaction Hamiltonian in a straightforward way and obtain

$$H = H_o + \hbar\, \Lambda_{SA}\, b^{+}_{q_1}\, b^{+}_{q_1}\, b_{q_2}\, b_{q_3} + \hbar\, \Lambda^{*}_{SA}\, b^{+}_{q_3}\, b_{q_2}\, b_{q_1}\, b_{q_1}$$

$$+ \hbar\, \Lambda_S\, b^{+}_{q_1}\, b_{q_2}\, b^{+}_{q_2}\, b_{q_1} + hc$$

$$+ \hbar\, \Lambda_A\, b^{+}_{q_3}\, b_{q_1}\, b^{+}_{q_1}\, b_{q_3} + hc \qquad (84)$$

and the effective coupling constant Λ_{sa} is proportional to the product of the stokes and antistokes coupling constant $\tilde{g}$ and $\hat{g}$ eq. (69) and eq. (83)

$$\Lambda_{SA} \sim \tilde{g}\, \hat{g}^{*}$$

This process can be interpreted as a two-step process, where in the first scattering event a phonon is created by stokes shifting the incoming laser field (b_{q_1}). The laser which is the highest intensity field in this problem, is scattered again and upshifted in frequency, while the phonon is anihilated.

Obviously this game can be played on and on, by allowing for more and more frequencies to interact. In this way higher order stokes and antistokes-shifted fields are produced. In intense laser fields scattered radiation can be observed up to about the 10th order[17]. It may be interesting to note that besides the similar appearance of the interaction terms Λ_S, Λ_A which describe the interaction of the laser field with the Stokes or the Anti-Stokes radiation, individually there exists an essential difference. While the interaction $\sim\Lambda_S$ includes the spontaneous emission of Stokes photons, $\sim\Lambda_A$ describes only stimulated processes because of the inverted order of the photon operators $b^+_{q_3}$, b_{q_3}. The physical reason for this difference is obvious.

With this example we want to close this chapter where we have given an idea of how in principle the different nonlinearities which are responsible for the most prominent effects in nonlinear optics arise in a microscopic picture. We have used rather simple models in order to clearify the basic ideas - we did not intend to derive quantitative expressions e.g. for the nonlinear susceptibilities.

D. FLUCTUATIONS

The problem of the interaction of light and matter in quantum mechanical formulation is in general a problem of many degrees of freedom, due to the macroscopic number of atoms involved on the one hand, and the denumberably infinite number of field modes in the optical cavity on the other hand. The simplicity and the success of the previous approach was brought about by restricting ourselves somewhat artificially and without detailed justification to a small number of variables which we are tempted to call the relevant ones. In a classical picture where no spontaneous processes are possible, this somewhat naive line of reasoning does not lead into any inconsistencies. In the quantum formulation, however, it is not possible in principle to choose e.g. just a single mode of the light field as the relevant variable, because it cannot be guaranteed that spontaneous emission will not also occur into other modes as well which we do not want to consider as relevant variables explicitly.

The simplification of the general many-body - many-mode problem therefore has to be performed in a quantum mechanically consistent way and not just by dropping the unwanted variables in order not to violate e.g. quantum mechanical commutation relations. The elimination of the bulk of the degrees of freedom in favor of a small number of relevant or macroscopic variables brings about dissipative as well as fluctuating correction terms. This can be achieved in many ways, and various methods have been developed to perform this elimination procedure (6), (18), (19).

a) Langevin Formalism

An elegant, but rather formal way to eliminate the irrelevant degrees of freedom from the description of a many-body problem, is the projection operator method developed by H. Mori (20). Here the equations of motion assume the formal appearance of the Langevin equation which has been used to describe the stochastic motion of Brownian Particles.

The generalized Langevin equation of Mori has been derived primarily to describe critical spin phenomena but can be used as well to formulate many different problems in solid state physics, when suitable projection operators are used[21].

The fundamental idea consists in the construction of a linear unitary operator space $\{\Omega_i\}$ in which the dynamical evolution of a complex system is described in terms of its Heisenberg equations of motion. At the time t=0, a separation is made of the fundamental set of operators into macroscopically relevant $\{x_i\}$ and irrelevant $\{y_i\}$ variables, using e.g. a time scale argument. An arbitrary operator can be split into relevant and irrelevant parts

$$\Omega = \sum_i \langle x_i \Omega \rangle x_i + \sum_j \langle y_j \Omega \rangle y_j \tag{85}$$

where the angular brackets indicate the scalar product of this space. By the time evolution of Ω

$$\dot{\Omega} = \frac{i}{\hbar} [H, \Omega] \tag{86}$$

relevant and irrelevant parts become mixed and a separation at a finite time will reveal that the projections in the relevant and irrelevant subspaces will have changed.
Formally we can separate these parts in the time evolution of Ω by introducing the projection operators Π and $\hat{\Pi} = 1-\Pi$ which allows us to write eq. (85) in the following form:

$$\Omega = \Pi\, \Omega + \hat{\Pi}\, \Omega \tag{87}$$

By formal integration of eq. (86) we obtain the wellknown generalized Langevin equation [20]

$$\dot{x}_i = iL_{ij}\, x_j - \int_0^t \Gamma_{ij}(t-t')\, x_j(t')\, dt' + F_i(t) \tag{88}$$

with

$$L_{ij} = \langle L\, x_i(0),\, x_j(0) \rangle \tag{89}$$

$$\Gamma_{ij} = \langle L \exp i\, \hat{\Pi}\, L(t-t')\, \Pi\, L\, x_i(0),\, x_j(0) \rangle \tag{90}$$

$$F_j(t) = i \exp i\, \hat{\Pi}\, Lt \;\; \hat{\Pi}\, L\, x_j(0) \tag{91}$$

For the special definition of a scalar product with respect to which L is a symmetric operator, we can give the integral kernel Γ_{ij} an intuitive interpretation [20], [21] - as the "correlation function" of the fluctuating forces F_i:

$$\Gamma_{ij}(t-t') = \langle F_i(t)\, F_j(t') \rangle \tag{92}$$

So far the result is exact but the kernel Γ_{ij} can be calculated exactly only for linear processes. If we evaluate e.g. in Born-Markoff approximation the integral kernel Γ, this term will lead to dissipation, while the "rest" $F_i(t)$, which has only components in the irrelevant subspace for all times t, is identified with the fluctuating force, giving eq. (88) the formal appearance of the classical Langevin equations.
The physical motivation behind this separation into relevant and irrelevant subspaces is based on the assumption that the time scale of the irrelevant variables is essentially shorter than for the relevant ones. Under this assumption, the fluctuating forces have no memory

and are assumed to be δ-correlated.

$$\langle F_i(t)\ F_j(t')\rangle = \Gamma_{ij}\ \delta(t-t') \qquad (93)$$

The matrix L_{ij} describes the coherent part of the interaction, i.e. the reversible time evolution described by the system Hamiltonian alone.
So far, we have essentially disregarded in our calculation the irrelevant degrees of freedom and have only dealt with this coherent term L_{ij}.

For a realistic formulation of the many particle problems described so far, however, we have to add the dissipative terms Γ to our previous description, and in order to guarantee the quantum mechanical consistency of the equations we have to add the fluctuating forces F. These forces, which are quantum mechanical in nature, prevent that a dissipative quantum system relaxes to a totally classical state. If the ensemble of the eliminated variables is assumed to be in thermal equilibrium with the zero temperature heat bath, the fluctuating forces describe pure quantum fluctuations. In general, however, they may contain the thermally imposed fluctuations as well. Macroscopic systems often are also subject to strong nonthermal noise sources which often can be the dominant contribution to the fluctuations.

In quantum optics there exists an important and elementary problem - the Resonance Fluorescence - which demonstrates the fundamental importance of fluctuations for the understanding of the light-matter interaction in a very elucidatory way. We will derive the general features of this phenomenon here, because it gives an intuitive and physical picture of the role of fluctuations.

The basic question of resonance fluorescence is: What are the spectral properties of the fluorescence light emitted by an ensemble of independent atoms which are continuously driven by a coherent external light source?

Under the approximation of a classical driving field, where the phenomenon is entirely linear, this problem has been solved first by Mollow[22], correcting the generally believed wrong picture of this effect[11]. In this first derivation the classical Bloch equations have been used in connection with the Quantum Regression Theorem[24]. Here we want to describe the problem in terms of the quantum mechanical Bloch equations which are of the form eq. (35-37), supplemented by fluctuating forces. This approach is physically equivalent with the previous one, but allows an intuitive understanding of the role of the fluctuating forces.

$$\dot{P}^+ = (i\Omega - \frac{1}{T_2})P^+ - 2ig\, W\, E^* + \Gamma^+ \tag{94}$$

$$\dot{W} = -\frac{1}{T_1}(W-W_o) - ig\,(P^+ E^- - E^* P^-) + \Gamma_o \tag{95}$$

where $E^* = E_o \exp i\omega_o t$ is the amplitude of the coherent external field, and $P^+ = P_{2,1}$, $W = P_{2,2} - P_{1,1}$, W_o is the equilibrium population difference in the absence of the external field E. The spatial phase factors have been dropped because we are interested in the incoherent fluorescence field which is a single atom effect. The operators $\Gamma^{\pm}$, Γ_o [6], [25] describe the fluctuations associated with spontaneous emission. The induced atomic polarization $P^{\pm}(t)$ is the source of the fluorescence light which is emitted from the sample in all directions. The spectral profile of the emitted field is proportional to the steady state polarization correlation function

$$S(\omega) = \frac{1}{2\Pi} \lim_{t\to\infty} \int_{-\infty}^{\infty} e^{i\omega\tau} \langle P^+(t+\tau)\, P^-(t)\rangle \, d\tau \tag{96}$$

where $P^{\pm}(t)$ is determined by the Langevin equations eq. (94), (95)[6]. As this system of evolution equations is linear, it can be handled in an elementary and straightforward way.

Before we formulate the exact solution of the problem, we would like to discuss for a moment the result which is obtained when the fluctuating forces in eq. (94) and eq. (95) are neglected in order to demonstrate the effect of the fluctuations.

Without the inhomogeneities, these equations are identical with the semiclassical eq. (35-37) and have formally the same solutions. The wellknown time evolution of the Bloch equation consists of an oscillatory relaxation from the given "initial condition" - which here are the operators in the Schrödinger picture - towards the unique steady state solution which oscillates at the frequency of the external field. Formally we can express this result in the following form:

$$P^{+}(t) = T(P^{\pm}(0),\ W(0),\ \{z_i\},\ t) + P^{+}_{\infty}\, e^{i\omega_0 t} \tag{97}$$

where T describes the transient relaxation

$$\lim_{t\to\infty} T(t) = 0$$

and P^{+}_{∞} is the steady state amplitude eq. (44), independent of the initial condition and therefore a pure c-number contribution. The spectral distribution obtained in this approximation solely reflects the δ-function distribution of the incoming classical laser field

$$S(\omega) = P^{+}_{\infty}\, P^{-}_{\infty}\, \delta(\omega-\omega_0) \tag{98}$$

and does not contain any specific information on the properties of the scattering atomic ensemble.

Already from a fundamental point of view it is obvious that this result has to be wrong, because by neglecting the fluctuating forces we allow the system to relax from a quantum mechanical initial state to a purely classical state, where the operators assume c-number values and therefore commute in the asymptotic limit $t\to\infty$.

$$\lim_{t\to\infty} [P^+(t),\ P^-(t)] = 0 \tag{99}$$

The fluctuating forces, however, prevent the system from settling down to the classical stationary state and conserve the quantum-mechanical properties.
If we interprete the noise operators $\Gamma^{\pm}$, Γ_o for a moment naively just as continuous random perturbations that act on the system, then we expect that the system is forced to undergo continuous transient relaxations (similar to those in"optical coherent transients"described in appendix A) and will emit a spectrum which is characterized by the eigenvalues z_i eq. (42) i=1,2,3.

The three eigenvalues - one real and two in general conjugate complex - then are expected to give rise to a three peaked spectrum with one unshifted peak at the frequency position of the external field and two side bands. All three peaks will have to be of finite width, reflecting the relaxation constants (T_1^{-1}, T_2^{-1}) of the atomic ensemble.

We will show now that these somewhat naive arguments already give us the correct picture of the phenomenon of resonance fluorescence.

The eq. (94) and eq. (95) are solved easily in terms of the Laplace transform, e.g.

$$P^+(z) = f^{-1}(z)\left(\left(z+\frac{1}{T_1}\right)\left(z+i\Delta+\frac{1}{T_2}\right)+\frac{1}{2}g^2E_o^2\right)\Gamma^+(z)$$

$$-\ i2gE_o\, f^{-1}(z)\left(z+i\Delta+\frac{1}{T_2}\right)\Gamma^o(z)$$

$$+\ 2g^2E_o^2\, f^{-1}(z)\,\Gamma^-(z)+\frac{1}{z}P_\infty^+ + \ldots \qquad (100)$$

where the dots stand for all the terms which are not of interest here because they approach zero in the limit $t\to\infty$. $P^+(z)$ is the Laplace transform of $P^+(t)\exp -i\omega_o t$, $f(z)$ is defined in eq. (42) - the zero's of f are the eigenvalues of the Bloch problem.

As the steady state correlation function eq. (96) is constructed in the limit $t\to\infty$, we have to collect only the contributions which survive this limiting procedure, and obtain the following result in a straightforward way:

$$S(\omega) = |P_\infty|^2\,\delta(\omega-\omega_o) + \sum_{l=1}^{3}\frac{D_l}{(\omega-\omega_o)^2+z_l^2} \qquad (101)$$

with

$$D_l = \frac{(2gE_o)^4}{8\Pi T_1}\;\frac{\left(\frac{1}{T_1}\right)^2+(gE_o)^2-z_l^2}{\Delta^2+\left(\frac{1}{T_2}\right)^2+(gE_o)^2}\,(z_l^2-z_{l+1}^2)^{-1}(z_l^2-z_{l-1}^2)^{-1} \qquad (102)$$

The only additional informations used are the correlation functions of the fluctuating forces which can be taken from the literature [6], [25].

In addition to the coherent δ peak we find a sum of three Lorenzian lines which are caused by the fluctuation. In order to show that this correction to the simple minded calculation does not describe at all a negligible feature, we will simplify eq. (101) for the case of a strong driving field:

$$gE_0 \gg T_1^{-1},\ T_2^{-1},\ \Delta \tag{103}$$

and obtain

$$\begin{aligned} S(\omega) = {} & (2gE_0)^{-2}\left(\Delta^2 + \left(\frac{1}{T_2}\right)^2\right)\delta(\omega-\omega_0) \\ & + \frac{1}{4\Pi}\,\frac{T_2^{-1}}{(\omega-\omega_0)^2 + \left(\frac{1}{T_2}\right)^2} \\ & + \frac{3}{16\Pi}\,\frac{T_2^{-1}}{(\omega-\omega_0-2gE_0)^2+\left(\frac{3}{2T_2}\right)^2} + \frac{3}{16\Pi}\,\frac{T_2^{-1}}{(\omega-\omega_0+2gE_0)^2+\left(\frac{3}{2T_2}\right)^2} \end{aligned} \tag{104}$$

At first, this may be a somewhat surprising result because it is just the coherent part of the spectrum which disappears in the limit eq. (103) of a strong external field. However, when we go back to the semiclassical Bloch equations we find that this tendency is obviously due to the familiar saturation behaviour of the two-level system.

We feel that this is an excellent example to demonstrate in a very transparent way the importance of fluctuation because

1. the problem is linear and can therefore be solved exactly by analytical methods;

2. the inclusion of fluctuations leads to a qualitatively different prediction from the semiclassical result;

3. there exists a conceptionally important limiting case, at least theoretically, where the fluctuation-induced part dominates the result;

4. the theoretical predictions have been verified experimentally[26].

b) Fokker-Planck Equation

The Langevin picture is only one formalism used in quantum statistics to describe the dynamics of small systems in contact with reservoirs. Other methods to eliminate the irrelevant variables are based e.g. on the density operator ρ and derive for its relevant part an effective irreversible equation of motion[(6),(18),(19)]. Depending on the special representations used for the density operator, the formal appearance of these equations may be quite different. The use of a generalized Wigner distribution e.g. allows to formulate the time evolution of the relevant variables in terms of a generalized Fokker-Planck equation[(6)].

In the following chapters we will restrict our discussion primarily on the classical nonlinear dynamics of the fields, subject to classical fluctuations. We will therefore interprete the Heisenberg equation of motion in the previous chapters in terms of classical field equations, and will add classical noise terms and dissipation. In this way we obtain coupled nonlinear Langevin equations of the following structure:

$$\dot{x}_i = K_i \left(\{x_j\}\right) + G_{ij} \left(\{x_l\}\right) \quad F_j(t) \tag{105}$$

where x_i stands for an individual field amplitude and $F_j(t)$ for the fluctuating force. Under the assumption that eq. (105) describes a continuous Markoff process with delta-correlated Gaussian forces F_j

$$\langle F_j(t) \; F_l(t') \rangle = \delta_{jl} \quad \delta(t-t')$$

this process can be formulated in the stochastically equivalent picture of the classical Fokker-Planck equation[(27)]

$$\frac{\partial}{\partial t} P(\{x_j\},t) = - \sum_i \frac{\partial}{\partial x_i} (K_i + \frac{1}{2} G_{lj} \frac{\partial G_{ij}}{\partial x_l}) P(\{x_j\})$$

$$+ \frac{1}{2} \sum_{ij} \frac{\partial^2}{\partial x_i \partial x_j} G_{il} G_{jl} P(\{x_j\},t) \qquad (106)$$

for the probability density $P(\{x_j\})$. The formal appearance of this equation is that of a continuity equation for a conserved quantity - the total probability.

The Langevin equation eq. (105) is in general a nonlinear ordinary differential equation for the field variables x_i, while the concept of the Fokker-Planck equation is a linear one.

No general methods are known to treat nonlinear equations of the Langevin type, while linear partial differential equations of the form eq. (106) are among the standard problems in mathematical physics.

For the description of nonlinear phenomena, in the following chapters we will therefore only start from the Langevin picture with its intuitive physical interpretation, but will then resort to the Fokker-Planck description when stochastic properties of nonlinear models are to be derived.

If the coefficients of the Langevin equation (105) do not depend explicitely on time, the corresponding Fokker-Planck equation can be written in the form of a nonhermitian eigenvalue problem:

$$L P_n = - \lambda_n P_n \qquad (107)$$

with

$$P(\{x_i\},t) = P(\{x_i\}) e^{-\lambda t}$$

subject in general to natural boundary conditions.
The eigenvalues λ which can form a partly discrete, partly continuous spectrum, characterize the transient relaxation to the steady state P_o, $\lambda_o=0$.

For one-dimensional problems and problems which satisfy the condition of detailed balance[28], the steady state distribution can be calculated in a straightforward way:
e.g. for a one-dimensional process we obtain

$$P_o(x) = N\, G^{-1} \exp 2 \int^{x} \frac{K(x')}{G^2(x')}\, dx' \tag{108}$$

where the constant N is introduced for normalization.

The condition of detailed balance is always guaranteed for systems in equilibrium[29], but there are fortunate cases where this condition is also satisfied "accidently" for nonequilibrium systems. This allows to solve for the steady state of some multidimensional nonequilibrium problems as well.

One example is the absorptive optical bistability in adiabatic approximation[30], an example which we will discuss in the next chapter. If detailed balance is not satisfied, one has to resort to approximation schemes even for the solution of the stationary case. We will give an example of an approximation strategy, using as an explicit example the model of dispersive optical bistability. Some basic properties of processes without detailed balance and the mathematical consequences are presented in appendix B, together with an outline of a perturbation approach for the weak fluctuation limit.

The general time dependent solution can be expanded in terms of the eigenfunction P_n which satisfy the normalization condition

$$\int P_n(x)\, P_m(x)\, P_o^{-1}(x)\, dx = \delta_{n,m} \tag{109}$$

in one dimensional problems.

If the functions P_n form a complete basis set, we can write in general

$$P(x,t) = \sum_n \int dx' \, P_o^{-1}(x') \, P_n(x') \, P(x',t=0) \, P_n(x) \, e^{-\lambda_n t}$$

where $P(x,t=0)$ is the given initial distribution. In order to calculate steady state correlation functions we also need multi-time probability distributions subject to the special initial condition $P(x,t=0) = P_o$. We find e.g. for the stationary two-time probability distribution

$$P(x_2,t_2; x_1,t_1) = \sum_n P_n(x_2) \, P_n(x_1) \, e^{-\lambda_n(t_2-t_1)} \tag{110}$$

which allows us e.g. to calculate the correlation function

$$G(\tau) = \lim_{t\to\infty} \langle x(t+\tau) \, x(t) \rangle = \sum_n g_n^2 \, e^{-\lambda_n \tau} \tag{111}$$

with $g_n = \int_{-\infty}^{\infty} x \, P_n(x) dx$ and the power spectrum as the Fourier transform of $G(\tau)$.

The eigenvalues λ_n characterize the time scales of the nonlinear diffusion process. If the spectrum is discrete and has well separated eigenvalues $\Delta\lambda=\lambda_n-\lambda_{n-1}$, we expect that for time $t \gg \Delta\lambda^{-1}$ the lowest nontrivial eigenvalue already gives a satisfactory estimate of the relaxation time of the stochastic process.

Unfortunately, in many physical important cases the time dependent Fokker-Planck equation resists an exact analytical solution, and approximate solutions have to be found. Problems subject to weak fluctuations away from any critical regions which contain only a single stationary point can well be approximated by a linearized model. However, for situations close to critical points or for multistable systems, i.e. systems which contain metastable points, the local analysis of a linearized approximation does not give a satisfactory answer to the physical questions, and better approximation methods have to be applied.

For stochastic processes which satisfy the potential condition[27] the eigenvalue problem eq. (107) can be rewritten in terms of a selfadjoint problem for which various approximation schemes have been developed. One of the methods wellknown from the elementary course of quantum mechanics is the variational method of Ritz. This method is quite generally applicable, as it does not rely on additional assumptions on the parameters of the model, as e.g. the WKB Method. Another advantage is that the variational ansatz gives a rigorous upper bound for the eigenvalue λ_n.

For an eigenvalue problem of the Fokker-Planck form we can derive a variational expression which assumes the compact and convenient form[31]

$$\lambda_n \leq \frac{\int P_o(\{x_j\}) \frac{\partial S_n}{\partial x_i} G_{ij} G_{lj} \frac{\partial S_n}{\partial x_l} dV}{\int P_o(\{x_j\}) S_n^2 \; dV} \tag{112}$$

where S_n is a suitably chosen variational test function which satisfies additional orthogonality relations as e.g.

$$\int S_1(\{x_i\}) \; P_o \; (\{x_i\}) \; dV = 0 \tag{113}$$

The summation convention is implied.

For a one-dimensional problem with a constant diffusion coefficient $G^2 = \frac{1}{2}$ Q we can rewrite the eq. (112) for the first variational eigenvalue in the rather compact form

$$\lambda_1 \leq \frac{Q}{2} \langle S_1'^2 \rangle / \langle S_1^2 \rangle \tag{114}$$

where the angular brackets denote the average over the stationary distribution P_o.

For a one-dimensional problem with the steady state $P_o(x)$ the relation eq. (113) can be satisfied e.g. by using $S_1(x)=x-\langle x\rangle$.

With this simple ansatz for S_n we obtain the following result:

$$\lambda_1 \leq \frac{Q}{2} K_2^{-1} \tag{115}$$

where K_2 is the second cumulant of the process defined by

$$K_2 = \langle x^2\rangle - \langle x \rangle^2$$

This result seems to be quite reasonable from an intuitive physical point of view.

If the Langevin equation of motion is controlled by a given external field amplitude y through the following linear dependence

$$x = K(x) + g(x)y + F(t) \tag{116}$$

we can derive with the help of eq. (108) and (115) some practically helpful relations.

The deterministic approximation of the process eq. (116) has the stationary solution $x=x_o$ defined by

$$K(x_o) + g(x_o)y = 0, \qquad x_o=x_o(y) \tag{117}$$

the local stability of which is characterized by the sign of the curvature of the effective potential

$$u(x) = \int (K(x') + g(x')y)\, dx' = u_o + y \cdot f$$

$$-\lambda = \frac{dK}{dx} + y \left.\frac{dg}{dx}\right|_{x=x_o} \tag{118}$$

taken at the stationary point ($\lambda>0$ locally stable, $\lambda<0$ locally in-stable).

A linearized solution of the stochastic eq. (116) reveals λ as the slowest relaxation rate for the local decay. Inserting (117) into (118) we find

$$\lambda = g(x_o)\, \frac{\partial y}{\partial x_o} \tag{119}$$

or with the relation $f'(x)=g(x)$

$$\lambda^{-1} = g^{-2}(x_o)\, \frac{df(x_o)}{dy} \tag{120}$$

If for the variational principle eq. (114) we choose the testfunction

$$S = f(x) - \langle f \rangle$$

we obtain

$$\lambda_1^{-1} \geq \frac{\langle f^2 \rangle - \langle f \rangle^2}{\langle g^2 \rangle} = \frac{1}{\langle g^2 \rangle}\, \frac{d\langle f \rangle}{dy} \tag{121}$$

where we used the evident relation

$$\frac{d}{dy} \langle f \rangle = \langle f^2 \rangle - \langle f \rangle^2 \qquad (122)$$

The formal resemblance of the two results is obvious:

deterministic time constant $\tau_{det} = \lambda^{-1} = \frac{1}{g^2} \frac{df}{dy}$ (123)

variational lower bound for the relaxation time $\tau_{var} = \lambda_1^{-1} \geq \frac{1}{\langle g^2 \rangle} \frac{d\langle f \rangle}{dy}$ (124)

In the limit of weak fluctuations $Q \to 0$, we are tempted to expect that the two results coincides:

$$\lim_{Q \to 0} P_o(x) \to \delta(x - x_o), \qquad \tau_{det} = \tau_{var}$$

But this is the case only for problems away from critical points or with only a single stationary point x_o. For multistable problems the definition eq (119) is not unique, and $P_o(x)$ does not necessarly collapse into a single δ-function peak. In these cases we expect even in the weak fluctuation limit considerable differences between the deterministic and the stochastic approach. An example of this behaviour is the optical bistability which we will discuss in chapter E.

With the help of these formal results we will describe in the following chapter various aspects of nonlinear phenomena under the special aspect of fluctuations.

E. FLUCTUATIONS IN NONLINEAR OPTICS

We will now combine the concepts of the two preceeding paragraphs in order to discuss nonlinear optical phenomena under the influence of internal and external noise. For macroscopic systems one might expect that fluctuations are extremely weak and not essential for the description of physical processes. If the system under consideration is in a globally stable state, the statistical nature of the macroscopic dynamics is indeed of minor importance and can safely be neglected. When, however, by changing e.g. some external parameters the state of the system approaches its limits of stability, large excursions about the deterministically described values may occur, and fluctuations are enhanced up to a degree where they play an essential role for the understanding of the macroscopic evolution. Equilibrium phase transitions are one class of examples - another class of phenomena consists of the various phase transition analogs that have been found in nonequilibrium systems. Nonlinear quantum optical processes are some of the most widely discussed examples of nonequilibrium systems that undergo phase transition[(33),(34)]. The most wellknown example among the optical processes is the single-mode laser[(6)]. Here we will, however, restrict ourselves to the special processes in nonlinear optics. Most instabilities that have been found in this field are bearing a strong resemblance to the critical behaviour of a second order phase transition[(33),(34)]. Quite recently, however, instabilities have been found which contain regions of metastability and exhibit hysteresis cycles somewhat analogous to first order phase transitions[(35-38)].

Multistable optical devices may prove to be practically important in the field of information processing, due to their fast time response. On the other hand, multistable or metastable systems are of special interest in connection with fluctuations.

In this chapter, we will now discuss the most important nonlinear processes separately under the influence of classical and quantum noise.

1) Parametric Oscillators

We consider a nonlinear crystalline medium with a strong second order susceptibility responsible for three-wave mixing, contained in an optical cavity and pumped from the outside by a coherent external field. The finite lifetime of the photons in the cavity is governed by the index of reflection $R(\omega)$.

The quantum mechanical Langevin equations for these processes can be written in the form

$$\dot{b}_1^+ = (i\omega_1 - \varkappa_1) b_1^+ + ig\, b_3^+ b_2 + F_1^+ \qquad (125a)$$

$$\dot{b}_2^+ = (i\omega_2 - \varkappa_2) b_2^+ + ig\, b_3^+ b_1 + F_2^+ \qquad (125b)$$

$$\dot{b}_3^+ = (i\omega_3 - \varkappa_3) b_3^+ + ig^* b_1^+ b_2^+ + F_3^+ + \varkappa_3 P^+ e^{i\omega_3 t} \qquad (125c)$$

where we supplemented the eq. (57) and eq. (62) by the photon damping constant $\varkappa_i = \frac{c}{L}(1 - R(\omega_i))$ and the corresponding fluctuating forces:

$$\langle F_i(t)\, F_j^+(t') \rangle = 2\varkappa_i\, \delta_{ij}\, (1 + n_{th}(\omega_i))\, \delta(t-t')$$

The field b_3 is coupled to an external coherent laser source. This coupling is simulated by the force term P in order to circumvent the delicate boundary value problem. A finite pump amplitude P will prevent the field b_3 from decaying to zero. To simplify the calculation, we will assume energy conservation:

$$\hbar\omega_3 = \hbar\omega_1 + \hbar\omega_2$$

We will find that this system eq. (125) possesses a critical threshold region. Below threshold only the laser field b_3 has the properties of a quasi-coherent field while the subharmonic fields b_1 and b_2 can be characterized by low intensity noise. In this region, we will describe the field b_3 as a classical variable but use a quantum formulation for b_1 and b_2 in order to account for the quantum noise. The quantum nature of the relevant fluctuations will show some unexpected peculiarities.

Above the threshold all three fields are partially coherent, and the process is essentially classical.

a) Quantum Fluctuations Below Threshold

We eliminate the rapid oscillations by the transformation

$$b_j^+ = \tilde{b}_j^+ \exp i\omega_j t, \qquad F_j^+ = \tilde{F}_j^+ \exp i\omega_j t$$

(no confusion will arise,when we drop the tilde again for simplicity)

Below threshold the so called signal b_1 and idler b_2 fields remain weak, and the field b_3 is entirely controlled from the outside source. This condition has to be confirmed selfconsistently at the end of the calculation. We replace

$$b_3^+ \sim P^+$$

and obtain the following closed linear system of equations for the Heisenberg operators $b_1^+(t)$, $b_2(t)$ with fluctuations.

$$\dot{b}_1^+ = -\varkappa_1 b_1^+ + i g P^+ b_2 + F_1^+$$

$$\dot{b}_2 = -\varkappa_2 b_2 - i g P^- b_1^+ + F_2 \qquad (126)$$

As these equations are ordinary linear differential equations, they can be solved by the standard analytical methods for c-numbers, bearing in mind only that the initial values are given by the operators b_1^+, b_2 in the Schrödinger picture.

The eigenvalues of this homogeneous system are given by[39]

$$\lambda_{1,2} = -\frac{1}{2}(\varkappa_1+\varkappa_2) \pm \frac{1}{2}\left((\varkappa_1-\varkappa_2)^2 + 4\varkappa_1\varkappa_2\frac{|P|^2}{P_c^2}\right)^{1/2} \tag{127}$$

where we introducedthe threshold power $P_c^2 = g^{-2}\varkappa_1\varkappa_2$. If $|P|$ approaches P_c from below, one of the eigenvalues approaches zero, marking the limit of stability of the low intensity solution, i.e. the linearization approximation.

The stationary correlation function characterizing the power spectrum

$$G(\tau) = \lim_{t\to\infty} \langle b_1^+(t+\tau)\, b_1(t)\rangle \tag{128}$$

is a linear functional of the correlation function of the random forces F_1^+ and F_2 and their hermitian conjugates

$$G(\tau) = F\,\{\langle F_i^+(t')\, F_i(t'')\rangle\} \tag{129}$$

At optical frequencies the thermal noise can safely be neglected, and all normal order correlation functions of the fluctuating forces vanish. Obviously the quantum noise of the idler field b_2, b_2^+ is therefore responsible for the fluctuations of the signal field b_1, b_1^+, while the fluctuations of this field itself have no effect on normally ordered correlations.

The solution of the eq. (126) is most easily obtained by Laplace transform and yields

$$G(\tau) = \frac{\varkappa_1 \varkappa_2^2 q^2}{\lambda_2^2 - \lambda_1^2} \left(\frac{1}{\lambda_1} e^{\lambda_1 \tau} - \frac{1}{\lambda_2} e^{\lambda_2 \tau} \right) \tag{130}$$

which leads to a spectral distribution of the following form

$$G(\omega) = \frac{2}{\sqrt{2\Pi}} \varkappa_1 \varkappa_2^2 q^2 (\omega^2 + \lambda_1^2)^{-1} (\omega^2 + \lambda_2^2)^{-1} \tag{131}$$

where $q = \frac{|P|}{P_c}$ is the external pump power normalized on the threshold value.

Close to the transition point q=1 the spectrum consists only of a single narrow Lorenzian line with the width $\Delta = |\lambda_1|$, which vanishes at the threshold.

The intensity of the random signal field is given by

$$\langle b_1^+ b_1 \rangle = \frac{2\varkappa_2}{\varkappa_1 + \varkappa_2} (q^{-2} - 1)^{-1}, \quad q < 1 \tag{132}$$

and diverges at the critical pump power q=1. For the degenerate case $\varkappa_1 = \varkappa_2 = \varkappa$ we obtain the simplified results

linewidth $$\Delta = \frac{\varkappa}{1 + \langle b_1^+ b_1 \rangle} \left(1 + \sqrt{\frac{\langle b_1^+ b_1 \rangle}{1 + \langle b_1^+ b_1 \rangle}} \right) \tag{133}$$

photon number $$\langle b_1^+ b_1 \rangle = q^2 (1 - q^2)^{-1} \tag{134}$$

Well below threshold the signal - and idler fields are qualitatively similar to thermal radiation, but mutually triggered by quantum fluctuation. For a strongly pumped system, the above solutions do no longer hold but give a qualitative understanding of how quantum fluctuation initiate the transient evolution of an instable system. The exponential growth will not go on forever, and saturation in terms of the depletion of the pumpfield b_3, b_3^+ will become important and will have to be taken into account. This behaviour is reminiscent of the laser in the threshold region, and indeed a strong analogy between these two effects exists. The laser theory can therefore be reformulated in terms of the parametric oscillator and can give a complete quantum mechanical description as long as the cavity model with the mode picture is used [39]. When, however, spatial degrees of freedom play a role, and one is concerned with the phenomena of light propagation, this analogy breaks down [16], due to the different boundary conditions of the two problems.

b) Nonlinear Classical Model

We will not repeat the laser theory here but focus our interest on a special aspect - the role of external fluctuations brought about by the fluctuations of the external driving field. This will be done by using an entirely classical picture for the field amplitude $b_i = A_i \exp i\omega_i t$, and the fluctuations. In order to simplify the nonlinear problem, we will assume that the laser field A_3 is strongly damped $\varkappa_3 >> \varkappa_1, \varkappa_2$ and follows adiabatically the slow variables A_1, A_2. By neglecting the time derivative in eq. (125c) we obtain the algebraic relation

$$A_3^* = \frac{1}{\varkappa_3} (i\, g\, A_1^*\, A_2^* + F_3^*) + P_o^* \tag{135}$$

This assumption reduces the general problem to the four-dimensional problem for the signal - and idler fields alone:

$$\dot{A}_1^* = -\varkappa_1 A_1^* - \frac{g^2}{\varkappa_3} |A_2|^2\, A_1^* + i\, g\, A_2 (P_o^* + \frac{F_3^*}{\varkappa_3}) + F_1^* \tag{136}$$

$$\dot{A}_2^* = -\varkappa_2 A_2^* - \frac{g^2}{\varkappa_3} |A_1|^2\, A_2^* + i\, g\, A_1\, (P_o^* + \frac{F_3^*}{\varkappa_3}) + F_2^* \tag{137}$$

We neglect for a moment the fluctuations in the driving field $F_3=0$ and assume equal damping of the fields $A_1, A_2, \varkappa_1 = \varkappa_2$.

$$\langle F_i(t)\, F_i(t')\rangle = Q\delta(t-t')$$

In this case the stationary Fokker-Planck equation corresponding to the Langevin equation (136), (137) is solved in a straightforward way, and we obtain

$$P(\{A_i, A_i^*\}) = \exp - \frac{1}{Q}\, u(\{A_i, A_i^*\}) \tag{138}$$

with the potential[40]

$$u = \varkappa \sum_{i=1}^{2} |A_i|^2 + ig\,(P_o^* A_1 A_2 - c.c.) + \frac{g^2}{\varkappa_3} |A_1|^2 \; |A_2|^2 \tag{139}$$

which governs at the same time the deterministic equations of motion in the following way

$$\dot{A}_i^* = - \frac{\partial u}{\partial A_i} \tag{140}$$

For the pump intensity $|P|^2$ below threshold eq. (138) has a single peak at the origin describing the noisy signals below threshold. Above threshold the stationary distribution has an extremum in the four-dimensional space at

$$|A_i|^2 = \frac{\varkappa\varkappa_3}{g^2}\,(q-1) \tag{141}$$

and

$$\arg\,(P_o^* + A_1 + A_2) = \Pi/2$$

which is continuously degenerate with respect to the phase difference arg (A_1-A_2). This is interpreted as phaselocking for the sum of the field phases to the external field, while the phase difference diffuses freely analogous to a one-dimensional random walk process. This phenomenon is well known from the phase diffusion of the single mode laser above threshold.

The description in the four-dimensional phase space of signal and idler does not give immediately an intuitive and simple picture of the process. We will therefore go one step further and eliminate also the idler field adiabatically by assuming that the idler field is strongly damped as well $\varkappa_2 >> \varkappa_1$, and arrive at the following two-dimensional model:

$$\dot{A}_1^* = dA_1^* - b\,|A_1|^2\,A_1^* + \tilde{F}_1^* + A_1^*\,\tilde{F}_2^* \qquad (142)$$

where

$$d = \varkappa_1\,(q^2-1) \qquad \text{and} \qquad b = g^2\,\frac{\varkappa_1}{\varkappa_2\varkappa_3}\,q^2 \qquad (143)$$

The $\tilde{F}_i$ are collections of the various fluctuating forces F_i. If we neglect the multiplicative noise source $\tilde{F}_2$, the model eq. (142) is identical with the model of the single mode laser in adiabatic approximation, and all the fundamental results from the laser theory apply here as well[(6),(28),(40)]. In contrast to this approach we now want to assume that the additive noise source is of negligible strength compared to the multiplicative one. This can be substantiated by the fact that the multiplicative force F_2 contains the fluctuations of the driving laser field, the strength of which can in principle be controlled externally.

By these arguments we are motivated to discuss the properties of the following multiplicative stochastic model[(41),(42)].

$$\dot{A} = dA - b\,|A|^2\,A + AF \qquad (144)$$

and assume for the fluctuating forces:

$$\langle \mathrm{Re}\, F(t)\ \mathrm{Re}\, F(t') \rangle = Q\delta(t-t')$$

$$\langle \mathrm{Im}\, F(t)\ \mathrm{Im}\, F(t') \rangle = Q\delta(t-t') \qquad (145)$$

It is possible to solve this problem exactly by means of analytical methods when we express this process by the stochastically equivalent Fokker-Planck equation[(41),(42)].

$$\frac{\partial P}{\partial t} = \frac{1}{r}\frac{\partial}{\partial r}\left[r(dr-br^3-\frac{Q}{2}\frac{\partial}{\partial r}r^2)P\right] + \frac{Q}{2}\frac{\partial^2}{\partial\varphi^2}P \qquad (146)$$

where polar coordinates have been used: $A = r \exp - i\varphi$

The general result

$$P(r,\varphi,t) = \sum_{n,m} C_n^m P_n^m(r)\, e^{im\varphi}\, e^{-\lambda_n^m t} \qquad (147)$$

subject to natural boundary conditions, i.e. $\lim_{r\to\infty} P(r)=0$ is obtained by standard methods.

For the discrete branch of eigenfunctions we find

$$P_n^m(r) = N\, r^{-2+2(d/Q\ -n)}\, e^{-\frac{b}{Q}r^2}\ {}_1F_1(-n,\frac{d}{Q} - 2n+1, \frac{br^2}{Q}) \qquad (148)$$

with the corresponding eigenvalues

$$\lambda_n^m = \frac{1}{2}m^2Q + 2nQ\,(\frac{d}{Q} - n) \qquad (149)$$

subject to the restriction $\frac{d}{Q} > 2n$. N is determined by the normalization condition. In addition we find a continuous branch in the spectrum of decay rates λ.

An experimentally observable quantity is the two-time correlation function which - with the general expression of chapter D, eq. (110) (111) - can be written in the following form:

$$\lim_{t\to\infty} \langle A^*(t+\tau)\, A(t)\rangle = \sum_n (g_n^1)^2 \quad e^{-\lambda_n^1 \tau} \tag{150}$$

with

$$g_n^1 = \int P_n^1(r)\, r^2\, dr \tag{151}$$

We notice a drastic difference between the multiplicative process here and the analogous additive one like e.g. the single mode laser. While with increasing pump rate the phase diffusion in the case of the laser slows down continuously, leading to the well known line narrowing effect above threshold, here the corresponding relaxation rate remains essentially constant. In other words: The fluctuations of the external driving field set a lower bound to the line width of the parametric process. With increasing pump power the average field amplitude increases, and so does the effective strength of the fluctuations which then balances exactly the line narrowing effect, well known for cases of phase diffusion with additive noise.

2) Subharmonic Generation

The effect of subharmonic generation is the special case of a degenerate parametric oscillator with identical signal and idler frequencies $\omega_1 = \omega_2$. Therefore it would be superfluous to add here a chapter about this effect, because not very much new could be learned on top of what we already know from the previous paragraph.

Here we would like to emphasize however, a new aspect of subharmonic generation which comes about when both modes are pumped simultaneously by coherent or partially coherent external laser fields. This device can operate in two different states and can be switched between these states by external means. The occurence of metastable states and hysteresis is new in connection with parametric processes described so far, and bears a strong resemblance to first order phase transitions [43], [44].

We consider the nonlinear coupling of two waves of frequency ω_1 and ω_2 via a nonlinear medium in an optical cavity. The coherent pump field P_2 at the fundamental frequency ω_2 provides the pump necessary to establish subharmonic oscillations, while the additional pump P_1 is used to control the subharmonic bistability. We eliminate the fundamental field A_2 by an adiabatic assumption $\varkappa_2 \gg \varkappa_1$ and obtain the following Langevin equation for the subharmonic field:

$$\dot{A}_1 = -\varkappa_1 A_1 - \frac{g^2}{2\varkappa_2} |A_1|^2 A_1 + \left(\frac{g}{\varkappa_2}\right) P_2 A_1^* + P_1$$

$$+ F_1 + \left(\frac{g}{\varkappa_2}\right) A_1^* F_2 \qquad (152)$$

where the noise forces F_1, F_2 are assumed to be statistically independent and δ-correlated

$$\langle F_i(t)\, F_j(t')\rangle = 2Q_i\, \delta_{ij}\, \delta(t-t') \qquad (153)$$

As the process eq. (152) cannot be solved in full generality, we will restrict ourselves to two special cases:

a) Additive Fluctuations ($F_2 = 0$)

In this case, the process eq. (152) satisfies the condition of detailed balance, and the steady state distribution can be given immediately[44]

$$P_o(A_1, A_1^*) = N \exp - \frac{1}{Q_1} \phi(A_1, A_2^*) \qquad (154)$$

where the potential is given by

$$\phi = \kappa_1 |A_1|^2 - \left(\frac{g}{2\kappa_2}\right) (P_2 A_1^{*2} + P_2^* A_1^2)$$

$$- (P_1^* A_1 + P_1 A_1^*) + \left(\frac{g^2}{4\kappa_2}\right) |A_1|^4 \qquad (155)$$

In order to discuss the properties of the potential ϕ we introduce polar coordinates:

$$A_1 = r\, e^{i\varphi}, \qquad P_1 = r_1\, e^{i\varphi_1}, \qquad P_2 = r_2\, e^{i\varphi_2}$$

Without limiting the generality of the calculations we can take the pump P_2 to be real. For the phase of the pump field P_1 we discuss two special cases, $\varphi_1 = 0$ and $\varphi_1 = \frac{\Pi}{2}$.

i)

In case $\varphi_1 = 0$ we find two minima of the potential Φ, separated by a saddle point, if the following relations are satisfied:

$$r_1 = r_{1,o}^c = \left(\frac{8}{27}\, g (r_2 - r_2^c)^3\, \kappa_2^{-3}\right)^{1/2} \qquad (156)$$

and

$$r_2 > r_2^c = \varkappa_1 \varkappa_2 / g^2$$

The intensities of the two minima are roughly the same, while the phases are $\varphi=0$, $\varphi=\Pi$. The probability of the inphase component $\varphi=\varphi_1=0$ increases with increasing pump amplitude r_1. If r_1 is increased beyond the critical value $r^c_{1,0}$, the potential has only a single minimum. This result reflects the tendency of the subharmonic field to adjust to the phase of the external pump field P_1.

ii)

When we choose $\varphi_1 = \frac{\Pi}{2}$, the potential has two minima located at the positions

$$r^2 = \frac{1}{g}(r_2 - r_2^c), \qquad \sin\varphi = \frac{r_1}{r_2} \frac{\varkappa_2}{\sqrt{8g}} (r_2 - r_2^c)^{-1/2} \tag{157}$$

provided that the following condition is met:

$$r_1^2 < (r^2_{1,\Pi/2})^2 = \frac{8g}{\varkappa_2^2} (r_2 - r_2^c)\, r_2^2$$

The dependence of these minima on the pump power P_1 is most easily followed in the fig. 1a-1f given on the next page:
We have plotted $P(x,y)$, $x=\mathrm{Re}A_1$; $y=\mathrm{Im}A_1$ for fixed values of r_2 and for different values of r_1, φ_1.

Fig. 1a - 1c $\qquad 0 < r_1 < r^c_{''0}$, $\quad \varphi_1 = 0$

Fig. 1d - 1f $\qquad 0 < r_1 < r^c_{''\Pi/2}$, $\quad \varphi_1 = \Pi/2$

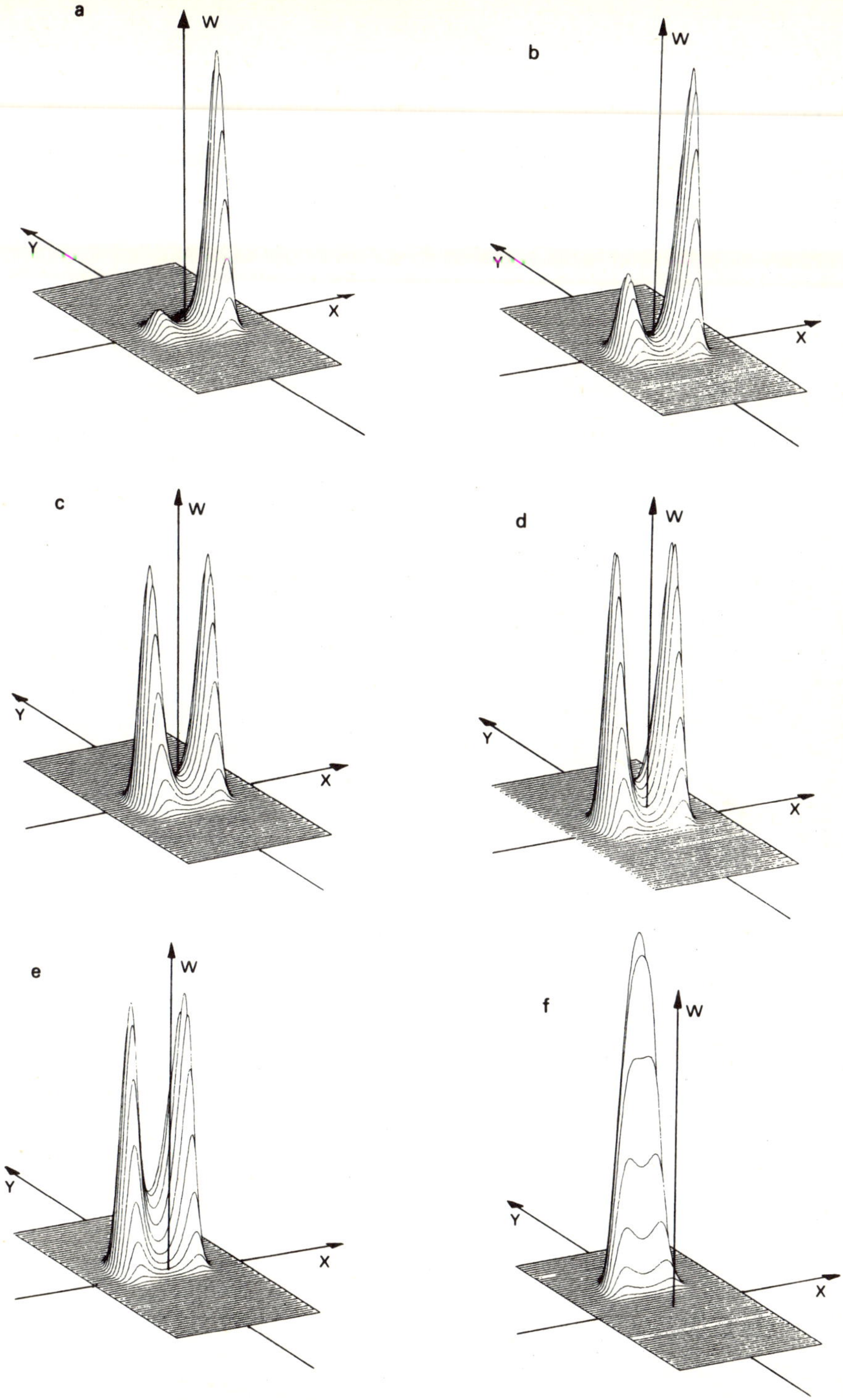
a
W
Y
X
b
W
Y
X
c
W
Y
X
d
W
Y
X
e
W
Y
X
f
W
Y
X

b) Additive and Multiplicative Fluctuations

A general solution for the steady state P_o has not been found so far. When we recognize, however, the phase locking tendency in these coherent interactions, we may as well discuss the one-dimensional approximation disregarding the phase fluctuations. In this case both pump fields have to be taken in phase.

The potential obtained under this assumption

$$\phi = \left(\frac{\varkappa_2^2\varkappa_1}{g^2Q_2} + \frac{1}{2} - \frac{Q_1\varkappa_2^3}{2Q_2^2g^2} - \frac{P_2\varkappa_2}{gQ_2}\right)\ln\left|A_1^2 + \frac{Q_1\varkappa_2^2}{Q_2g^2}\right|$$
$$+ \frac{\varkappa_2A_1^2}{2Q_2} - \frac{2P_1\varkappa_2}{2\sqrt{Q_1Q_2}}\arctan\left(A_1\frac{g}{\varkappa_2}\sqrt{\frac{Q_2}{Q_1}}\right) \tag{158}$$

is a model for the amplitude diffusion under the simultaneous influence of additive and multiplicative noise sources. This potential describes the inphase bistable behaviour, which is qualitatively similar to eq. (155) as long as the multiplicative fluctuations remain weak. For increasing Q_2, however, the subharmonic oscillations disappear as the multiplicative fluctuations increase the thresholds pump power:

$$P_2 \geq \varkappa_1\varkappa_2/g^2 + \frac{g^2}{2\varkappa_2}Q_2 \tag{159}$$

This is easily understood in physical terms, when we remember that the noise Q_2 contains, besides other sources of fluctuations, the noise of the external pump field P_2. For a purely incoherent pumpfield, however, subharmonic oscillations cannot be expected, and the result eq. (158) interpolates between the coherent and the incoherent limiting cases.

3) Absorptive Optical Bistability

An ensemble of two-level atoms contained in a partially reflecting optical cavity, driven on resonance by an external laser field, exhibits a bistable behaviour with respect to the transmitted field intensity[37]. The deterministic equations contain regions with only a single stationary point for small and large pump field intensities, and a domain of multistable stationary points in the region of intermediate intensities. Utilizing the linear stability analysis, the deterministic equations allow to differentiate between e.g. stable and instable stationary points. As the linearization procedure can only make predictions about the local properties, stable points may still turn out to be only metastable when the global dynamics is considered. The inclusion of fluctuations will reveal the global stability properties in a quantitative way, because they provide a mechanism for the relaxation towards a unique equilibrium or steady state distribution even when potential barriers have to be overcome.

How important this distinction between metastable and stable really is from a practical or experimental point of view, depends on the time scale of relaxation.

We will present here a stochastic theory of optical bistability in terms of a simple model[38] which still contains the basic features of the effect.

The steady state solution of the corresponding Fokker-Planck equation will allow us to differentiate between stable and metastable states, while the timedependent solution will give a quantitative estimate on how longlived the metastable states really are.

We start from the Langevin equation (94) and eq. (95) for the dynamics of the atomic system, and include explicitly the Maxwell equation for a single cavity mode, driven externally by the laser field E_o

$$E^* = (i\omega-\varkappa)\, E^* + i\, g\, P^+ + \varkappa\, E_o^+\, e^{i\omega_o t} + F^+ \qquad (160)$$

$F^{\pm}$ describes the fluctuation corresponding to the dissipative term $\varkappa E^{+}$ as well as the fluctuations of the external source field E_{o}^{+}. $\bar{g} \sim g \frac{N}{V}$ For the discussion in this paragraph we will make the simplifying assumption that the three fields P, E, E_o are in resonance, $\omega = \omega_o = \Omega$. This restriction will be lifted in the next section when we include dispersive effects as well.

The entire five-dimensional problem resists all attempts for an exact and general solution, and we will use the adiabatic principle again to eliminate the atomic degrees of freedom under the assumption $T_2\varkappa << 1$, and obtain, by collecting all terms systematically in lowest order in $\varkappa T_2$ (30):

$$\frac{d\tilde{E}}{dt} + \frac{1}{2}\tilde{E}\frac{d}{dt}|\tilde{E}|^2 = -\tilde{E}|\tilde{E}|^2 + \frac{1}{2}\tilde{E}(\tilde{E}_o\tilde{E}^* - \tilde{E}_o^*\tilde{E})$$

$$+ (1+\Gamma^2)\tilde{E} + \tilde{E}_o \quad + \text{fluctuations} \tag{161}$$

where we used the rescaled variables

$$\tilde{E}^{\pm} = E^{\pm}(4g^2T_1T_2)^{1/2}$$

$$\tilde{E}_o^{\pm} = E_o^{\pm}(4g^2T_1T_2)^{1/2}$$

$$\tilde{t} = t\varkappa$$

and

$$\Gamma^2 = -2g\bar{g}\,T_2\,\varkappa^{-1}\,w_o \tag{162}$$

We will drop the tilde again to simplify the notation. The fluctuations contain the following contributions (30):

i) Fluctuation of the polarization

$$\sim -i\frac{\Gamma^2}{w_o}(T_1T_2)^{1/2}\,\Gamma^{\pm} \tag{163}$$

ii) Fluctuations of the inversion, which are multiplicative in nature

$$\frac{\Gamma^2}{w_o} T_1 E \Gamma^o \tag{164}$$

iii) Fluctuations of the pump field $E_o^{\pm} \rightarrow E_o^{\pm} + F_o^{\pm}$ and $\langle F_o^+ F_o^- \rangle = 2Q\delta(t-t')$

$$F_o^- (1+\frac{1}{2} |E|^2) + \frac{1}{2} F_o^+ E^2 \tag{165}$$

Models which contain each of these fluctuating terms separately have been discussed with respect to their stationary behaviour[30].

The only model, however, which allows a solution including amplitude as well as phase fluctuations is the third case (iii), and we will restrict ourselves to this model here.

The Fokker-Planck equation corresponding to this process eq. (161) with the fluctuations eq. (165) can be written in terms of polar coordinates:

$$\frac{\partial}{\partial t} P(r,\varphi,t) = \frac{1}{r}\frac{\partial}{\partial r} r \left(r-r_o \cos(\varphi-\varphi_o) + \frac{\Gamma^2 r}{1+r^2} + \frac{Q}{2}\frac{\partial}{\partial r}\right)P$$
$$+ \frac{1}{r}\frac{\partial}{\partial \varphi}\left(r_o \sin(\varphi-\varphi_o) + \frac{Q}{2}\frac{1}{r}\frac{\partial}{\partial \varphi}\right)P \tag{166}$$

with $E = r \exp -i\varphi$, $E_o = r_o \exp -i\varphi_o$

a) <u>Stationary Properties</u>

In spite of the fact that eq. (166) describes a process which can be far from thermal equilibrium, this process nevertheless satisfies the condition of detailed balance, and the steady state distribution is obtained in a straightforward way:

$$P_o(r,\varphi) = N\,(1+r^2)^{-\Gamma^2/Q} \exp -\frac{1}{Q}\,(r^2-2rr_o\cos(\varphi-\varphi_o)) \tag{167}$$

where N is included for normalization.

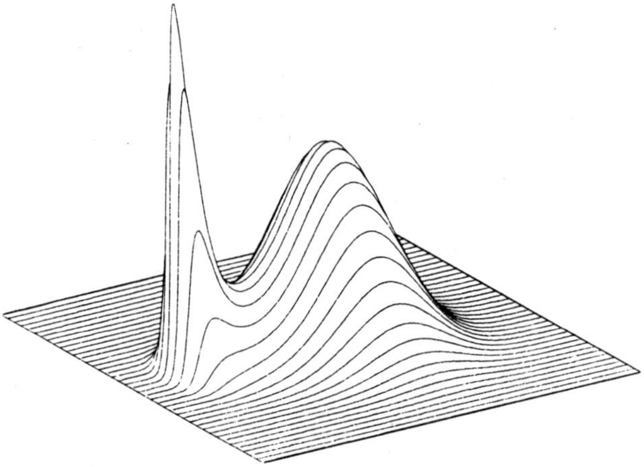

Fig. 2: Steady state probability $P_o(r,\varphi)$ for absorptive optical bistability (eq. 167)

In fig. 2 the probability distribution is plotted for a special choice of parameters in the region of bistability. The peak characterizing the low intensity stationary point, i.e. the absorptive solution, is narrower than the peak describing the saturated or bleached state of the device.

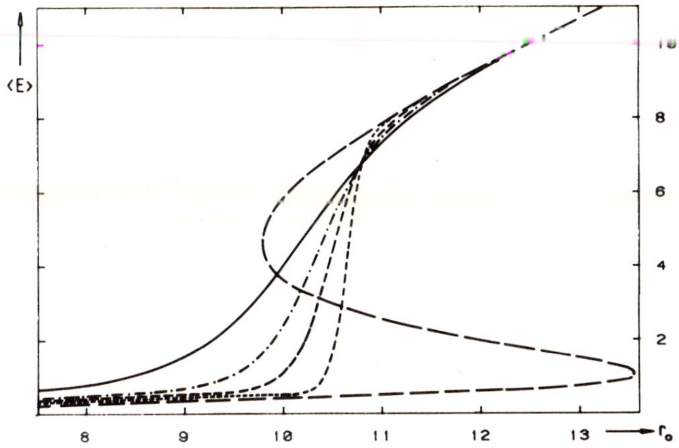

Fig. 3: Averaged amplitude <E> for absorptive bistability. Compared with the deterministic steady states (dashed curve).

With the exact solution eq. (167) we are in a position to calculate all steady state expectation values. The average field amplitude e.g. $\langle E\rangle = \int rP_o \, rdrd\varphi$ is plotted in fig. 3 as a function of the average amplitude E_o of the driving field. The dashed line indicates the hysteresis cycle of the deterministic theory. In the limit $Q \to 0$ the averaged field <E> will coincide with the most probable value in the monostable regimes. As <E> is a single valued function of E_o, it will have to decide which branch it is going to follow inside of the bistable domain. As expected, we find that a sudden transition occurs from one branch to the other. The point where the two branches exchange *global* stability, corresponds e.g. to the coexistence vapor pressure of the van der Waals gas, which can be characterized by the Maxwell construction. One can show that the Maxwell construction for this nonequilibrium but detailed balance case still holds[46]. In contrast to this, the boundaries of the bistable domain are defined by the limits of *local* stability.

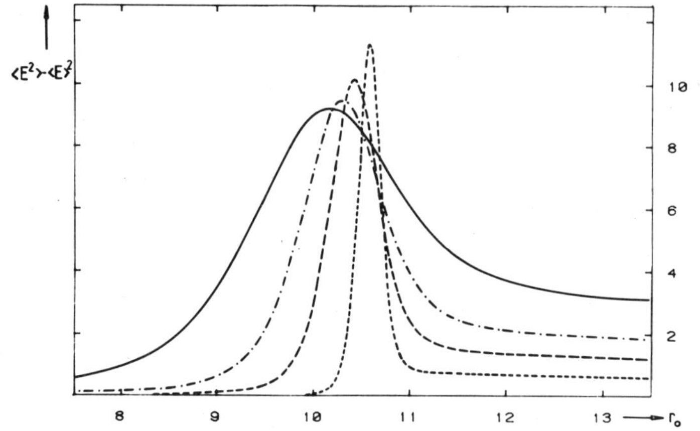

Fig. 4: Second cumulant for absorptive bistability.

Using the relations of eq. (122), the second cumulant which characterizes the fluctuations of the field, is easily obtained merely by differentiation - the result is plotted in fig. 4.

It clearly exhibits enhanced fluctuations at the transition point mentioned before.

Here we want to emphasize that the point characterized by enhanced fluctuations is the point of stability exchange of the two branches and not the boundaries of the bistable regime which have recently been called erroneously 'critical points' in a number of publications.

It should also be mentioned that this model includes the single mode laser if we set $E_o = 0$; $\Gamma^2 = -1$ defines the laser threshold. The classical laser model is only obtained after expanding the logarithmic term in the exponent up to second order in the field intensity. A difference between the two models, however, is only observed in the asymptotic dependence $P(r)$, $r \to \infty$ of the probability density.

b) Dynamic Properties

In order to discuss the dynamical properties of optical bistability we would need a general solution of the time dependent Fokker-Planck equation eq. (166). So far, no exact solution has been found and we have to resort to approximation methods in order to determine what metastability really means quantitatively for the physical process here. When we look for an approximation strategy, we have to keep in mind that we are dealing with a rather complicated model:

i) Linearization can only be used in the monostable regimes well away from the boundaries of the bistable domain.

ii) A "mean first passage time" calculation will deliver reasonable results well inside of the bistable domain but deteriorates, when the boundaries are reached.

iii) In order to demonstrate the role of fluctuation we would like to compare results for strong as well as for weak fluctuations.

A promising method which may overcome all these problems is the variational principle as described in chapter D eq. (112).

This principle does apply here because the problem under consideration satisfies the potential condition(27) and an exact analytical solution for the steady state is already known. The only troublesome point in the variational approach may lie in the fact that one may not have enough physical or mathematical motivation for guessing a proper variational function. The simple appearance of the variational expression eq. (112), however, is very helpful for choosing a suitable ansatz.

Before we will proceed in this way, it may be interesting to stop for a moment and see if we cannot somehow guess the result qualitatively.

If the fluctuations are very weak, we expect that already the local stability consideration will give satisfactory results. This concept has been outlined in chapter D, eq. (123) and eq. (124), where from the deterministic equations we define the relaxation rate simply by differentiating the relation $\hat{E} = \hat{E}\,(E_o)$, where $\hat{E}$ is the deterministic or most probable field amplitude. This means that we have to differentiate only the dashed curve in fig. 3 and invert the result. The corresponding curve is plotted in fig. 5 as a dashed line marked $Q = 0$. This curve is not single valued in the bistable domain and therefore cannot be the proper approximation for the global relaxation in the weak noise limit, but may still approximate satisfactorily the relaxation rates in the monostable regimes. In the limit $Q \to 0$, when the potential in P contains two minima, there exists no physical process which would allow the system to approach the steady state. The relaxation rate λ_1 will therefore have to vanish inside the bistable domain in this limit.

In fig. 5 we compare the two cases $Q = 0$ and $Q \to 0$ and see that they do not agree when bistability occurs.

A better approximation is given by the relation eq. (124) where we derived a general approximate upper bound from the variational principle which is determined simply by differentiating the averaged field amplitude $\langle E \rangle$ and not as above the most probable value, with respect to the pump field amplitude E_o. This, however, has already been done in fig. 4. The only thing left for us to do is to invert this picture and compare it with our estimates above. This result has been plotted in fig. 6 for different strengths of the fluctuations.

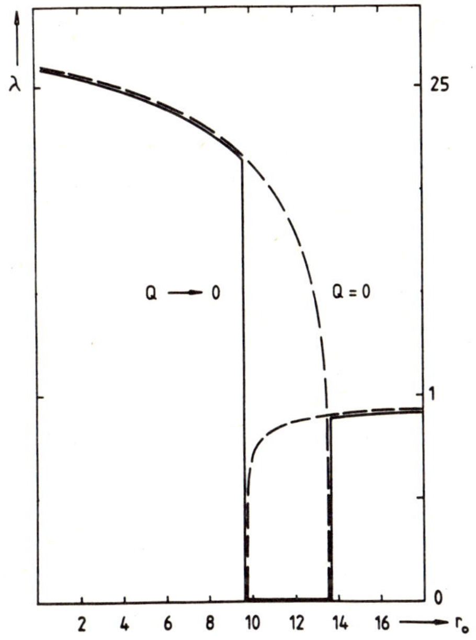

Fig. 5: Comparison of the determinstic time constants (Q = O) with the Fokker-Planck limit Q → O.

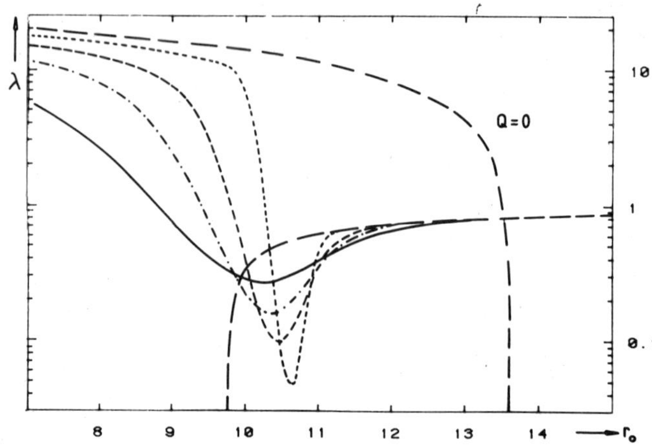

Fig. 6: A simple variational estimate of the eigenvalue λ_1 (eq. 115)

We expect that the correct variational eigenvalue λ_1 (E_o, Q) will follow qualitatively the deterministic predictions in the monostable region regime, but will go through a deep minimum inside of the bistable domain. For decreasing fluctuations, we expect that the curve $Q \to O$ will be approximated better and better, the smaller the fluctuations are. No singularity or irregularity, however, is expected to occur at the boundaries of the metastable domain.

The variational calculation has been performed, using the following ansatz(31):

$$S_1(r,\varphi) = \frac{1-\exp-a(r-r_1)}{1+\exp-a(r-r_2)} \tag{168}$$

and the results are plotted for different values of the fluctuations Q on the linear scale in fig. 7 and on a log scale in fig. 8.

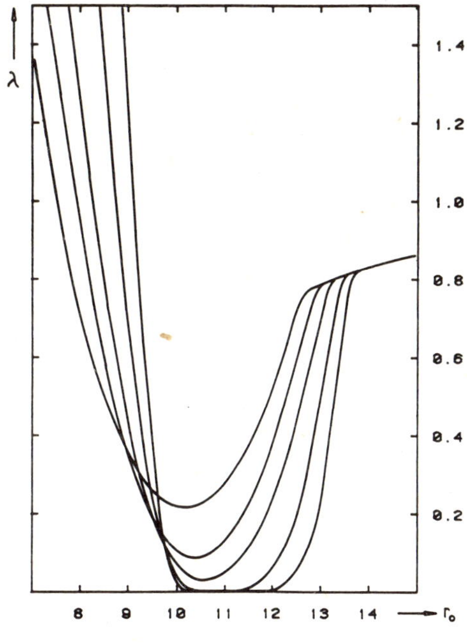

Fig. 7: Variational eigenvalue λ_1 using the testfunction eq. 168

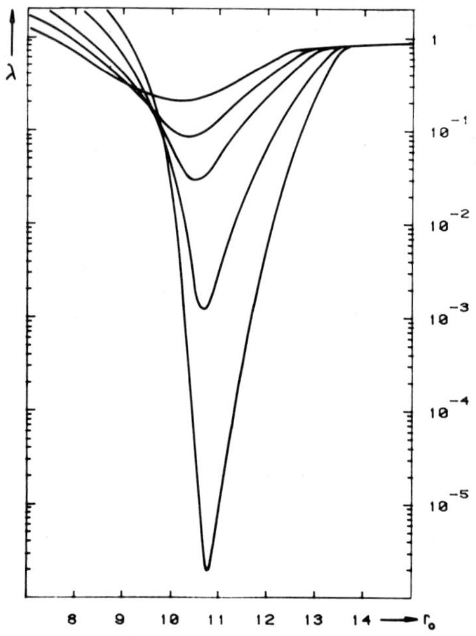

Fig. 8: The same result as fig. 7 but on a log scale.

From the linear plot it is rather obvious that the $Q \to 0$ prediction of fig. 5 is approached while the logarithmic scale plot exhibits the drastic variation inside the bistable domain, but a continuous, smooth behaviour at the boundaries of bistability.

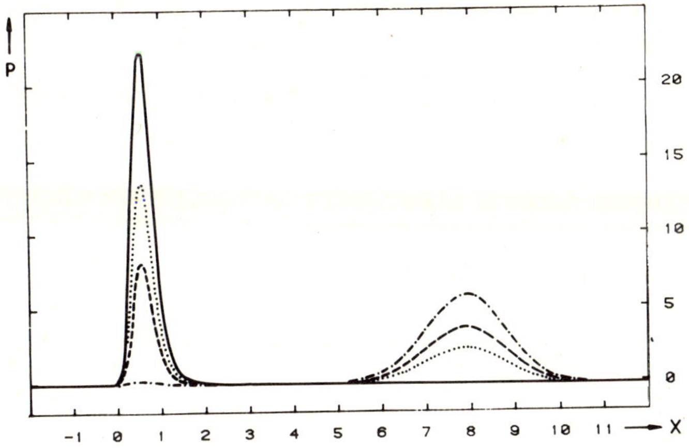

Fig. 9: Time dependent probability P(x,t) according to eq. 169.

In order to get an intuitive picture of the transition behaviour of the probability distribution P(x,t) which can be described by using only the steady state eigenfunction P_o and the variational approximation $P_o \cdot S$ for the first "excited" state, we have plotted in fig. 9 the time evolution of the distribution which has been localized initially around the metastable low intensity state. To simplify the plot, we have chosen the one-dimensional limit of the previous model

$$P(x,t) = P_o(x)\ (1-S_1(x)\ e^{-\lambda_1 t}) \qquad (169)$$

4) Dispersive Optical Bistability

The assumption of perfect resonance between the cavity mode ω and the atomic transition frequency Ω on the one hand and zero detuning between the cavity mode and the external laser field ω_o on the other hand was the essential assumption to establish detailed balance, and to provide us with an exactly solvable model. If one of the two resonance conditions is violated, the condition of detailed balance is violated as well, and there is no systematic way to derive an exact solution even for the steady state. Introducing the detuning parameter $\delta = (\omega - \omega_o)/\varkappa$ and $\Delta = (\omega - \Omega) T_2$ we can reformulate the Fokker-Planck equation (166) for this generalized case:[45]

$$\frac{\partial P}{\partial t} = -\frac{1}{r}\frac{\partial}{\partial r} r \left(\frac{r}{1+r^2}(1+r^2+\Gamma^2) - r_o \cos(\varphi - \varphi_o) + \frac{Q}{2}\frac{\partial}{\partial r}\right) P$$

$$+ \frac{1}{r}\frac{\partial}{\partial \varphi}\left(\delta r + \Delta\Gamma^2 \frac{r}{1+r^2} + r_o \sin(\varphi - \varphi_o) + \frac{Q}{2}\frac{1}{r}\frac{\partial}{\partial \varphi}\right) P \qquad (170)$$

An exact general solution for the steady state of this process is not known.

In this paragraph we will present an exact solution of a special case of finite but symmetric detuning $\Delta = \delta$ and an approximate solution for the general case $\Delta \neq \delta$ in the limit of weak fluctuations.

a) The special case of equal detuning $\Delta = \delta$

In spite of the fact that detailed balance is not satisfied for this model as long as $\delta \neq 0$, $\Delta \neq 0$, we can derive an exact steady state solution in this special case for equal detuning $\delta = \Delta$ in the following form :[45,46]

$$P_o(r,\varphi) = N \exp - \frac{1}{Q}\left(r^2 - 2\frac{rr_o}{\sqrt{1+\Delta^2}}\cos(\varphi-\varphi_o-\psi_o) + \Gamma^2 \ln(1+r^2)\right) \qquad (171)$$

where

$$\tan\psi_o = -\Delta = -\delta = \frac{\omega_o-\omega}{\varkappa}$$

The new features of dispersive optical bistability in this model compared with the absorptive case eq. (167) are the rescaling of the external field amplitude $r_o \rightarrow r_o(1+\Delta^2)^{1/2}$ and the rotation of the entire distribution by the angle ψ_o. The most probable phase shift between the cavity mode and the external field therefore is neither zero nor $\Pi/2$, but intermediate to pure dissipation and pure dispersion.

According to the general discussion of appendix B we are now in the position to decompose the drift vector into the gradient part and the residual part r which governs the steady state probability current j_i. We can write r_i as a superposition of a reversible and an irreversible part:

$$r_i = r_i^{rev} + r_i^{irr} \qquad (172)$$

Using the potential (171) we find explicitly in polar coordinates:

$$r^{rev} = \frac{\Delta r_o}{1+\Delta^2}\begin{pmatrix}\sin\varphi \\ \cos\varphi\end{pmatrix} - \Delta r\begin{pmatrix} 0 \\ \frac{1+\Gamma^2+r^2}{1+r^2}\end{pmatrix} \qquad (173)$$

$$r^{irr} = -\frac{\Delta^2 r_o}{1+\Delta^2}\begin{pmatrix}-\cos\varphi \\ +\sin\varphi\end{pmatrix} \qquad (174)$$

If no external field r_o is applied, the system relaxes to equilibrium and r_i transforms like a reversible current depending on the sign of Δ in accordance with our general consideration. A non vanishing external field r_o drives the system away from thermal equilibrium changing the reversible part (173) and introducing a non zero irreversible part (174). Notice that r_i^{irr} exists only for a detuned system far from thermal equilibrium i.e.

$$r^{irr} \sim r_o \Delta^2$$

and that r_i^{irr} does not change sign with the frequency mismatch Δ.

b) Approximate solution in the limit of weak fluctuations

For the general case we will utilize now the perturbation scheme outlined in appendix B based on the weak fluctuation limit $\varepsilon \to 0$ which allows to approximate the Fokker-Planck equation (170) formally by a Hamilton Jacobi equation. In accordance with the arguments there we will use the amplitude r_o of the external field as the expansion parameter. Setting $\Phi_n = 1/Q\ \varphi_n$ we obtain in zero order

$$\varphi_o = (r^2 + \Gamma^2 \ln (1+r^2)) \tag{175}$$

For the correction in first order in the field r_o we obtain the following linear inhomogeneous partial differential equation of first order:

$$\frac{\partial \varphi_1}{\partial r} \left(-\frac{r}{1+r^2}(1+\Gamma^2+r^2)\right) + \frac{1}{r}\frac{\partial \varphi_1}{\partial \varphi} r\left(\delta + \frac{\Delta \Gamma^2}{1+r^2}\right) = - \frac{r r_o}{1+r^2}(1+\Gamma^2+r^2) \cos \varphi \tag{176}$$

This equation can be solved exactly by the standard method of characteristics and we obtain[45,46]

$$\Phi = \frac{1}{Q}\left(r^2 - 2\frac{rr_o}{\sqrt{1+\delta^2}}\cos(\varphi-\psi_o) + \Gamma^2 \ln(1+r^2)\right)$$

$$- 2\frac{\Gamma^2}{Q}(\delta-\Delta)\frac{rr_o}{\sqrt{1+\delta^2}}\,\mathrm{Im}\,\frac{e^{i(\psi-\psi_o)}}{1+\Gamma^2-i(\Delta\Gamma^2+\delta)}\,F\left(-\frac{r^2}{1+\Gamma^2}\right) \qquad (177)$$

where F(z) is a hypergeometric function in the notation of

$$F(z) = {}_2F_1\left(1, \frac{1}{2} - \frac{i\delta}{2}; \frac{3}{2} - \frac{i}{2}\frac{\delta+\Gamma^2\Delta}{1+\Gamma^2}; z\right) \qquad (178)$$

The potential Φ is a unique solution of the linearized Hamilton Jacobi equation under the requirement of single-valuedness, and approximates the exact result rigorously to first order in r_o and $1/Q$. This is the most general result we have derived for this problem for finite Δ, δ, $\Delta \neq \delta$, and r_o by analytical methods. It illustrates the scope and the power of the general methods described in appendix B and allows us to describe the stationary statistical properties of dispersive optical bistability in detail.

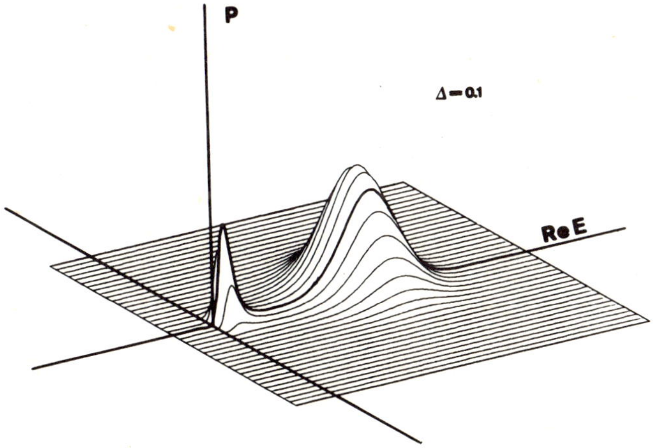

Fig. 10: Steady state probability $P_o(r,\varphi)$ for dispersive optical bistability eq. (179).

The expansion eq. (177) is rather involved and does not give an immediate and intuitive picture of the process. A simplified result, however, which still contains most of the essential features of dispersive optical bistability is obtained for $\delta=0$ by expanding the result to first-order in Δ:

$$P(r,\varphi) = N \exp -\frac{1}{Q} \quad r^2-2rr_o \cos(\varphi-\varphi_o)+\Gamma^2 \ln(1+r^2)$$

$$+ 2\Delta\Gamma^2 \frac{r_o}{\sqrt{1+\Gamma^2}} \sin(\varphi-\varphi_o) \arctan \frac{r}{\sqrt{1+\Gamma^2}} \qquad (179)$$

This distribution is plotted in fig. 10.

One new feature of the dispersive effect is that in addition to the hysteresis cycle for the field intensities we also obtain a hysteresis for the phases of the field. In fig. 11 and fig. 12 we compare the phases of the most probable values and of the deterministic steady state as a function of the pump field amplitude E_o and the detuning parameter Δ; δ has been set equal to zero for simplicity. This comparison gives us an idea of the value of the approximation scheme, because for an exact solution of the Hamilton-Jacobi equation the two results would have to coincide. The difference therefore has to be attributed to the additional expansion with respect to the field intensity.

Knowing the location of the most probable values of the stationary distribution we can calculate with the help of eqs. (177) the relative stability of the two branches. In this way we can determine the field strength r_o for which both branches are simultaneously stable in order to generalize the idea of the Maxwell construction to processes far from thermal equilibrium.

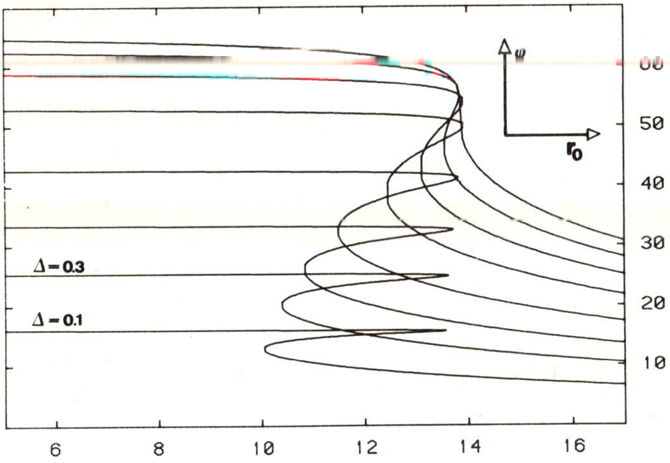

Fig. 11: Most probable phase shift between the cavity mode and the driving field (eq. 177).

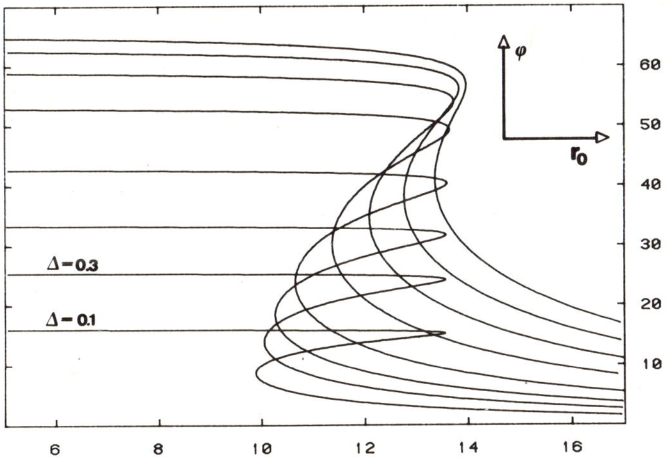

Fig. 12: Phase at the deterministic steady state of dispersive optical bistability.

In fig. 13 we have indicated by dashed lines the special values of r_o where global stability is exchanged by the two branches. By rotating this plot by 90 degrees we get a picture which qualitatively resembles the typical picture of the coexistence curve of a van der Waal gas. For $\Delta=0$, $\delta=0$, it is easy to prove that the Maxwell construction is still valid [46].

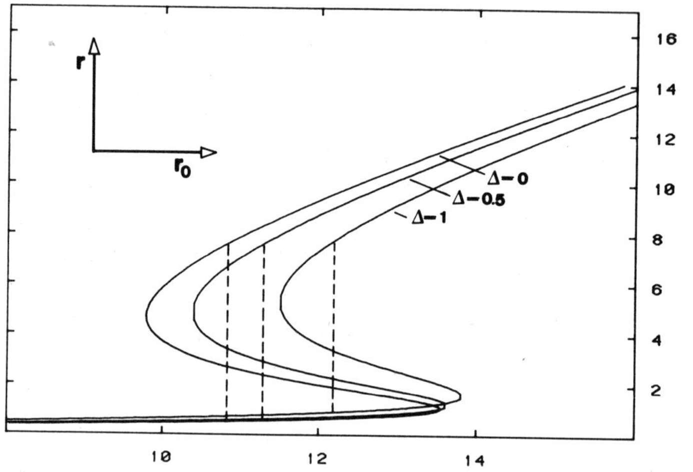

Fig. 13: Hysteresis of dispersive bistability. The dashed line indicates the exchange of global stability.

APPENDIX A

1. Theorem on Optical Coherent Transients

A typical experiment in the optical thin sample regime consists of a gas filled absorption cell close to resonance with an external light field. For the emission following the pulsed preparation of finite width τ we can proove that the following general statement holds:

When a pulse of finite duration τ (interval $0 \leq t \leq \tau$) excites an optical thin sample, the coherent emission which follows lasts no longer than an additional time τ ($\tau \leq t \leq 2\tau$).

This statement can be proven rigorously (47) for a travelling wave of any puls shape and for arbitrary atomic relaxation parameters T_1, T_2, when the assumption of an infinite inhomogeneous line width is made; i.e. $D(\Delta)$ = const. In the optical regime this is often an excellent approximation.

The formal integration of eq. (39) for the initial condition $P_q^*(\Delta,t=0) = 0$ leads to

$$\int P_q^*(\Delta,t)\, D(\Delta) d\Delta = -i \int_{-\infty}^{\infty} d\Delta \int_0^t gW(\Delta,t')E^*(t')\, e^{(i\omega - i\Delta - \frac{1}{T_2})(t-t') + i\omega t'} dt' \tag{A.1}$$

with

$$E^*(t) = \begin{cases} E^*(t) & 0 \leq t \leq \tau \\ 0 & t > \tau\,,\ t < 0 \end{cases} \tag{A.2}$$

Interchanging the order of integration, the Δ-integral can be evaluated by contour integration for time $t > 2\tau$ in the upper half plane. Realizing that $W(\Delta)$ has no singularities in the finite upper Δ-plane, the integral vanishes and so does the polarization which is the source, of the emitted field:

$$\int D(\Delta)\, P_q^*(\Delta,t)\,d\Delta \quad \equiv \quad 0 \qquad \text{for} \qquad t > 2\tau \quad . \tag{A.3}$$

No assumption has been made on the time dependence of the field as long as it can be considered to be slowly varying. This is a rather general result which is important for many spectroscopic experiments when the pulse width becomes comparable to the decay times of the sample.

A similar statement can be made for multilevel systems using the same mathematical arguments, as long as we consider travelling field modes. Under somewhat technical restrictions we find that the coherent emission of a multilevel system excited by a finite pulse width τ, will be zero for times larger than

$$t \quad \geq \quad \left(1 + \frac{\Omega\ max}{\Omega\ min}\right)\ \tau$$

Ω max and Ω min are the largest and the smallest dipol allowed transition frequency in the multilevel ensemble.

For a single two-level system this result reduces to our previous statement.

2. Oscillatory Free Induction Decay

While in the previous paragraph we defined the time regime where no coherent radiation can be expected, it is certainly interesting to see how the signal will decay during the time interval immediately after the pulse. This is also an interesting question from the mathematical point of view, because the polarization is not an analytic function in time.

For the simplified pulse envelope

$$E(t) = \begin{cases} 0 & t < 0,\ t > \tau \\ E_o & 0 \leq t \leq \tau \end{cases}$$

we can derive by analytical methods that the polarization of the radiative source - upto some unimportant factors - can be written in the following from [48]:

$$\int D(\Delta)\, P_q^*(\Delta,t)\, d\Delta = \begin{cases} 0 & t \geq 2\tau \\ 2\pi\, D(0)\, g\, E_o \sum_{m=1}^{\infty} (2\,\frac{\tau}{t} - 1)^m\, J_{2m}\,(g\, E_o \sqrt{t(2\tau - t)}) & t \leq 2\tau \end{cases} \tag{A.4}$$

in the limit of large field amplitudes we may use the asymptotic expansion of the Bessel functions and obtain the simpler expression

$$\frac{(2\tau - t)^{3/4}}{t^{1/4}} \cos\,(g\, E_o \sqrt{t(2\tau - t)} - \frac{\pi}{4}) \tag{A.5}$$

This oscillatory decay of the pulsed free induction decay has been rather unexpected, but closer inspection reveals that an experimental indication of the nonexponential decay has been seen before. Recently the oscillatory behaviour has been verified in great detail in a spin resonance experiment over a wide range of parameters [11], and excellent agreement between theory and experiments has been demonstrated.

In order to obtain the result (A.4) we have made the simplifying but not necessary assumption that the relaxation times are infinitely long. To demonstrate the influence of the finite width of the pulse, we compare this result with the free induction decay signal after steady state preparation, which is assumed to terminate at $t = 0$ [49]:

$$\int D(\Delta)\, P(\Delta,t)\, d\Delta \;=\; \left(\left(4g^2 E_o{}^2 T_1 T_2 + 1\right)^{-1/2} - 1\right) \times$$

$$\exp - \frac{t}{T_2}\left(1 + \left(4g^2 E_o{}^2 T_1 T_2 + 1\right)^{1/2}\right) \qquad (A.6)$$

This is a pure exponential decay with the "power broadened" relaxation constant

$$\frac{1}{T_2}\left(1 + \left(1 + 4g^2 E_o{}^2 T_1 T_2 \right)^{1/2}\right) \qquad (A.7)$$

which is 180° out of phase with the steady state preparation

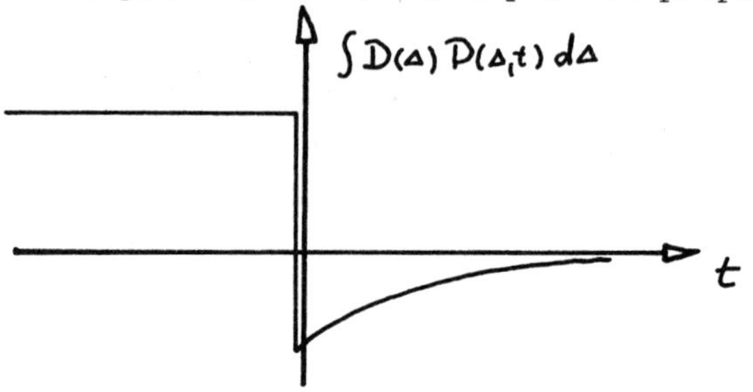

The steady state polarization to lowest order in gE_o is linear while for the emitted transient field the linear terms vanish leaving a cubic response.

3. Optical Coherent Transients by Phase Switching

Optical coherent transients can be triggered by various methods. Every sudden change in the properties of the external driving field E(x,t) will cause transient relaxations in the atomic ensemble by which the system relaxes in its own time scale towards the new steady state. The most obvious experiment consists in applying sudden changes in the field intensity. A sudden change in the field frequency is followed by a transient response of the atoms as well [50]. A rather subtle method consists of an instantaneous change in the phase of the driving field. For the two previous methods it is rather obvious how the atomic system would respond; however, the response on a mere phase shift is not immediately obvious. From an experimental point of view, this is an interesting question, because fast phase changes can be produced easily.

We want to include this example here because it also reveals a basic property of the Bloch equations - the sensitivity of the solution to the phase relation between the atomic polarization and the driving field.

As there exists no general solution of the Bloch equations we[8] present an approximate solution for the weak intensity limit $g^2 E_o^2 T_1 T_2 \ll 1$. For the driving field amplitude E_o we assume that it suffers a sudden phase shift at time t = 0

$$E^*(t) = E_o \begin{cases} 1 & t < 0 \\ \exp i\varphi & t > 0 \end{cases} \tag{A.8}$$

The response of the atomic ensemble is a transient evolution of the polarization which emits a coherent field. This field will be detected as a superposition with the driving laser field of the same frequency, because the coherent radiation is emitted in the foreward direction. The transient signal observed is then given by[49],[50]:

$$S = E(t)\,E(t) \sim |g\,E_o|^4\, T_1 T_2 \sin^2 \varphi/2 \cdot \cdot\left(\frac{\exp -\frac{2t}{T_2}}{1 - \varepsilon/2} - \frac{\varepsilon}{1-\varepsilon}\, e^{-\frac{t}{T_2}} + \frac{\varepsilon}{(1-\varepsilon)(2-\varepsilon)}\, e^{-\frac{t}{T_1}}\right) \tag{A.9}$$

where $\varepsilon = T_2/T_1$.

Due to the disturbed phase relation between the driving field and the atomic polarization, a transient excursion of the polarization is detected which reestablishes the absorptive phase relation again. As the atomic ensemble stays at resonance during the entire process, polarization - as well as energy relaxation plays a role. Only in the limit $T_2 \ll T_1$ a single exponential decay remains and the experiment can serve as a versatile spectroscopic method.

A rather elegant experiment which allows to determine the phase decay constant T_2 in the presence of a finite energy decay constant T_1 can be achieved by a double phase shift in the driving field [50]. When the phase of the field is shifted suddenly by an angle $\pi/2$ and after a time T back to its original value in general a discontinuous response is observed at $t = T$. For the special time separation $T = T_2 \frac{1}{2} \ln 2$ the discontinuity vanishes and allows thereby a direct observation of the phase decay time T_2. A qualitative picture is given in the following figures.

NULL METHOD FOR MEASUREMENT OF T_2

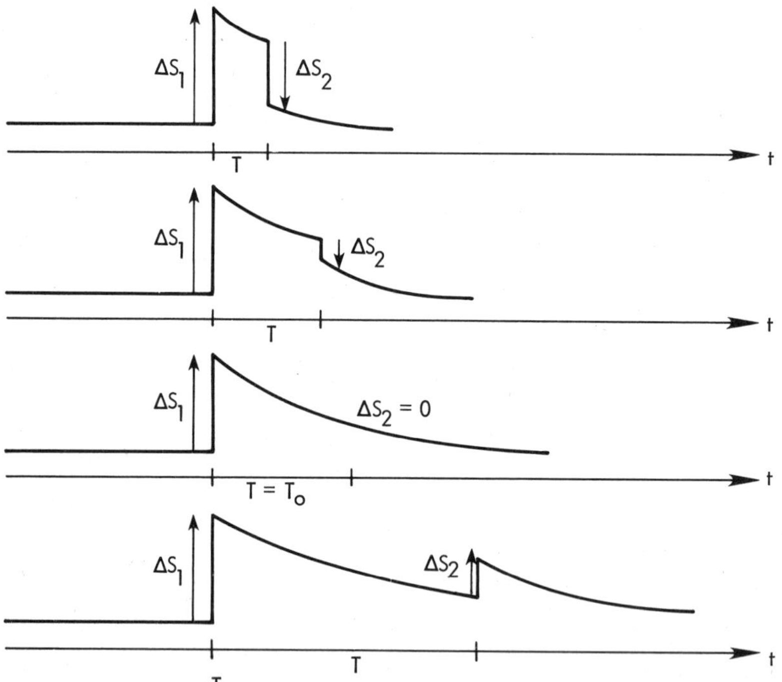

FOR $\Phi = \pi/2$, $T_o = \frac{T_2}{2} \ln 2$

APPENDIX B

Steady States without Detailed Balance

We assume, that a multidimensional stochastic process can be described by the following Fokker-Planck equation with constant diffusion:

$$\frac{\partial P}{\partial t} = - \frac{\partial}{\partial x_i} K_i(\{x_j\}) P + \frac{1}{2} \varepsilon Q_{ij} \frac{\partial^2}{\partial x_i \partial x_j} P \tag{B.1}$$

where $P(\{x_i\},t)$ is the conditional probability density which reduces to the n-dimensional δ-function in the limit $t \to 0$, K_i is the drift vector determined by the deterministic equations of motion $\dot{x}_i = K_i$ and $Q_{ij} = \varepsilon$ is the positive definite diffusion matrix which characterizes the strength of the fluctuations. Q_{ij} is assumed to be independent of the variables x_j and will later be considered as small ($\varepsilon \to 0^+$). Natural boundary conditions are implied.

The stationary properties of the process are described by the steady state solution of (B.1)

$$\frac{\partial P_o}{\partial t} = 0 \qquad P_o = \exp(-\varepsilon^{-1} \Phi) \tag{B.2}$$

If the external applied forces allow the system to reach thermodynamic equilibrium, the potential Φ as described by eq. (B.2) is proportional to the corresponding thermodynamic potential. The drift vector K_i then can be separated into a dissipative part $d_i(\{x_j\})$ and a reversible part $r_i(\{x_j\})$:

$$K_i(\{x_j\}) = d_i(\{x_j\}) + r_i(\{x_j\}) \tag{B.3}$$

According to the physical meaning of the variables x_i we can distinguish between even and odd variables under time reversal. Denoting the time reversed of x_i by $\tilde{x}_i$

$$\tilde{x}_i = \varepsilon_i x_i \quad , \quad \varepsilon_i = \pm 1 \tag{B.4}$$

we can identify:

$$d_i(\{\tilde{x}_j\}) = \varepsilon_i \, d_i(\{x_j\}) \tag{B.5}$$

and

$$r_i(\{\tilde{x}_j\}) = -\varepsilon_i \, r_i(\{x_j\}) \tag{B.6}$$

In expressions containing ε_i the summation convention is dropped. The dissipative part then is of the form given by Onsager [29]

$$d_i = -\frac{1}{2} Q_{ij} \frac{\partial \Phi}{\partial x_j} \tag{B.7}$$

The reversible part of r_i is not determined by Φ, but leaves the thermodynamic potential and the volume element in phase space invariant:

$$r_i \frac{\partial \Phi}{\partial x_i} = 0 \qquad \text{and} \qquad \frac{\partial r_i}{\partial x_i} = 0 \tag{B.8}$$

Defining the reversible drift through (B.3) - (B.6), we can derive from the Fokker Planck equation a more general form of (B.8) which makes only use of the detailed balance condition of the assumed equilibrium case.

$$\frac{1}{\varepsilon} r_i \frac{\partial \Phi}{\partial x_i} - \frac{\partial r_i}{\partial x_i} = 0 \tag{B.9}$$

For steady states lacking detailed balance we use (B.3) and (B.7) instead of (B.3) - (B.6) to define r_i by

$$K_i = -\frac{1}{2} Q_{ij} \frac{\partial \Phi}{\partial x_j} + r_i \tag{B.10}$$

Inserting (B.10) into the time independent Fokker Planck equation, we obtain the analogy of (B.9)

$$\frac{1}{\varepsilon} r_i \frac{\partial \Phi}{\partial x_i} - \frac{\partial r_i}{\partial x_i} = 0 \tag{B.11}$$

Note, however, that here we did not assume, that r_i is the reversible part of K_i; we merely define it through (B.10). Therefore we can no longer conclude like in equilibrium that the stationary probability current j_i

$$j_i = (\frac{1}{2} Q_{ij} \frac{\partial \Phi}{\partial x_j} + K_i) P_o = r_i P_o \tag{B.12}$$

is reversible under time reversal transformations.

However, the formal structure of the equilibrium theory expressed in terms of a thermodynamic potential carries over to the non-equilibrium theory expressed in terms of a generalized potential Φ. This gives us the key for a direct comparison of equilibrium and non-equilibrium steady states lacking detailed balance.

While for equilibrium states Φ can easily be given in terms of quadratures, in the absence of detailed balance we have to deal with an elliptic nonhermitian boundary value problem for which no general method of solution exists.

Recognizing, however, that in many relevant physical examples fluctuations are extremely weak we may obtain the leading contribution to Φ in the limit $\varepsilon \to 0^+$ by solving the first order problem

$$\frac{1}{2} Q_{ij} \frac{\partial \Phi}{\partial x_i} \frac{\partial \Phi}{\partial x_j} + K_i \frac{\partial \Phi}{\partial x_i} = 0 \tag{B.13}$$

It should be emphasized that in this limit Φ becomes independent of ε, while P still depends on ε in the form

$$P = \exp(- \frac{1}{\varepsilon} \Phi)$$

The relation (B.9) between r_i and Φ reduces, for $\varepsilon \to 0^+$, to the orthogonality relation

$$r_i \frac{\partial \Phi}{\partial x_i} = 0 \tag{B.14}$$

Equation (B.13) is a first order partial differential equation of the form of a time independent Hamilton Jacobi equation in classical mechanics for vanishing total energy. From a mathematical point of view we have replaced the second order boundary value problem of the Fokker Planck equation by a first order problem with Φ given on a hypersurface intersecting the field of characteristics. Mathematical literature exists, where these two problems are related in the limit of small ε for the corresponding time dependent equation. As these theorems are restricted to finite time intervals, it is not clear in which sense the solution of (B.13) approximates the solution of the full stationary Fokker Planck equation.

In the absence of a well developed mathematical framework for our procedure we have adopted a pragmatic attitude: If from a general solution of eq. (B.13) we can derive a unique single-valued one with the property $\Phi \to \infty$ for $\{x_j\} \to \infty$ necessary for normalizability, it will be taken as an approximation of the corresponding solution of the time independent Fokker Planck equation (B.1).

Apart from some elementary examples, unfortunately, a solution of (B.13) can not be obtained in general in a systematic way and we have to resort to further approximations based on the existence of a suitable small parameter λ in the drift vector K_i:

$$K_i = K_i^{1} + \lambda K_i^{2} \tag{B.15}$$

An approximate solution of eq. (B.13) can then be derived systematically in the form

$$\Phi = \sum_{n=0}^{\infty} \lambda^n \Phi_n \tag{B.16}$$

satisfying the hierarchy of equations:

$$\sum_{l=0}^{n} Q_{ij} \frac{\partial \Phi_{n-l}}{\partial x_i} \frac{\partial \Phi_l}{\partial x_j} + K_i^{1} \frac{\partial \Phi_n}{\partial x_i} + K_i^{2} \frac{\partial \Phi_{n-1}}{\partial x_i} = 0 \tag{B.17}$$

The potential Φ, as defined by (B.2), (B.3), (B.7), and (B.14), is a Lyapunoff function of the deterministic equations. This may be seen as follows: Due to eq. (B.2) Φ is a positive function. Due to eq. (B.14) Φ can only decrease if it evolves under the deterministic equations of motion

$$\dot{\Phi} = \frac{\partial\Phi}{\partial x_i}\dot{x}_i = \frac{\partial\Phi}{\partial x_i}K_i = -\frac{1}{2}Q_{ij}\frac{\partial\Phi}{\partial x_i}\frac{\partial\Phi}{\partial x_i} \leq 0 \tag{B.18}$$

Thus the locally (globally) stable attractors of the deterministic equations are identified with the local (global) minima of Φ.

We summarize what actually has been achieved once a single-valued solution of (B.13) has been obtained one or the other way:

(i) The steady state probability is known in the limit of small fluctuations;

(ii) A Lyapunoff function of the deterministic equations of motion is available, characterizing the locally and globally stable attractors;

(iii) The condition replacing the Maxwell construction for a first order type phase transition far from thermal equilibrium can be given;

(iv) The deterministic drift K_i can be decomposed into a gradient part which stabilizes the attractors and a remaining part r_i containing Φ as a constant of motion. This separation is a generalization of the separation into reversible and irreversible part in thermodynamic equilibrium.

This perturbation approach has been tested for a variety of simple mathematical models as well as for more involved physical processes. Dispersive optical bistability as discussed in chapter E is an example of a two-dimensional highly nonlinear process lacking detailed balance.

REFERENCES

1. S. Schweber, *An Introduction to Relativistic Quantum Field Theory*, Evanstone, III 1961
2. E. Henley and W. Thirring, *Elementary Quantum Field Theory*, New York 1962
3. H. Haken, *Quantenfeldtheorie des Festkörpers*, Teubner, Stuttgart 1973
4. P.L. Taylor, *A Quantum Approach to the Solid State*, Englewood Cliffs, N.Y., Prentice Hall 1970
5. N.H. March, W.H. Young and S. Sampanthar, *The Many-Body Problem in Quantum Mechanics*, Cambridge 1967
6. H. Haken, *Encyclopedia of Physics*, Vol. XXV/2c, Springer, Berlin
7. R.G. Brewer, in Les Houches Lectures: *Frontiers in Laser Spectroscopy*, eds. R. Balian, S. Haroche and S. Liberman (North Holland, 1977)
8. H.C. Torrey, Phys.Rev. 76, 1059 (1949)
9. A. Schenzle, N.C. Wong, and R.G. Brewer, Phys.Rev. A21, 887 (1980)
10. A. Schenzle, N.C. Wong, and R.G. Brewer, Phys.Rev. A22, 635 (1980)
11. M. Kunitomo, T. Endo, S. Nakanishi, and T. Hashi, Phys.Lett. preprint
12. A. Yariv, J. Opt.Soc. Amer. 66, 301 (1976)
13. I.A. Armstrong, N. Bloembergen, J. Ducuing, and P.S. Pershan, Phys.Rev. 127, 1918 (1962)
14. N. Bloembergen, *Nonlinear Optics*, New York, Benjamin 1965
15. H. Cheng, P.B. Miller, Phys. Rev. 134, 683 (1964); N. Bloembergen, and Y.R. Shen, Phys.Rev. 133, 37 (1964)
16. A. Schenzle, Thesis, Stuttgart 1970
17. V. Wilke and W. Schmidt, Appl.Phys. 18, 177 (1979)
18. S. Nakajma, Progr.Theor.Phys. 20, 948 (1958)
19. R. Zwanzig, J.Chem.Phys. 33, 1338 (1960)
20. H. Mori, Progr.Theor.Phys. 33, 423 (1965); 34, 394 (1965)
21. A. Schenzle, Thesis, Stuttgart 1974
22. B.R. Mollow, Phys.Rev. 188, 1969 (1969)
23. W. Heitler, *Quantum Theory of Radiation*, Clarendon Press, Oxford (1953)
24. M. Lax, Phys.Rev. 129, 2342 (1963)

25. H. Risken, *Progress in Optics*, Vol. VIII, ed. by E. Wolf, North Holland (1970)

26. W. Hartig, W. Rasmussen, R. Schieder, H. Walther, Z.Physik 278, 205 (1976)

27. R.L. Stratonovich, *Topics in the Theory of Random Noise* (Gordon and Breach, N.Y. 1963)

28. R. Graham and H. Haken, Z.Physik 243, 289 (1971); H. Risken, Z. Physik 251, 231 (1972)

29. L. Onsager, Phys.Rev. 37, 407 (1931) and 38, 2265 (1931)

30. A. Schenzle and H. Brand, Opt.Comm. 27, 485 (1978)

31. A. Schenzle and H. Brand, Opt.Comm. 31, 401 (1979)

32. H. Brand, A. Schenzle, and G. Schröder, to be published

33. R. Graham and H. Haken, Z.Physik 237, 31 (1970)

34. V. De Giorgio and M.O. Scully, Phys.Rev. A2, 1170 (1970)

35. J.F. Scott, M. Sargent III, and C.D. Cantrell, Opt.Comm. 15, 13 (1975)

36. R.B. Schaefer and C.R. Willis, Phys.Lett. 58A, 53 (1976)

37. S.L. Mc Call, Phys.Rev. A9, 1515 (1974); H.M. Gibbs, S.L. Mc Call, and T.N.C. Venkatesan, Phys.Rev.Lett. 36, 1135 (1976)

38. R. Bonifacio and L. Lugiato, Opt.Comm. 19, 172 (1975)

39. R. Graham, Z. Physik 210, 319 (1968)

40. R. Graham, *Springer Tracts in Modern Physics*, Vol. 66, Springer, Berlin (1973)

41. A. Schenzle and H. Brand, Phys.Lett. A69, 313 (1979)

42. A. Schenzle and H. Brand, Phys.Rev. A20, 1628 (1979)

43. P.D. Drummond, K.J. Mc Neil, and D.F. Walls, Opt.Comm. 28, 255 (1979)

44. H. Brand, R. Graham, and A. Schenzle, Opt.Comm. 32, 359 (1980)

45. R. Graham and A. Schenzle, *Proceedings of the Conference on Optical Bistability*, Asheville 1980, ed. by H.R. Robl and C.M. Bowden

46. R. Graham and A. Schenzle, Phys.Rev. A23, in press

47. F.A. Hopf, R.F. Shea and M.O. Scully, Phys.Rev. A7, 2105 (1973)

48. R.G. Brewer and A.Z. Genack, Phys.Rev.Lett. 36, 959 (1976)

49. A.Z. Genack, D.A. Weitz, R.M. Macfarlane, R.M. Shelby, and A. Schenzle, Phys.Rev.Lett. 45, 438 (1980)

50. A. Schenzle and A.Z. Genack, to be published

NON-LINEAR OPTICS

J.D. Hey, Department of Physics, University of Cape Town, Rondebosch 7700, South Africa

F.A. Hopf, Optical Sciences Center, University of Arizona, Tucson, Arizona 85721, USA

INTRODUCTORY CHAPTER

INTRODUCTION TO NON-LINEAR OPTICS

The field of non-linear optics originated from three classic experiments related to the generation {1,2} and absorption {3} of optical harmonics of an incident laser beam by a crystalline medium. In the experiment of Franken et al. {1}, a ruby laser producing approximately 3J of 6943 Å light per one-millisecond pulse was projected through crystalline (α-) quartz. The emergent beam was found upon spectrometric analysis to contain the second harmonic (3472 Å) of the fundamental. The existence of the second harmonic wave was confirmed in two ways: it exhibited the expected energy dependence upon polarization and orientation of the incoming beam, and disappeared upon replacement of the quartz by an isotropic medium (glass).

Whereas the existence of the second harmonic in this experiment {1} could be ascribed to the lack of a centre of inversion in a particular member of the trigonal crystal system {4}, the later experiment of Terhune et al. {2} employed the crystal calcite, which possesses a centre of inversion. In addition, a d.c. electric field of up to 250 kV/cm was imposed on the crystal, which was illuminated with the output beam of a 1J pulsed ruby laser. In this case, the second harmonic generation could be ascribed to an induced electric quadrupole moment in the crystal, as well as a dipole polarization term associated with gradients in the impressed (time-dependent) electric field. In addition, third harmonic generation could be observed at suitable orientations of the laser beam to the optic axis of the crystal.

Kaiser and Garrett {3} investigated the generation of blue fluorescent light around 4250 Å by two photon absorption in $CaF_2:Eu^{2+}$ crystals illuminated with the red light (λ_r = 6943 Å) of a ruby laser. The crystal CaF_2 has a highly symmetric (cubic) structure, the absorption of laser radiation at $2\nu_r$ being due to the presence of 0.1% Eu^{2+} ions substituted for Ca^{2+}. The blue fluorescent light was interpreted as arising through non-radiative decay to intermediate (virtual) atomic states (charge transfer bands), together with a radiative transition to the ground state. The observed magnitude of the effect was in satisfactory agreement with a theoretical discussion of the process {5}, which differs significantly in principle from second-harmonic production {1,2}, so that the restrictions imposed by symmetry considerations are entirely different {6}.

From this brief historical introduction, we proceed to introduce the subject more formally through the Maxwell equations of electrodynamics.

CHAPTER 1

THE MAXWELL EQUATIONS

The Maxwell equations may be written (c g s units) as:

$$\nabla \cdot \underset{\sim}{E} = 4\pi(\rho + \rho_b) \tag{1.1}$$

$$\nabla \cdot \underset{\sim}{B} = 0 \tag{1.2}$$

$$\nabla \times \underset{\sim}{E} = -\frac{1}{c}\frac{\partial B}{\partial t} \tag{1.3}$$

$$\nabla \times \underset{\sim}{B} = \frac{4\pi}{c}(\underset{\sim}{J} + \underset{\sim}{J}_b) + \frac{1}{c}\frac{\partial E}{\partial t} \tag{1.4}$$

where ρ and $\underset{\sim}{J}$ denote the free charge and current sources, and ρ_b and $\underset{\sim}{J}_b$ the additional source terms arising from induced molecular polarization and magnetization within the medium under consideration.

For present purposes we shall usually consider the situations where the free source terms vanish. The above equations will be used in their macroscopic form, in which case one may expand {7,8}:

$$\rho_b = -\nabla \cdot \underset{\sim}{P} + \nabla \cdot (\nabla \cdot \underset{\approx}{Q}) - \cdots\cdots \tag{1.5}$$

$$J_b = \frac{\partial}{\partial t}(\underset{\sim}{P} - \nabla \cdot \underset{\approx}{Q} + \cdots\cdots) + (c\nabla \times \underset{\sim}{M} + \cdots\cdots)\,. \tag{1.6}$$

Here $\underset{\sim}{P}$, $\underset{\approx}{Q}$ .... denote the dipole, quadrupole (dyadic) and higher multipole orders in the electric polarization of the medium; $\underset{\sim}{M}$, ... denote the dipole and higher multipole orders in the magnetization.

It is convenient to introduce at this point the secondary (macroscopic) field quantities:

$$\underset{\sim}{D} = \underset{\sim}{E} + 4\pi(\underset{\sim}{P} - \nabla \cdot \underset{\approx}{Q} + \cdots\cdots) \tag{1.7}$$

$$\underset{\sim}{H} = \underset{\sim}{B} - 4\pi(\underset{\sim}{M} + \cdots\cdots) \tag{1.8}$$

in which case the Maxwell equations assume the form:

$$\nabla \cdot \underset{\sim}{D} = 4\pi\rho \tag{1.9}$$

$$\nabla \cdot \underset{\sim}{B} = 0 \tag{1.10}$$

$$\nabla \times \underset{\sim}{E} = -\frac{1}{c}\frac{\partial B}{\partial t} \tag{1.11}$$

$$\nabla \times \underset{\sim}{H} = \frac{4\pi}{c}\underset{\sim}{J} + \frac{1}{c}\frac{\partial \underset{\sim}{D}}{\partial t} \ . \tag{1.12}$$

Equations (1.9)-(1.12) are however merely a concise re-formulation of (1.1)-(1.8); for practical purposes one requires, in addition, constitutive equations relating $\underset{\sim}{P}$, $\underset{\approx}{Q}$, ... to $\underset{\sim}{E}$, and $\underset{\sim}{M}$ .... to $\underset{\sim}{B}$. Since we shall be ignoring magnetic properties of the materials under consideration, the latter are of no further interest for these purposes. In the former case we shall write for homogeneous media phenomenological equations in the form:

$$\underset{\sim}{P} = \chi^{(1)}\underset{\sim}{E} + \chi^{(2)}\underset{\sim}{E}\underset{\sim}{E} + \chi^{(3)}\underset{\sim}{E}\underset{\sim}{E}\underset{\sim}{E} + \cdots\cdots \tag{1.13}$$

$$\underset{\approx}{Q} = \eta^{(2)}\underset{\sim}{E}\underset{\sim}{E} + \eta^{(3)}\underset{\sim}{E}\underset{\sim}{E}\underset{\sim}{E} + \cdots\cdots \tag{1.14}$$

where $\chi^{(1)}$, $\chi^{(2)}$, $\chi^{(3)}$ are electric susceptibility tensors of the second, third, fourth .... ranks, respectively, and $\eta^{(2)}$, $\eta^{(3)}$, ... tensors of the fourth, fifth, ... ranks, respectively. For clarity, the corresponding Cartesian component forms of the above equations are:

$$P_i = \chi^{(1)}_{ij} E_j + \chi^{(2)}_{ijk} E_j E_k + \chi^{(3)}_{ijkl} E_j E_k E_l + \cdots\cdots \tag{1.15}$$

$$Q_{ij} = \eta^{(2)}_{ijkl} E_k E_l + \eta^{(3)}_{ijklm} E_k E_l E_m + \cdots\cdots \tag{1.16}$$

summation over the repeated indices being implied.

Such relations are extremely complicated, in view of the large number of coefficients of the susceptibility required for the evaluation of individual field components of the induced polarization. However, examination of the symmetry properties of the seven crystal systems, comprising thirty-two crystal classes, drastically reduces the effective number of coefficients, particularly when terms in equations (1.13)-(1.16) are retained to leading order only {4}. For example, in any crystal with inversion symmetry, $\chi^{(2)}_{ijk}$ is identically zero {6}, as in the experiment of Terhune et al. {2}. Here the (third) harmonic generation could be ascribed to the terms containing $\chi^{(3)}$ and $\eta^{(2)}$ and, in addition, a term of the type

$$\chi^{(3)'}_{ijk\ell} E_j \nabla_k E_\ell$$

introduced to account for inhomogeneity (which we shall not consider further).

Taking the curl on both sides of equation (1.3) together with (1.4) and (1.6) yields the wave equation for the electric field:

$$\nabla(\nabla \cdot \underset{\sim}{E}) - \nabla^2 \underset{\sim}{E} + \frac{1}{c^2}\frac{\partial^2 \underset{\sim}{E}}{\partial t^2} = -\frac{4\pi}{c^2}\left(\frac{\partial^2 \underset{\sim}{P}}{\partial t^2} - \nabla \cdot \frac{\partial^2 \underset{\approx}{Q}}{\partial t^2} + \cdots\right). \tag{1.17}$$

In the simplest case of a linear susceptibility only, this reduces to:

$$\nabla(\nabla \cdot \underset{\sim}{E}) - \nabla^2 \underset{\sim}{E} + \frac{1}{c^2}\frac{\partial^2}{\partial t^2}\left[\underset{\approx}{\epsilon} \cdot \underset{\sim}{E}\right] = 0 \tag{1.18}$$

where the dielectric tensor (dyadic) $\underset{\approx}{\epsilon}$ is related to the susceptibility $\chi^{(1)}$ by:

$$\epsilon_{ij} = \delta_{ij} + 4\pi\chi^{(1)}_{ij} . \tag{1.19}$$

From simple considerations of energy flux {9}, or by means of standard thermodynamic relations {10}, one readily shows that the dielectric tensor is symmetrical ($\epsilon_{ij} = \epsilon_{ji}$), and therefore possesses only six independent components. By a suitable choice of coordinate axes, $\underset{\approx}{\epsilon}$ can be represented by a diagonal matrix; these are called <u>principal axes</u> of the crystal, and in this system the diagonal elements of $\underset{\approx}{\epsilon}$ will be denoted simply by ϵ_j (j = 1,2,3) {11}. The components of $\underset{\sim}{D}$ and $\underset{\sim}{E}$ are then related by:

$$D_j = \epsilon_j E_j . \tag{1.20}$$

CHAPTER 2

ELECTROMAGNETIC WAVE PROPAGATION IN A LINEAR ANISOTROPIC MEDIUM

Consider a monochromatic plane wave

$$\underset{\sim}{E}(\underset{\sim}{r},t) = \underset{\sim}{E}_o \exp[i(\underset{\sim}{k}\cdot\underset{\sim}{r} - \omega t)] \tag{2.1}$$

$$\underset{\sim}{H}(\underset{\sim}{r},t) = \underset{\sim}{H}_o \exp[i(\underset{\sim}{k}\cdot\underset{\sim}{r} - \omega t)] \tag{2.2}$$

propagating in a linear, anisotropic medium. In such a medium, the amplitudes $\underset{\sim}{E}_o$ and $\underset{\sim}{H}_o$ are constant. With $\nabla \rightarrow i\underset{\sim}{k}$, the wave equation (1.18) immediately yields:

$$\left(-|\underset{\sim}{k}|^2 \underset{\approx}{I} + \underset{\sim}{k}\,\underset{\sim}{k} + \frac{\omega^2}{c^2}\underset{\approx}{\epsilon}\right)\cdot\underset{\sim}{E} = 0 \tag{2.3}$$

where $\underset{\approx}{I}$ denotes the unit dyadic. Now express the wave vector $\underset{\sim}{k}$ as:

$$\underset{\sim}{k} = k\underset{\sim}{n} = \frac{\omega}{v}\underset{\sim}{n} \tag{2.4}$$

in terms of the unit vector ($\underset{\sim}{n}$) in the direction of the wave normal (also called here the direction of propagation), and v the phase velocity in the medium (which will turn out to depend both upon the direction of propagation and plane of polarization of the wave). We define also the principal velocities in the crystal as:

$$v_j = \frac{c}{\sqrt{\epsilon_j}} \tag{2.5}$$

in terms of the principal dielectric constants ϵ_j, the subscript j = 1,2,3 again denoting the principal axis (xyz) system, relative to which the direction cosines of the wave normal will be denoted by n_j. Thereby, equation (2.3) can be recast in the form of the homogeneous system (j = 1,2,3):

$$n_j \sum_{i=1}^{3} n_i E_i + \left(\frac{v^2}{v_j^2} - 1\right) E_j = 0. \tag{2.6}$$

Now from equations (1.1), (1.5), (1.9) and (1.19), we have for a linear medium:

n = unit vector in the direction of propagation

$$\underset{\sim}{n}\cdot\underset{\sim}{D} = \underset{\sim}{n}\cdot\underset{\approx}{\epsilon}\cdot\underset{\sim}{E} = 0 , \qquad (2.7)$$

which implies that whereas the displacement will always be perpendicular to the direction of propagation, the electric field will not be, in general. Only in certain special cases will $\underset{\sim}{n}\cdot\underset{\sim}{E} = 0$, these being: (i) isotropic crystals (i.e. cubic in structure); (ii) propagation along one of the principal axes of any crystal; (iii) the ordinary wave in a uniaxial crystal (see below); (iv) propagation along or at right angles to the optic axis in a uniaxial crystal (see below).

Now direct substitution of D_j in terms of $\sum_i n_i E_i$ from (2.6) in (2.7), together with (1.20), yields the celebrated Fresnel equation {10-13}:

$$\sum_{j=1}^{3} \frac{n_j^2}{v^2 - v_j^2} = 0. \qquad (2.8)$$

An analysis of this equation, which is quadratic in v^2, for the general case $\epsilon_1 < \epsilon_2 < \epsilon_3$ shows that for every direction of the wave normal, there are two distinct phase velocities v' and v" (the phenomenon of birefringence), which will not coincide unless $n_2 = 0$ {11}. In that case ($v' = v'' = v_2$), there are two possible directions of the wave normal given by

$$v_2^2 - v_1^2 - n_1^2(v_3^2 - v_1^2) = 0. \qquad (2.9)$$

These two directions in which there is only one phase velocity are known as the optic axes of a biaxial crystal. With the above labelling of the principal axes, the optic axes lie in the (x,z) plane and are symmetrical with respect to the x-axis.

It should be pointed out that the derivation of equation (2.8) shows it to be inapplicable in the special case of propagation along a principal axis. Here, however, two phase velocities are again obtained {13}, namely the principal velocities

$$v' = \frac{c}{\sqrt{\epsilon_i}} \quad ; \quad v'' = \frac{c}{\sqrt{\epsilon_k}}$$

corresponding to $n_j = 1$, a result which follows immediately from equations (2.6).

The special case of the uniaxial crystal (only one optic axis) occurs if $\epsilon_1 = \epsilon_2$, equation (2.9) predicting that this axis lies in the z direction. With θ denoting the angle between $\underset{\sim}{n}$ and the z axis, Fresnel's equation (2.8) for propagation in the direction (θ,ϕ) with respect to the z axis yields the following two solutions for the phase velocity, in terms of $v_o = v_1 = v_2$ and $v_e = v_3$:

$$v^2 = v_o^2 \qquad \text{Ordinary wave} \tag{2.10}$$

$$v^2 = v_o^2 \cos^2\theta + v_e^2 \sin^2\theta \qquad \text{Extraordinary wave} \tag{2.11}$$

θ = angle from optic axis (z-axis)

The wave travelling with velocity v_o independent of the direction of propagation is termed the <u>ordinary wave</u>. The other wave is termed the <u>extraordinary wave</u>, its velocity depending upon the direction of propagation. The latter will not obey Snell's Law when refracted at the surface of a uniaxial crystal. Only for propagation along the z axis ($\theta = 0$) are the two velocities the same. In a direction perpendicular to the z axis ($\theta = \pi/2$), the velocity of the extraordinary wave is v_e. The corresponding <u>ordinary</u> (μ_o) and <u>extraordinary</u> (μ_e) refractive indices of the crystal may be defined by:

Refractive indices:

$$\mu_o = \sqrt{\epsilon_1} = \sqrt{\epsilon_2} \quad ; \quad \mu_e = \sqrt{\epsilon_3} \ .$$

Equation (2.11) may now be recast as an expression for the refractive index ($\mu = c/v$) applicable to propagation of the <u>extraordinary</u> wave:

$$\mu_e(\theta) = \frac{\mu_o \mu_e}{\sqrt{\mu_o^2 \sin^2\theta + \mu_e^2 \cos^2\theta}} \tag{2.12}$$

(Note therefore that $\mu_e = \mu_e(\pi/2)$, from (2.12).)

If $\mu_e > \mu_o$ (i.e. $\epsilon_3 > \epsilon_1 = \epsilon_2$), the birefringence is said to be positive, whereas if $\mu_e < \mu_o$ (i.e. $\epsilon_3 < \epsilon_1 = \epsilon_2$), the birefringence is said to be negative. The corresponding crystals are called <u>positive</u> or <u>negative uniaxial</u>, respectively {9}. In the former case, the ordinary wave is faster, in the latter case it is slower.

A supplementary relation to (2.8) for v^2 in terms of the <u>direction cosines (γ_j) of $\tilde{D}$</u> with respect to the principal axes may also be derived {11} from equation (2.6):

$$v^2 = \sum_{j=1}^{3} \gamma_j^2 v_j^2 \ . \tag{2.13}$$

γ_j = direction cosines of D w.r.t. principal axes.

We now proceed to examine the inter-relationships of the various fields in more detail. From equation (1.12) it follows that:

$$\underset{\sim}{D} = -\mu\, \underset{\sim}{n} \times \underset{\sim}{H} \tag{2.14}$$

(with $\mu = {}^{c}/_{v}$ as before). Thus, $\underset{\sim}{D}$ is perpendicular to both $\underset{\sim}{n}$ and $\underset{\sim}{H}$. From equation (1.3) it follows in addition that:

$$\underset{\sim}{H} = \mu\, \underset{\sim}{n} \times \underset{\sim}{E} \tag{2.15}$$

while equations (1.11) and (1.12) together yield:

$$\underset{\sim}{D} = \mu^2 [\underset{\sim}{E} - (\underset{\sim}{n} \cdot \underset{\sim}{E})\underset{\sim}{n}] . \tag{2.16}$$

Thus, while $\underset{\sim}{H}$ is perpendicular to both $\underset{\sim}{n}$ and $\underset{\sim}{E}$, the latter is coplanar with $\underset{\sim}{n}$ and $\underset{\sim}{D}$ but not in general perpendicular to $\underset{\sim}{n}$ unless the crystal is cubic.

Therefore, the direction of energy flow associated with the wave, given by the Poynting vector:

$$\underset{\sim}{S} = \frac{c}{8\pi} \underset{\sim}{E} \times \underset{\sim}{H}^* \tag{2.17}$$

is not along the normal to the surfaces of constant phase in an anisotropic crystal. From (2.15), one finds that:

$$\underset{\sim}{S} = \frac{\mu c}{8\pi} [|\underset{\sim}{E}|^2 \underset{\sim}{n} - (\underset{\sim}{E} \cdot \underset{\sim}{n})\underset{\sim}{E}^*] \tag{2.18}$$

from which a <u>velocity of energy propagation</u> $\underset{\sim}{u}$ may be obtained from the definition:

$$\underset{\sim}{S} = U \underset{\sim}{u} \tag{2.19}$$

in terms of the electromagnetic energy density:

$$U = \frac{1}{16\pi} (\underset{\sim}{E} \cdot \underset{\sim}{D}^* + \underset{\sim}{B} \cdot H^*) \tag{2.20}$$

$$= \frac{\mu^2}{8\pi} [|\underset{\sim}{E}|^2 - (\underset{\sim}{n} \cdot \underset{\sim}{E})(\underset{\sim}{n} \cdot E^*)] \tag{2.21}$$

(note that the electric and magnetic contributions to U remain equal for anisotropic crystals). Equations (2.18) to (2.21) reveal that $\underset{\sim}{u}$

and v are related by

$$v = \underset{\sim}{u}\cdot\underset{\sim}{n} \qquad (2.22)$$

; $\underset{\sim}{u}$ = energy velocity
$\underset{\sim}{n}$ = unit vector in direction of propagation.

or the phase velocity is equal to the projection of the energy velocity on the wave normal.

Now the fields corresponding to the two distinct solutions (v' and v") of the Fresnel equation (2.8) may be shown by (2.6) to have the property that {11}:

$$\underset{\sim}{D}'\cdot\underset{\sim}{D}'' = 0 \,. \qquad (2.23)$$

Hence, since $\underset{\sim}{D}'$, $\underset{\sim}{E}'$, $\underset{\sim}{n}$ and $\underset{\sim}{D}''$, $\underset{\sim}{E}''$, $\underset{\sim}{n}$ form two sets of coplanar vectors,

$$\underset{\sim}{D}'\cdot\underset{\sim}{E}'' = \underset{\sim}{D}''\cdot\underset{\sim}{E}' = 0 \,. \qquad (2.24)$$

However, $\underset{\sim}{E}'\cdot\underset{\sim}{E}'' \neq 0$ for the two solutions of the Fresnel equation corresponding to a particular direction of propagation ($\underset{\sim}{n}$).

From equation (2.6) together with (1.20), it follows that {11}:

$$D_x (v^2 - v_1^2)/n_1 = D_y (v^2 - v_2^2)/n_2 = D_z (v^2 - v_3^2)/n_3 \qquad (2.25)$$

and

$$E_x(1 - v^2/v_1^2)/n_1 = E_y(1 - v^2/v_2^2)/n_2 = E_z(1 - v^2/v_3^2)/n_3 \,. \qquad (2.26)$$

Hence the directions of $\underset{\sim}{D}$, $\underset{\sim}{E}$ and $\underset{\sim}{H}$ (from (2.15)) are known provided that v is known. Thus, if $\underset{\sim}{n}$ is given, v' and v" are determined from (2.8) and thereafter the directions of the components of the two waves are specified uniquely, except for the special case discussed below. On the other hand if $\underset{\sim}{D}$ is given $\underset{\sim}{E}$ follows at once and the direction of $\underset{\sim}{n}$ is known from the fact that it is coplanar with $\underset{\sim}{D}$ and $\underset{\sim}{E}$, and perpendicular to $\underset{\sim}{D}$. In this case, v is derived from (2.13) and $\underset{\sim}{H}$ from (2.15).

In the case of propagation along the optic axis of a crystal (i.e. when v' = v" = v_2), however, we see from (2.25) and (2.26) that the directions of $\underset{\sim}{D}$ and $\underset{\sim}{E}$ are no longer uniquely determined. Therefore, plane waves can propagate along an optic axis with any arbitrary direction of polarization.

A special case of interest is the propagation of the ordinary wave in a uniaxial crystal. Equation (2.25) implies in this case the vanishing of D_z, and hence the condition (2.7) requires that $\underset{\sim}{D}$ be perpendicular to the plane containing $\underset{\sim}{n}$ and the optic axis. Similarly, equation (2.26) implies the vanishing of E_z, in which case equation (1.20) for the principal axis system implies that $\underset{\sim}{n}\cdot\underset{\sim}{E} = 0$ for the ordinary wave. Thus, <u>the ordinary wave in a uniaxial crystal is polarized perpendicularly to the optic axis</u>.

Conversely, (2.23) now requires that $\underset{\sim}{D}$ for the <u>extraordinary</u> wave lie in the plane containing $\underset{\sim}{n}$ and the optic axis. Consequently, (1.20) and (2.26) together mean that $\underset{\sim}{n}\cdot\underset{\sim}{E}$ cannot be zero except for the special cases of $\underset{\sim}{n}$ along or perpendicular to the optic axis.

Referring to equation (2.18), we therefore see that, whereas the direction of energy propagation (the <u>ray direction</u>) always coincides with the wave normal in the case of the ordinary wave in a uniaxial crystal, the two directions will not coincide in the case of the extraordinary wave unless the wave normal is either along or perpendicular to the optic axis.

The above comments are confirmed by the following useful expressions for the angles involved in wave propagation in a uniaxial crystal. We define as before the angle between the <u>wave normal</u> (direction of propagation) and the optic axis as θ. Equation (2.12) yields:

$$\tan\theta = \pm \frac{\mu_e}{\mu_o}\left(\frac{\mu_o^2-\mu^2}{\mu^2-\mu_e^2}\right)^{\frac{1}{2}} . \tag{2.27}$$

Let β denote the angle between the electric field direction and the optic axis. With

$$v_1 = v_2 = v_o = c/\mu_o$$

$$v_3 = v_e = c/\mu_e$$

$$D_x = \mu_o^2 E_x \; ; \; D_y = \mu_o^2 E_y \; ; \; D_z = \mu_e^2 E_z$$

equations (2.25) and (2.26) both reduce to:

$$\frac{E_x}{n_1} = \frac{E_y}{n_2} = \left(\frac{\mu^2-\mu_e^2}{\mu^2-\mu_o^2}\right)\frac{E_z}{n_3} \tag{2.28}$$

whence one obtains

$$\tan\beta = \left(\frac{\mu^2-\mu_e^2}{\mu^2-\mu_o^2}\right)\tan\theta . \tag{2.29}$$

Let $\varphi = \beta \mp \frac{\pi}{2}$ denote the angle between the direction of energy propagation (ray direction) and the optic axis. (The relationship between the appropriate quadrants for θ and β, from (2.27) and (2.29), must be carefully considered here.) From equations (2.27) and (2.29) one therefore has:

$$\tan\varphi = \pm\frac{\mu_o}{\mu_e}\left(\frac{\mu_o^2-\mu^2}{\mu^2-\mu_e^2}\right)^{\frac{1}{2}} = \left(\frac{\mu_o}{\mu_e}\right)^2 \tan\theta . \qquad (2.30)$$

Therefore in positive uniaxial crystals ($\mu_o < \mu_e$), $\varphi < \theta$, whereas in negative uniaxial crystals ($\mu_o > \mu_e$), $\varphi > \theta$.

The angle (α) between $\underset{\sim}{D}$ and $\underset{\sim}{E}$, which is also the angle (α) between the wave normal ($\underset{\sim}{n}$) and ray ($\underset{\sim}{S}$) directions, is given by $\alpha = |\theta-\varphi|$, and may therefore be calculated from equations (2.27) and (2.30):

$$\tan\alpha = \frac{1}{\mu_o\mu_e}\left[(\mu_o^2-\mu^2)(\mu^2-\mu_e^2)\right]^{\frac{1}{2}} . \qquad (2.31)$$

For an alternative presentation of the various geometrical relationships for extraordinary wave propagation expressed by equations (2.27), (2.29), (2.30) and (2.31), the reader is referred to Appendix IV.

CHAPTER 3

OPTICAL HARMONIC GENERATION IN A NON-LINEAR MEDIUM

INTRODUCTION

In this section we shall consider the general phenomenon of optical harmonic generation and related processes, and thereafter the restrictions imposed by crystal symmetry. In order to facilitate the understanding of the problem a simple physical model will be described for the polarization of the medium. Now, whereas a complete treatment of the problem can in general be given only within the framework of the quantum theory, many polarization properties can be analyzed on the basis of the classical anharmonic oscillator {4}. This in turn presupposes that the frequencies under consideration are sufficiently far removed from absorption bands, i.e. fall in regions of optical transparency for the particular crystal {4,9}.

Below, we shall consider the effect of the first non-linear term $\underset{\sim}{\chi}^{(2)}\underset{\sim}{E}\underset{\sim}{E}$ in equation (1.13) when two fields:

$$\underset{\sim}{\mathcal{E}}_1 \exp[i(\underset{\sim}{k}_1\cdot\underset{\sim}{r} - \omega_1 t)] \quad , \quad \underset{\sim}{\mathcal{E}}_2 \exp[i(\underset{\sim}{k}_2\cdot\underset{\sim}{r} - \omega_2 t)]$$

are impressed upon a non-linear medium. Evaluation of the second-order polarization shows that four types of interactions arise (called three-frequency {4} or three-wave {14} interactions), one of which corresponds to the generation of waves at the sum frequency:

$$\omega_3 = \omega_1 + \omega_2 \quad .$$

For this process, a particular component of the non-linear (NL) polarization can be written:

$$P_i^{NL}(\omega_3) = \sum_{jk} \chi^{(2)}_{ijk}(\omega_3;\omega_1,\omega_2)\,\mathcal{E}_{1j}\,\mathcal{E}_{2k}\,\exp i[(\underset{\sim}{k}_1+\underset{\sim}{k}_2)\cdot\underset{\sim}{r} - (\omega_1+\omega_2)t] \,. \tag{3.1}$$

Initially, it will be necessary to retain explicitly the frequency dependence associated with each of the three principal axes (ijk) as implied {14} by equation (3.1), and thereafter consideration will be given to situations which permit one to simplify this detailed form of description {6,15}.

One important general property of the susceptibility tensor $\chi^{(n)}_{ijk\ldots q}$ is immediately apparent on consideration of the reality of the physical fields $\underset{\sim}{P}$ and $\underset{\sim}{E}$: from a basic theorem of Fourier analysis, it follows that {16}:

$$\chi^{(n)}_{ijk\ldots q}(-\omega;-\omega_1,\ldots,-\omega_n) = \chi^{(n)*}_{ijk\ldots q}(\omega;\omega_1,\ldots,\omega_n)\ . \tag{3.2}$$

Since the susceptibility itself must be a real quantity, it follows that each tensor component of the type (3.2) must be an <u>even function of frequency</u>. (For simplicity, we shall however employ complex electric fields in this chapter. In that case, allowance for dissipation in the medium would yield an imaginary part to $\chi^{(n)}$ which would, from (3.2), be an odd function of frequency {16}.)

As a preliminary to this investigation, however, an important question needs to be answered, namely the rôle of the <u>local</u> fields in determining the non-linear susceptibilities in equation (1.13).

(a) THE RELATIONSHIP BETWEEN MACROSCOPIC AND LOCAL FIELD QUANTITIES

Whereas the fields appearing in the Maxwell equations (1.1)-(1.4), (1.9)-(1.12) are macroscopic quantities {7} (i.e. averages over macroscopic volume elements containing many atoms, in which microscopic fluctuations are smoothed out), a dynamical model for the polarization of the medium would require one to take into account the <u>local field</u> at the position of a particular molecule or atom in the crystal lattice.

In a linear (L), isotropic medium (a fluid or crystal with cubic symmetry), the local field $\underset{\sim}{\tilde{E}}$ is related to the macroscopic field $\underset{\sim}{E}$ by the Lorentz formula {7,8}:

$$\underset{\sim}{\tilde{E}} = \underset{\sim}{E} + \frac{4\pi}{3}\underset{\sim}{P}^{L} \tag{3.3}$$

$$\therefore\ \underset{\sim}{\tilde{E}} = \left(\frac{\epsilon+2}{3}\right)\underset{\sim}{E} \tag{3.4}$$

by (1.13) and (1.19).

Now certain crystals like ZnS are isotropic (cubic, class T_d) but lack inversion symmetry {4,17}. Thus $\chi^{(2)}_{ijk}$ is non-zero for all $i \neq j \neq k$

which implies that second harmonic generation can readily take place (see below). The corresponding form of (3.3) now becomes:

$$\underset{\sim}{\tilde{E}} = \underset{\sim}{E} + \frac{4\pi}{3}\underset{\sim}{P}^{L} + \frac{4\pi}{3}\underset{\sim}{P}^{NL} \tag{3.5}$$

while the displacement may be written:

$$\underset{\sim}{D} = \underset{\sim}{E} + 4\pi\underset{\sim}{P}^{L} + 4\pi\underset{\sim}{P}^{NL} . \tag{3.6}$$

In terms of the local electric field and susceptibility,

$$\underset{\sim}{P}^{L} = \tilde{\chi}^{(1)}\underset{\sim}{\tilde{E}} \tag{3.7}$$

and we find by (3.5):

$$\underset{\sim}{P}^{L} = \left[\frac{\tilde{\chi}^{(1)}}{1 - \frac{4\pi}{3}\tilde{\chi}^{(1)}}\right]\left\{\underset{\sim}{E} + \frac{4\pi}{3}\underset{\sim}{P}^{NL}\right\} \tag{3.8}$$

$$= \left(\frac{\epsilon - 1}{4\pi}\right)\underset{\sim}{E} + \left(\frac{\epsilon - 1}{3}\right)\underset{\sim}{P}^{NL} \tag{3.9}$$

where the permittivity ϵ of the medium has been obtained from $\tilde{\chi}^{(1)}$ by the Clausius-Mossotti equation {7,8}:

$$\tilde{\chi}^{(1)} = \frac{3}{4\pi}\left(\frac{\epsilon - 1}{\epsilon + 2}\right) . \tag{3.10}$$

Thus by (3.6):

$$\underset{\sim}{D} = \epsilon\underset{\sim}{E} + 4\pi\left(\frac{\epsilon + 2}{3}\right)\underset{\sim}{P}^{NL} . \tag{3.11}$$

Equation (3.11) shows that the effective macroscopic non-linear polarization is $\frac{1}{3}(\epsilon + 2)$ times the value calculated from the local fields by:

$$\underset{\sim}{P}^{NL} = \tilde{\chi}^{(2)}\underset{\sim}{\tilde{E}}\,\underset{\sim}{\tilde{E}} . \tag{3.12}$$

Thus, from (3.1), (3.4), (3.11) and (3.12), i.e. considering the non-linear terms to be small perturbations on the linear terms, we obtain {14}:

$$\chi^{(2)}(\omega_3;\omega_1,\omega_2) = \left(\frac{\epsilon(\omega_1)+2}{3}\right)\left(\frac{\epsilon(\omega_2)+2}{3}\right)\left(\frac{\epsilon(\omega_3)+2}{3}\right)\tilde{\chi}^{(2)}(\omega_3;\omega_1,\omega_2) . \qquad (3.13)$$

This important relation, which may be generalized to higher order {4}, shows that the macroscopic quantities $\chi^{(2)}$ follow the same symmetry relations as the microscopic quantities $\tilde{\chi}^{(2)}$. Armstrong et al. {14} have shown that the same conclusion holds in the case of anisotropic crystals (see Appendix I).

This is the justification for our not making a careful distinction between macroscopic and local field quantities in equations (1.13)-(1.16), and we shall henceforth omit the tildes once more. When necessary, the distinction can readily be made, by means of the relations discussed in this section.

We point out, lastly, the physical interpretation {14} of the non-linearity in equation (3.9) leading to (3.11) and (3.13). Owing to the interaction between the non-linear dipole moment at one lattice site and the linear dipole moment at another site, the local field at the various sites and, therefore, the linear dipole moment, is affected by the existence of such a non-linearity. This in turn changes the permittivity (dielectric tensor $\underset{\approx}{\epsilon}$) of the medium. The main approximation introduced in (3.13) has been to ignore the interaction between non-linear dipole moments at different sites {4,14}.

(b) MEDIA WITH QUADRATIC SUSCEPTIBILITY

Evaluation of the second-order polarization $\chi^{(2)}\underset{\sim}{E}\underset{\sim}{E}$ in equation (1.13) shows that four types of interactions arise when two fields:

$$\underset{\sim}{\mathcal{E}}_\alpha \exp[i(\underset{\sim}{k}_\alpha\cdot\underset{\sim}{r} - \omega_\alpha t)] \qquad (\alpha = 1,2)$$

are impressed upon a non-linear medium {4}.

(i) Static, non-linear terms of the type:

$$\chi^{(2)}_{ijk}(0;\omega_1,-\omega_1)\,\mathcal{E}_{1j}\,\mathcal{E}^*_{1k} \quad ; \quad \chi^{(2)}_{ijk}(0;\omega_2,-\omega_2)\,\mathcal{E}_{2j}\,\mathcal{E}^*_{2k} .$$

This effect might be called rectification of high-frequency electromagnetic waves in a quadratic medium, but is not of great interest for present purposes.

(ii) Second harmonic generation, i.e. terms of the type:

$$\chi^{(2)}_{ijk}(2\omega_1;\omega_1,\omega_1)\,\mathcal{E}_{1j}\,\mathcal{E}_{1k}\exp[i(2\underset{\sim}{k}_1\cdot\underset{\sim}{r}-2\omega_1 t)]$$

$$\chi^{(2)}_{ijk}(2\omega_2;\omega_2,\omega_2)\,\mathcal{E}_{2j}\,\mathcal{E}_{2k}\exp[i(2\underset{\sim}{k}_2\cdot\underset{\sim}{r}-2\omega_2 t)]\ .$$

(iii) Frequency summation terms of the type:

$$\chi^{(2)}_{ijk}(\omega_3;\omega_1,\omega_2)\,\mathcal{E}_{1j}\,\mathcal{E}_{2k}\exp i[(\underset{\sim}{k}_1+\underset{\sim}{k}_2)\cdot\underset{\sim}{r}-(\omega_1+\omega_2)t]\ .$$

(iv) The generation of waves of the difference frequency by the interaction of two waves in a quadratic medium, i.e. terms of the type:

$$\chi^{(2)}_{ijk}(\omega_3;\omega_1,-\omega_2)\,\mathcal{E}_{1j}\,\mathcal{E}^{*}_{2k}\exp i[(\underset{\sim}{k}_1-\underset{\sim}{k}_2)\cdot\underset{\sim}{r}-(\omega_1-\omega_2)t]\ .$$

An interesting special case of these processes is termed the linear electro-optic (Pockels) effect {4,8}. The index of refraction for a high-frequency electromagnetic wave is changed when a static or low-frequency field $\underset{\sim}{E}_2$ is applied. The corresponding non-linear polarization contains terms of the type:

$$\chi^{(2)}_{ijk}(\omega_1;\omega_1,0)\,\mathcal{E}_{1j}\,\mathcal{E}_{2k}\exp[i(\underset{\sim}{k}_1\cdot\underset{\sim}{r}-\omega_1 t)]\ .$$

(c) MEDIA WITH CUBIC SUSCEPTIBILITY

The effects arising from the second non-linear term $\chi^{(3)}\underset{\sim}{E}\underset{\sim}{E}\underset{\sim}{E}$ in equation (1.13) when three fields:

$$\underset{\sim}{\mathcal{E}}_\beta\exp[i(\underset{\sim}{k}_\beta\cdot\underset{\sim}{r}-\omega_\beta t)]\qquad\qquad(\beta=1,2,3)$$

are impressed upon a non-linear medium, may be enumerated as follows {4}.

(i) Sum frequency generation, from:

$$\chi^{(3)}_{ijk\ell}(\omega_4;\omega_1,\omega_2,\omega_3)\,\mathcal{E}_{1j}\,\mathcal{E}_{2k}\,\mathcal{E}_{3\ell}\exp i[(\underset{\sim}{k}_1+\underset{\sim}{k}_2+\underset{\sim}{k}_3)\cdot\underset{\sim}{r}-(\omega_1+\omega_2+\omega_3)t]\ .$$

With $\omega_1 = 0$, this becomes sum frequency generation in the presence of a static field.

(ii) Sum and difference frequency generation:

$$\chi^{(3)}_{ijk\ell}(\omega_4;\omega_1,\omega_2,-\omega_3)\,\mathcal{E}_{1j}\,\mathcal{E}_{2k}\,\mathcal{E}^*_{3\ell}\exp i[(\underset{\sim}{k}_1+\underset{\sim}{k}_2-\underset{\sim}{k}_3)\cdot\underset{\sim}{r}-(\omega_1+\omega_2-\omega_3)t]$$

$$\chi^{(3)}_{ijk\ell}(-\omega_4;-\omega_1,-\omega_2,+\omega_3)\,\mathcal{E}^*_{1j}\,\mathcal{E}^*_{2k}\,\mathcal{E}_{3\ell}\exp -i[(\underset{\sim}{k}_1+\underset{\sim}{k}_2-\underset{\sim}{k}_3)\cdot\underset{\sim}{r}-(\omega_1+\omega_2-\omega_3)t].$$

Again, with $\omega_1 = 0$, this becomes difference frequency generation in the presence of a static field.

(iii) Frequency tripling of an electromagnetic wave:

$$\chi^{(3)}_{ijk\ell}(3\omega_1;\omega_1,\omega_1,\omega_1)\,\mathcal{E}_{1j}\,\mathcal{E}_{1k}\,\mathcal{E}_{1\ell}\exp i[3\underset{\sim}{k}_1\cdot\underset{\sim}{r}-3\omega_1 t].$$

(iv) Frequency doubling of an electromagnetic wave by application of a static field ($\omega_2 = 0$):

$$\chi^{(3)}_{ijk\ell}(2\omega_1;\omega_1,\omega_1,0)\,\mathcal{E}_{1j}\,\mathcal{E}_{1k}\,\mathcal{E}_{2\ell}\exp i[2\underset{\sim}{k}_1\cdot\underset{\sim}{r}-2\omega_1 t].$$

(v) Change of the index of refraction of a medium in the field of an intense electromagnetic wave, owing to polarization terms of the type:

$$\chi^{(3)}_{ijk\ell}(\omega_1;\omega_1,\omega_1,-\omega_1)\,\mathcal{E}_{1j}\,\mathcal{E}_{1k}\,\mathcal{E}^*_{1\ell}\exp i[\underset{\sim}{k}_1\cdot\underset{\sim}{r}-\omega_1 t].$$

(d) A DYNAMICAL MODEL FOR THE POLARIZATION

We consider firstly the situation in a linear medium. Denote by $\underset{\sim}{x}_i = x_i\,\underset{\sim}{\ell}_i$ the displacement of electrons or ions corresponding to the ith normal mode, which obeys {4}:

$$M_i \frac{d^2 x_i}{dt^2} + \Gamma_i \frac{dx_i}{dt} + k_i x_i = -e \, \underset{\sim}{\ell}_i \cdot \underset{\sim}{\tilde{E}} \qquad (3.14)$$

Here M_i, Γ_i and k_i are respectively the effective mass, damping constant and harmonic force constant of the mode x_i; the unit vector $\underset{\sim}{\ell}_i$ is directed along the electric dipole moment of the oscillation, while $\underset{\sim}{E}$ denotes the local electric field strength. The dipole moment per unit volume (linear polarization) is given by:

$$\underset{\sim}{P}^L = -eN \sum_i \underset{\sim}{\ell}_i x_i \qquad (3.15)$$

where N is the number of atoms or molecules per unit volume and the summation extends over all electronic and atomic normal oscillations with non-zero dipole moment, i.e. oscillations which are optically active in the absorption spectrum.

"The normal modes of electrons and atoms must conform to the symmetry properties of the molecule or, for a single crystal, of the crystal as a whole. Thus, for example, those modes which are optically active in absorption and in Raman scattering satisfy, first of all, the requirements of translational symmetry, according to which they must be invariant under displacement by any lattice vector. Accordingly, the only normal modes considered are those which correspond to in-phase displacement of all homologous atoms in all unit cells of the crystal. Thus the translational symmetry requirement enables us to restrict the discussion for crystals to a single 'crystal molecule' or unit cell. Besides translational symmetry of crystals, isolated molecules display various other types of symmetry. In many cases, a knowledge of the symmetry and structure of the molecules enables one to determine the possible normal modes and corresponding dipole moments. For a sufficiently high degree of rotational and reflection symmetry, some of the modes (fully symmetric oscillations) have zero moment and therefore do not contribute to the sum (3.15)" {4}.

Now with the normal angular frequency $\omega_{oi} = \sqrt{\frac{k_i}{M_i}}$, and considering the crystal to be a weakly absorbing dispersive medium, i.e. ignoring the Γ_i term in (3.14), the steady-state solution corresponding to:

$$\underset{\sim}{\tilde{E}} = \underset{\sim}{\mathcal{E}} \exp i(\underset{\sim}{k} \cdot \underset{\sim}{r} - \omega t) \qquad (3.16)$$

is simply

$$x_i = - \frac{e \, \underset{\sim}{\ell}_i \cdot \underset{\sim}{\mathcal{E}} \exp i(\underset{\sim}{k} \cdot \underset{\sim}{r} - \omega t)}{M_i(\omega_{oi}^2 - \omega^2)} \, . \qquad (3.17)$$

This yields a linear polarization:

$$\underset{\sim}{P}^{L} = \sum_{i} \frac{N e^2 \underset{\sim}{\ell}_i (\underset{\sim}{\ell}_i \cdot \underset{\sim}{\mathcal{E}}) \exp i(\underset{\sim}{k} \cdot \underset{\sim}{r} - \omega t)}{M_i (\omega_{oi}^2 - \omega^2)} \tag{3.18}$$

and thus the linear susceptibility (Cartesian components):

$$\tilde{\chi}^{(1)}_{ab} = \sum_{i} N e^2 \frac{\ell_{ia} \ell_{ib}}{M_i (\omega_{oi}^2 - \omega^2)} \quad . \tag{3.19}$$

Now for a medium with quadratic susceptibility (a quadratic medium), equation (3.14) may be modified by addition of the appropriate non-linear correction to the potential energy U of the molecule:

$$U = \frac{1}{2} \sum_{i} k_i x_i^2 + \Delta U \quad . \tag{3.20}$$

This correction term arises from the interaction (coupling) of the normal modes. The lowest order of these is cubic, which dominates higher order terms in weakly non-linear media:

$$\Delta U = \frac{1}{3} \sum_{ijk} \beta_{ijk} x_i x_j x_k \tag{3.21}$$

where the matrix β_{ijk} is invariant under any permutation of the indices. As shown below, such a correction term can only exist in a crystal which lacks a centre of inversion. (It is interesting to note that the same restriction applies in the case of piezoelectric crystals, which belong to exactly the same classes as those which admit quadratic susceptibility {10}.)

Now the resulting non-linear equation:

$$M_i \frac{d^2 x_i}{dt^2} + \Gamma_i \frac{dx_i}{dt} + k_i x_i + \sum_{jk} \beta_{ijk} x_j x_k = - e \underset{\sim}{\ell}_i \cdot \underset{\sim}{\tilde{E}} \tag{3.22}$$

cannot be solved exactly. However, we are interested in weakly absorbing dispersive media, for which Γ_i may be set equal to zero, and obtain an approximate solution to (3.22) by the perturbation method. With $\tilde{E}$ again given by (3.16), we write:

$$x_i = x_i^L + x_i^{NL} \tag{3.23}$$

where $|x_i^{NL}| \ll |x_i^L|$ by assumption (thus $|\frac{\beta_{ijk}}{k_i}| \ll |x_i^L|^{-1}$), and x_i^L is given by (3.17). Upon substitution of (3.23) into (3.22) and on collecting the first-order correction terms, one obtains for the real (i.e. physical) non-linear displacement:

$$x_i^{NL} = -\frac{1}{2}\sum_{jk} \frac{\beta_{ijk}\, e^2 (\tilde{\ell}_j \cdot \tilde{\mathcal{E}})(\tilde{\ell}_k \cdot \tilde{\mathcal{E}})}{M_i M_j M_k (\omega_{oj}^2 - \omega^2)(\omega_{ok}^2 - \omega^2)} \times \left[\frac{1}{\omega_{oi}^2} + \frac{\cos 2(\tilde{k}\cdot\tilde{r} - \omega t)}{(\omega_{oi}^2 - 4\omega^2)}\right]. \tag{3.24}$$

Thus,

$$\tilde{P}^{NL} = \tilde{P}^{NL}(0) + \tilde{P}^{NL}(2\omega)$$

where:

$$\tilde{P}^{NL}(2\omega) = \frac{Ne^3}{2}\sum_{ijk} \frac{\beta_{ijk}\, \tilde{\ell}_i (\tilde{\ell}_j \cdot \mathcal{E})(\tilde{\ell}_k \cdot \mathcal{E})}{M_i M_j M_k (\omega_{oj}^2 - \omega^2)(\omega_{ok}^2 - \omega^2)} \times \frac{\cos 2(\tilde{k}\cdot\tilde{r} - \omega t)}{(\omega_{oi}^2 - 4\omega^2)} \tag{3.25}$$

and therefore (in Cartesian components):

$$\tilde{\chi}^{(2)}_{abc}(2\omega; \omega, \omega) = \frac{Ne^3}{2}\sum_{ijk} \frac{\beta_{ijk}\ \ell_{ia}\, \ell_{jb}\, \ell_{kc}}{M_i M_j M_k (\omega_{oj}^2 - \omega^2)(\omega_{ok}^2 - \omega^2)(\omega_{oi}^2 - 4\omega^2)}. \tag{3.26}$$

Completely analogously, one finds on combining two wave fields $\tilde{\mathcal{E}}_1 \exp i(\tilde{k}_1\cdot\tilde{r} - \omega_1 t)$ and $\tilde{\mathcal{E}}_2 \exp i(\tilde{k}_2\cdot\tilde{r} - \omega_2 t)$ two quadratic susceptibility terms, corresponding to sum and difference frequencies $\omega_1 \pm \omega_2$, the one for $\omega_3 = \omega_1 + \omega_2$ being:

$$\tilde{\chi}^{(2)}_{abc}(\omega_3;\omega_1,\omega_2) = \frac{Ne^3}{2}\sum_{ijk}\frac{\beta_{ijk}\,\ell_{ia}\,\ell_{jb}\,\ell_{kc}}{M_iM_jM_k(\omega_{oi}^2-\omega_3^2)(\omega_{oj}^2-\omega_1^2)(\omega_{ok}^2-\omega_2^2)} \tag{3.27}$$

where it is understood that components a, b, c are associated respectively with angular frequencies ω_3, ω_1 and ω_2, respectively.

Examination of equation (3.27) reveals that it satisfies the important result {14} that:

$$\tilde{\chi}^{(2)}_{abc}(\omega_3;\omega_1,\omega_2) = \tilde{\chi}^{(2)}_{cab}(\omega_2;\omega_3,-\omega_1) = \tilde{\chi}^{(2)}_{bca}(\omega_1;-\omega_2,\omega_3)\,. \tag{3.28}$$

This, the most general permutation symmetry relation {14} states that the frequencies may be permuted at will provided that the Cartesian indices a, b and c are simultaneously permuted so that a given frequency is always associated with the same index.

However, (3.27) also satisfies the general relationship {4}:

$$\tilde{\chi}^{(2)}_{abc}(\omega_3;\omega_1,\omega_2) = \tilde{\chi}^{(2)}_{acb}(\omega_3;\omega_1,\omega_2) \tag{3.29}$$

i.e. that $\tilde{\chi}^{(2)}_{abc}$ is always symmetrical in the last two indices, as well as the weaker Kleinman symmetry condition {15} ("conjecture" {6}), that $\tilde{\chi}^{(2)}_{abc}$ is symmetrical in all three indices (i.e. without permutation of the corresponding frequencies):

$$\tilde{\chi}^{(2)}_{abc}(\omega_3;\omega_1,\omega_2) = \tilde{\chi}^{(2)}_{acb}(\omega_3;\omega_1,\omega_2) = \tilde{\chi}^{(2)}_{bac}(\omega_3;\omega_1,\omega_2)\,. \tag{3.30}$$

Now a more rigorous derivation of the susceptibilities than that presented here, i.e. with inclusion of energy dissipation, would show that whereas relations (3.28) and (3.29) are rigorously and generally valid, (3.30) would hold only in the limit of sufficiently weak dispersion, and thus provided that the crystal is effectively transparent throughout a spectral region that includes all the frequencies involved in the interaction {6,9}. While this very useful simplification {15} may be assumed to hold within experimental accuracy in many (if not most) cases of interest, it would of course be invalid were an absorption band to lie between any of the relevant frequencies. It therefore does not apply to the linear electro-optic (Pockels) effect, or to difference-frequency generation of far-infrared wavelengths {18}. An interesting case has been reported in the literature {19}, in which the crystal $LiIO_3$ (class C_6) was shown to possess a weak but non-vanishing second-order susceptibility $\chi^{(2)}_{123} = -\chi^{(2)}_{213}$, contrary to Kleinman's conjecture. The physical basis of Kleinman's arguments {15} has been re-examined by Franken and Ward {6}, who also present a validity criterion for judging the applicability of this approximation.

In the case of crystals possessing a centre of inversion, the quadratic susceptibility vanishes identically (see below). The predominant contribution to the non-linear susceptibility then arises from a quartic correction to the molecular potential energy U {4}:

$$\Delta U = \frac{1}{4}\sum_{ijkg} \alpha_{ijkg}\, x_i x_j x_k x_g \tag{3.31}$$

where the matrix α_{ijkg} is invariant under any permutation of the indices. Equation (3.22) is now replaced by:

$$M_i \frac{d^2 x_i}{dt^2} + \Gamma_i \frac{dx_i}{dt} + k_i x_i + \sum_{jkg} \alpha_{ijkg}\, x_j x_k x_g = - e\, \underset{\sim}{\ell}_i \cdot \underset{\sim}{E} \; . \tag{3.32}$$

Upon substitution of (3.23) into (3.32) and on collecting the first-order correction terms, one obtains for the real (i.e. physical) non-linear displacement:

$$x_i^{NL} = \frac{1}{4}\sum_{jkg} \frac{\alpha_{ijkg}\, e^3 (\underset{\sim}{\ell}_j \cdot \underset{\sim}{\mathcal{E}})(\underset{\sim}{\ell}_k \cdot \underset{\sim}{\mathcal{E}})(\underset{\sim}{\ell}_g \cdot \underset{\sim}{\mathcal{E}})}{M_i M_j M_k M_g (\omega_{oj}^2 - \omega^2)(\omega_{ok}^2 - \omega^2)(\omega_{og}^2 - \omega^2)}$$

$$\times \left\{ \frac{3\cos(\underset{\sim}{k}\cdot\underset{\sim}{r} - \omega t)}{(\omega_{oi}^2 - \omega^2)} + \frac{\cos 3(\underset{\sim}{k}\cdot\underset{\sim}{r} - \omega t)}{(\omega_{oi}^2 - 9\omega^2)} \right\} \; . \tag{3.33}$$

Thus,

$$\underset{\sim}{P}^{NL} = \underset{\sim}{P}^{NL}(\omega) + \underset{\sim}{P}^{NL}(3\omega)$$

where:

$$P^{NL}(3\omega) = -\frac{Ne^4}{4}\sum_{ijkg}\frac{\alpha_{ijkg}\,\underset{\sim}{\ell}_i(\underset{\sim}{\ell}_j\cdot\underset{\sim}{\mathcal{E}})(\ell_k\cdot\underset{\sim}{\mathcal{E}})(\ell_g\cdot\underset{\sim}{\mathcal{E}})}{M_iM_jM_kM_g}$$

$$\times\ \frac{\cos 3(\underset{\sim}{k}\cdot\underset{\sim}{r} - \omega t)}{(\omega_{oi}^2 - 9\omega^2)(\omega_{oj}^2 - \omega^2)(\omega_{ok}^2 - \omega^2)(\omega_{og}^2 - \omega^2)} \tag{3.34}$$

and therefore the Cartesian components of the third-order susceptibility are given by:

$$\tilde{\chi}^{(3)}_{abcd}(3\omega;\omega,\omega,\omega) = -\frac{Ne^4}{4}\sum_{ijkg}\frac{\alpha_{ijkg}\,\ell_{ia}\,\ell_{jb}\,\ell_{kc}\,\ell_{gd}}{M_iM_jM_kM_g}$$

$$\frac{1}{(\omega_{oi}^{2\prime} - 9\omega^2)(\omega_{oj}^2 - \omega^2)(\omega_{ok}^2 - \omega^2)(\omega_{og}^2 - \omega^2)}\,. \tag{3.35}$$

Completely analogously, one finds on combining three wave fields $\mathcal{E}_1 \exp i(\underset{\sim}{k}_1\cdot\underset{\sim}{r} - \omega_1 t)$, $\mathcal{E}_2 \exp i(\underset{\sim}{k}_2\cdot\underset{\sim}{r} - \omega_2 t)$ and $\underset{\sim}{\mathcal{E}}_3 \exp i(\underset{\sim}{k}_3\cdot\underset{\sim}{r} - \omega_3 t)$ four quadratic susceptibility terms corresponding to angular frequencies:

$$\omega_4 = \omega_1 \pm \omega_2 \pm \omega_3 \quad ; \quad \omega_4 = \omega_1 \pm \omega_2 \mp \omega_3$$

Such interactions are termed (counting the mode at the resultant frequency) <u>four-frequency</u> {4} or <u>four-wave</u> {14} <u>interactions</u>. The susceptibility corresponding to $\omega_4 = \omega_1 + \omega_2 + \omega_3$ may be written:

$$\tilde{\chi}^{(3)}_{abcd}(\omega_4;\omega_1,\omega_2,\omega_3) = -\frac{Ne^4}{4}\sum_{ijkg}\frac{\alpha_{ijkg}\,\ell_{ia}\,\ell_{jb}\,\ell_{kc}\,\ell_{gd}}{M_iM_jM_kM_g(\omega_{oi}^2-\omega_4^2)(\omega_{oj}^2-\omega_1^2)(\omega_{ok}^2-\omega_2^2)(\omega_{og}^2-\omega_3^2)} \tag{3.36}$$

where it is understood that components a, b, c, d are associated respectively with angular frequencies ω_4, ω_1, ω_2 and ω_3.

It is clear that (3.36), analogously to (3.27), satisfies the result that {14}:

$$\tilde{\chi}^{(3)}_{abcd}(\omega_4;\omega_1,\omega_2,\omega_3) = \tilde{\chi}^{(3)}_{dabc}(\omega_3;\omega_4,-\omega_1,-\omega_2)$$

$$= \tilde{\chi}^{(3)}_{cdab}(\omega_2;-\omega_3,\omega_4,-\omega_1) = \tilde{\chi}^{(3)}_{bcda}(\omega_1;-\omega_2,-\omega_3,\omega_4). \qquad (3.37)$$

As in the case of (3.28), this general permutation symmetry relation may be stated: the frequencies may be permuted at will provided that the Cartesian indices a, b, c and d are simultaneously permuted in such a way that a given frequency is always associated with the same index.

However, (3.36) also satisfies the general relationship {4}:

$$\tilde{\chi}^{(3)}_{abcd}(\omega_4;\omega_1,\omega_2,\omega_3) = \tilde{\chi}^{(3)}_{abdc}(\omega_4;\omega_1,\omega_2,\omega_3) = \tilde{\chi}^{(3)}_{adcb}(\omega_4;\omega_1,\omega_2,\omega_3) \qquad (3.38)$$

i.e. that $\tilde{\chi}^{(3)}_{abcd}$ is always symmetrical in the last three indices. In addition, it satisfies the weaker Kleinman symmetry condition {15} that $\tilde{\chi}^{(3)}_{abcd}$ is symmetrical in all three indices (i.e. without permutation of the corresponding frequencies):

$$\tilde{\chi}^{(3)}_{abcd}(\omega_4;\omega_1,\omega_2,\omega_3) = \tilde{\chi}^{(3)}_{dabc}(\omega_4;\omega_1,\omega_2,\omega_3)$$

$$= \tilde{\chi}^{(3)}_{cdab}(\omega_4;\omega_1,\omega_2,\omega_3) = \tilde{\chi}^{(3)}_{bcda}(\omega_4;\omega_1,\omega_2,\omega_3). \qquad (3.39)$$

As in the case of (3.30), a more rigorous derivation of the susceptibilities than that presented here, i.e. with inclusion of energy dissipation, would show that whereas relations (3.37) and (3.38) are rigorously and generally valid, (3.39) holds only in the limit of sufficiently weak dispersion, and thus provided that the crystal is effectively transparent throughout a spectral region that includes all the frequencies involved in the interaction {6,9}.

Comparison of equations (3.19), (3.27) and (3.36) suggests the validity of an empirical rule due to Miller {20}, implying that since the non-linear susceptibilities consist of products of terms which enter also in the corresponding linear (first-order) susceptibilities for the relevant frequencies, materials with high refractive index tend to be more strongly non-linear in their optical properties. We shall examine the basis for Miller's rule {20} in more detail below.

(e) EXPERIMENTAL NOTATION

Corresponding to the second-order non-linear susceptibility $\chi^{(2)}_{ijk}$ used above, one finds in the experimental literature a non-linear coefficient d_{ijk}. The relationship between the two is {9}:

$$d_{ijk} = \frac{1}{2}\chi^{(2)}_{ijk} \quad . \tag{3.40}$$

The reader should note carefully the ambiguity that exists in the literature concerning the omission or inclusion in relation (3.40) of the factor of two {9}. It is also common practice at this point to employ a parameter d with contracted indices in the notation of W. Voigt {16}. One sets:

$$d_{ijk} = d_{mn} \tag{3.41}$$

where $m = i$, and $n = j$ for $j = k$, $n = 4$ for $j + k = 5$, $n = 5$ for $j + k = 4$ with $j \neq k$, and $n = 6$ for $j + k = 3$. When employing these parameters with contracted indices in matrices, the reader should note the cautionary remarks in Section 2.9 of Ref. {9}.

A further grouping of parameters d_{mn} into an effective parameter d_{eff} (or simply d) is also employed {9} when summation is carried out over a number of interacting waves in various states of polarization.

(f) SYMMETRY TRANSFORMATIONS

Consider a particular physical situation described by equation (1.13), that is for given orientations of the medium, impressed electric fields and resultant polarization. One can transform from a description of the phenomenon in one chosen set of coordinate axes (e.g. the principal axes of the crystal, as in Chapter 2 above) to a new coordinate frame. The new axes are related to the old by a transformation $\underset{\approx}{A}$ which can be written as a (3×3) matrix representing an arbitrary combination of rotation and inversion. In the new frame one has:

$$\tilde{\chi}^{(2)'}_{\alpha\beta\gamma} = A_{\alpha i} A_{\beta j} A_{\gamma k} \tilde{\chi}^{(2)}_{ijk} \quad . \tag{3.42}$$

If $\underset{\approx}{A}$ is now restricted to be a symmetry transformation, then all properties of the material are identically described in both coordinate frames, so that:

$$\tilde{\chi}^{(2)}_{\alpha\beta\gamma} = A_{\alpha i} A_{\beta j} A_{\gamma k} \tilde{\chi}^{(2)}_{ijk} \quad . \tag{3.43}$$

These symmetry transformations correspond to so-called macroscopic symmetry elements in crystals {17}, which reduce to the following:

i) centre of symmetry (inversion),
ii) mirror plane,
iii) 1-, 2-, 3-, 4- or 6-fold rotation axes,
iv) 1-, 2-, 3-, 4- or 6-fold inversion axes.

Possible combinations of macroscopic symmetry elements are called point groups, of which there are just 32. Crystals are divided into 32 crystal classes according to the point-group symmetry they possess.

As an example of a symmetry transformation, choose $\underset{\approx}{A}$ to be the inversion transformation $A_{\alpha i} = -\delta_{\alpha i}$. Equation (3.43) then yields:

$$\tilde{\chi}^{(2)}_{\alpha\beta\gamma} = - \tilde{\chi}^{(2)}_{\alpha\beta\gamma} = 0 \tag{3.44}$$

from which one may conclude that second-harmonic generation cannot take place in crystals which possess a centre of inversion. By an extension of this argument, it may be shown that for crystals with a centre of inversion, all tensors of odd rank are zero.

Of the various crystal classes, there remain 21 without a centre of inversion. Of these, one (cubic, class O) is of no practical importance since all components of the second-order susceptibility vanish {16} on the grounds of other symmetry transformations which the crystal possesses {17}. The remaining 20 classes, which also exhibit piezoelectricity {10}, are listed in Appendix II.

(g) MILLER'S RULE

Consider again equations (3.19) and (3.27). In a crystal, the mode directions $\underset{\sim}{\ell}_i$ are tied to the crystallographic axes {4}, and therefore by choosing the principal axis system, one obtains a diagonal representation for $\tilde{\chi}^{(1)}$, as already seen in Chapters 1 and 2. Thus,

$$\tilde{\chi}^{(1)}_{aa} = Ne^2 \sum_i{}' \frac{1}{M_i(\omega_{oi}^2 - \omega^2)} \tag{3.45}$$

where the prime indicates that the sum includes only those modes which can be projected onto principal axis a. In this way, (3.27) may also be simplified as follows:

$$\tilde{\chi}^{(2)}_{abc}(\omega_3;\omega_1,\omega_2) = \sum_{ijk}{}' \frac{\frac{1}{2} Ne^3 \beta_{ijk}}{M_i M_j M_k(\omega_{oi}^2 - \omega_3^2)(\omega_{oj}^2 - \omega_1^2)(\omega_{ok}^2 - \omega_2^2)} \tag{3.46}$$

where the prime indicates that the sum includes only those normal modes lying along principal axes a (modes i), b (modes j) and c (modes k), respectively.

Then, a simple heuristic argument {9,20,21} suggests that one may approximate $\tilde{\chi}^{(2)}_{abc}$ by:

$$\tilde{\chi}^{(2)}_{abc}(\omega_3;\omega_1,\omega_2) = \tilde{\chi}^{(1)}_{aa}(\omega_3)\tilde{\chi}^{(1)}_{bb}(\omega_1)\tilde{\chi}^{(1)}_{cc}(\omega_2)\,\Delta_{abc} , \tag{3.47}$$

where the third-rank tensor Δ_{abc} may also be expected to obey the Kleinman symmetry condition {15}. Now it is clear from (3.46) that relation (3.47) cannot be derived rigorously from the preceding equations. However, Miller {20} has found that whereas experimental values for $\tilde{\chi}^{(2)}_{abc}$ vary over a range of about 600 for the various crystal types he considered (all suitable for optical second harmonic generation), the corresponding values of Δ_{abc} calculated by (3.47) from the experimental data were always within a factor of two of their average value. Moreover, Δ_{abc} for a particular coefficient for crystals in a given symmetry class were found to be equal, to within ± 50%.

Experiments by Patel {22,23} verified that Miller's phenomenological rule can be applied to infra-red fundamental wavelengths as well. In his work, a 10.6 μ CO_2 laser was employed, together with the following crystals: Te and Se (Class D_3); ZnS, CdS and CdSe (Class C_{6v}); InAs, GaAs, ZnS, CdTe, ZnSe and ZnTe (Class T_d). In addition to good qualitative and quantitative agreement with the paper of Miller {20}, the Kleinman symmetry relation {15} was verified within experimental accuracy for ZnS, CdS and CdSe.

An attempt to account for these observations on both classical and quantum-mechanical grounds led to the conclusion that the physical origin of the non-linearity could be ascribed to the shape of the potential at a lattice site {21}, and therefore that the magnitude of the non-linear polarization is determined primarily by geometrical factors rather than by the detailed structure of the atomic wave functions. This conclusion in fact provides a justification for the essentially classical description of non-linear optics in the present review.

(h) ORDERS OF MAGNITUDE

A simple order-of-magnitude estimate of the ratio between the first-order non-linear polarization and the linear polarization (or equivalently between non-linear polarizations of successive orders) may be obtained as follows {21,24}. Restoring the neglected damping terms in (3.17) et seq., and using the abbreviated notation of Bloembergen {24}:

$$D(\omega) = \omega_0^2 - \omega^2 - i\frac{\Gamma}{M}\omega$$

we have:

$$\left|\frac{P^{NL}(2\omega)}{P^{L}(\omega)}\right| \sim \frac{\beta e}{m^2}\,\frac{|\mathcal{E}|}{D(\omega)D(2\omega)}$$

using simply the electron mass for each M_i, and ignoring the three-dimensional aspects of the problem.

"Now it may be expected from the physical nature of electronic binding that if the deviation x is of the order of the radius a of the equilibrium orbital of the electron, the non-linear force βx^2 is of the same order as the linear force $m\omega_0^2 a = e|E_{at}|$, where E_{at} is the intra-atomic electric field binding the electron" {24}. Thus with both and 2ω well off resonance, we may estimate:

$$\left|\frac{P^{NL}(2\omega)}{P^{L}(\omega)}\right| \sim \frac{e}{ma}\,\frac{|\mathcal{E}|}{\omega_0^2} \sim \frac{|\mathcal{E}|}{|E_{at}|}$$

The electric field amplitude of the light wave must therefore be compared with the electric field inside the atom, which is typically of the order of 3×10^{8} volts/cm.

It may thus be concluded {24} that even for the high power flux densities ($\sim 10^{10}$ watts/cm^2), e.g. in the focus of a Q-switched laser, the non-linear response can still be treated as a small perturbation, since $|\mathcal{E}/E_{at}| \sim 3 \times 10^{-3}$ in this case.

With a resonance occurring in one of the D factors, this ratio would be enhanced, however, by the substantial factor $\frac{m\omega_0}{\Gamma}$. It should be pointed out, however, that even if the magnitude of the non-linear effect is small, its detectability is due to the excellent discrimination in the various experiments discussed here. Of primary importance in this regard is the subject of the following chapter.

CHAPTER 4

PHASE MATCHING IN CRYSTALS

(a) THE THREE-WAVE INTERACTION

Consider a medium possessing quadratic susceptibility, on which the wave fields:

$$\underset{\sim}{\mathcal{E}}_\alpha \exp[i(\underset{\sim}{k}_\alpha \cdot \underset{\sim}{r} - \omega_\alpha t)] \qquad (\alpha = 1, 2)$$

are impressed. As discussed in Chapter 3, a wave of angular frequency:

$$\omega_3 = \omega_1 + \omega_2$$

will be generated in the medium, and this can in turn interact with the wave at angular frequency ω_2 to produce a non-linear polarization of the medium at angular frequency ω_1, by the three-wave interaction {9}. Consider for example wave propagation in the z direction.

Then for the ith component of the non-linear polarization at frequency ω_1, equation (1.13) yields the summation:

$$P_{1i}^{NL}(z,t) = \sum_{jk} \Big[\chi_{ijk}^{(2)}(\omega_1; -\omega_2, \omega_3)\, \mathcal{E}_{2j}^{*}(z)\, \mathcal{E}_{3k}(z) + \chi_{ijk}^{(2)}(\omega_1; \omega_3, -\omega_2)\, \mathcal{E}_{3j}(z)\, \mathcal{E}_{2k}^{*}(z) \Big] \times \exp i[(k_3 - k_2)z - (\omega_3 - \omega_2)t] \qquad (4.1)$$

where explicit allowance has been made for amplitude variation with distance in a non-linear medium.

Now since by (3.28) and (3.29):

$$\chi_{ijk}^{(2)}(\omega_1; \omega_3, -\omega_2) = \chi_{ikj}^{(2)}(\omega_1; -\omega_2, \omega_3) = \chi_{ijk}^{(2)}(\omega_1; -\omega_2, \omega_3) \qquad (4.2)$$

and introducing the experimental parameter from (3.40):

$$\chi^{(2)}_{ijk} = 2\,d_{ijk} \tag{4.3}$$

we may simplify the notation further by contracting the summations over the indices jk by means of the effective parameter d (see Chapter 3 and Ref. {9}). "The advantage of this notation is that it reduces the problem to one dimension. All further derivations are made using this effective non-linearity, yielding in the end a simple, universally valid expression for the generated power. For given experimental conditions, the appropriate equation for d is substituted in this equation, restoring again the full three-dimensional aspect of the problem" {9}. It may be pointed out that at this stage in the text just cited, the approximation has already been made of neglecting all frequency dependences of the non-linear susceptibility, i.e. in effect Kleinman symmetry has been assumed.

With the above notational changes and simplifications, equation (4.1) and similar relations for P^{NL}_{2j}, P^{NL}_{3k} yield for the non-linear polarizations at the three frequencies:

$$P^{NL}_1(z,t) = 2d\,\mathcal{E}^*_2(z)\,\mathcal{E}_3(z)\exp i[(k_3-k_2)z-(\omega_3-\omega_2)t] \tag{4.4}$$

$$P^{NL}_2(z,t) = 2d\,\mathcal{E}_3(z)\,\mathcal{E}^*_1(z)\exp i[(k_3-k_1)z-(\omega_3-\omega_1)t] \tag{4.5}$$

$$P^{NL}_3(z,t) = 2d\,\mathcal{E}_1(z)\,\mathcal{E}_2(z)\exp i[(k_1+k_2)z-(\omega_1+\omega_2)t]\,. \tag{4.6}$$

These expressions may now be used in conjunction with the Maxwell equations to relate the spatial variations of the electric fields associated with the three frequencies. From equation (1.17) one has:

$$\nabla(\nabla\cdot\underset{\sim}{E}) - \nabla^2\underset{\sim}{E} + \frac{1}{c^2}\frac{\partial^2}{\partial t^2}[\underset{\approx}{\epsilon}\cdot\underset{\sim}{E}] = -\frac{4\pi}{c^2}\frac{\partial^2 \underset{\sim}{P}^{NL}}{\partial t^2} \tag{4.7}$$

where

$$\underset{\sim}{D} = \underset{\approx}{\epsilon}\cdot\underset{\sim}{E} + 4\pi\,\underset{\sim}{P}^{NL} \tag{4.8}$$

(with $\underset{\sim}{P}^{NL}$ now taken to be the effective <u>macroscopic</u> non-linear polarization, as distinct from the value calculated from the local fields in the medium as in (3.11) and (3.12)).

Now since $\frac{\partial}{\partial x} = \frac{\partial}{\partial y} = 0$, we have in the principal axis system of the crystal:

$$\frac{\partial^2 E_1}{\partial z^2} + \frac{\omega_1^2 \epsilon_1}{c^2} E_1 = -\frac{4\pi}{c^2} \omega_1^2 P_1^{NL} \qquad (4.9)$$

using also equation (4.4). Direct calculation of the first term in (4.9) yields:

$$\frac{\partial^2 E_1}{\partial z^2} = \left(\frac{d^2 \mathcal{E}_1}{d z^2} - k_1^2 \mathcal{E}_1 + 2 i k_1 \frac{d \mathcal{E}_1}{d z} \right) \exp i(k_1 z - \omega_1 t) . \qquad (4.10)$$

Now if the scale lengths L for significant changes of amplitude or phase of $\mathcal{E}_1$ with distance are such that:

$$L \gg \frac{2\pi}{k_1}$$

i.e. many wavelengths would be required for significant variation of the envelope function $\mathcal{E}_1(z)$, then the first term in parentheses in (4.10) may be neglected, and:

$$\frac{\partial^2 E_1}{\partial z^2} \simeq - \left(k_1^2 \mathcal{E}_1 - 2 i k_1 \frac{d \mathcal{E}_1}{d z} \right) \exp i(k_1 z - \omega_1 t) . \qquad (4.11)$$

Henceforth we shall consistently adopt the <u>slowly varying amplitude and phase approximation</u>. From equations (4.4) - (4.6), (4.9) and (4.11) or their equivalents, we obtain directly {9}:

$$\frac{d \mathcal{E}_1}{d z} = \frac{4\pi i \omega_1^2}{k_1 c^2} d \; \mathcal{E}_2^*(z) \mathcal{E}_3(z) \; \exp[i(k_3 - k_2 - k_1) z] \qquad (4.12)$$

$$\frac{d \mathcal{E}_2}{d z} = \frac{4\pi i \omega_2^2}{k_2 c^2} d \; \mathcal{E}_1^*(z) \mathcal{E}_3(z) \; \exp[i(k_3 - k_2 - k_1) z] \qquad (4.13)$$

$$\frac{d \mathcal{E}_3}{d z} = \frac{4\pi i \omega_3^2}{k_3 c^2} d \; \mathcal{E}_1(z) \mathcal{E}_2(z) \; \exp[i(k_1 + k_2 - k_3) z] \qquad (4.14)$$

where explicit use has been made of the substitution $k_i^2 = \epsilon_i \, \omega_i^2/c^2$ $(i = 1,2,3)$. (Actually, this substitution requires a little thought:

equation (4.9) in fact represents the propagation of both E_{1x} and E_{1y} in the z direction. From Chapter 2 we know that the corresponding phase velocities are in fact different, being given by $\frac{c}{\sqrt{\epsilon_x}}$ and $\frac{c}{\sqrt{\epsilon_y}}$, respectively. In our condensed notation, both of these modes are represented by a single wave vector k_1.)

Each of the three coupled equations (4.12) - (4.14) gives the rate of change with distance of the amplitude at one frequency as a function of the amplitudes at the two other frequencies and of the phase difference between the non-linear polarization wave and the electro-magnetic wave. We shall substitute {14}:

$$\Delta k = k_1 + k_2 - k_3 \tag{4.15}$$

and first consider the simple case in which the amount of power generated at the sum frequency (ω_3) is small enough for the amplitudes at the two input frequencies (ω_1 and ω_2) to be considered constant. Then, (4.12) - (4.15) reduce to the single equation:

$$\frac{d\mathcal{E}_3}{dz} = \frac{4\pi i \omega_3^2}{k_3 c^2} d\, \mathcal{E}_1 \mathcal{E}_2 \exp(i \Delta k z) \tag{4.16}$$

whence, integrating over a homogeneous slab of thickness L,

$$\mathcal{E}_3(L) = \frac{4\pi \omega_3^2}{k_3 c^2} \frac{d\,\mathcal{E}_1\mathcal{E}_2}{\Delta k}\left[\exp(i\Delta k L) - 1\right] . \tag{4.17}$$

In terms of the corresponding Poynting vectors (by equation (2.18)):

$$S_3 = \frac{\mu_3 c}{8\pi} |\mathcal{E}_3|^2$$

$$= \frac{512\, \pi^5 L^2 d^2 S_1 S_2}{\mu_1 \mu_2 \mu_3 \lambda_3^2 c} \left(\frac{\sin x}{x}\right)^2 \tag{4.18}$$

where $x = \Delta k\, L/2$, λ_3 is the wavelength <u>in vacuo</u> corresponding to ω_3, S_1 and S_2 are the Poynting vectors corresponding to the waves of angular frequency ω_1 and ω_2, respectively, and the refractive indices μ_1, μ_2 and μ_3 are those appropriate to the particular polarization states under consideration (or the effective values appropriate to the compositions of the beams ω_1 and ω_2).

The form of equation (4.18) illustrates three important points {9}:

i) For $\Delta k \neq 0$, the output power generated by the three-wave interaction varies as $\left(\frac{\sin x}{x}\right)^2$.

ii) For $\Delta k = 0$, the output power is proportional to the square of the number of output wavelengths corresponding to the length of the crystal.

iii) The output power is proportional to the product of the input powers. It should be borne in mind that (4.18) has been derived strictly within the <u>small-signal approximation</u>.

(b) THE MANLEY-ROWE RELATIONS

Equations (4.12)-(4.14), with $\frac{\omega}{k} = \frac{c}{\mu}$, may be written in the form:

$$\frac{\mu_1 c}{\omega_1} \mathcal{E}_1^* \frac{d\mathcal{E}_1}{dz} = 4\pi i d\, \mathcal{E}_1^* \mathcal{E}_2^* \mathcal{E}_3\, e^{-i\Delta k z} \tag{4.19}$$

$$\frac{\mu_2 c}{\omega_2} \mathcal{E}_2^* \frac{d\mathcal{E}_2}{dz} = 4\pi i d\, \mathcal{E}_2^* \mathcal{E}_1^* \mathcal{E}_3\, e^{-i\Delta k z} \tag{4.20}$$

$$\frac{\mu_3 c}{\omega_3} \mathcal{E}_3^* \frac{d\mathcal{E}_3}{dz} = 4\pi i d\, \mathcal{E}_3^* \mathcal{E}_1 \mathcal{E}_2\, e^{+i\Delta k z} \tag{4.21}$$

which may be combined as:

$$\frac{\mu_1 c}{\omega_1} \frac{d}{dz}(\mathcal{E}_1 \mathcal{E}_1^*) = \frac{\mu_2 c}{\omega_2} \frac{d}{dz}(\mathcal{E}_2 \mathcal{E}_2^*) = -\frac{\mu_3 c}{\omega_3} \frac{d}{dz}(\mathcal{E}_3 \mathcal{E}_3^*) \tag{4.22}$$

or, from equation (2.18):

$$\frac{1}{\omega_1} \frac{d}{dz} S(z,\omega_1) = \frac{1}{\omega_2} \frac{d}{dz} S(z,\omega_2) = -\frac{1}{\omega_3} \frac{d}{dz} S(z,\omega_3)\,. \tag{4.23}$$

These are the celebrated Manley-Rowe relations, originally derived {25-27} for application to non-linear electrical circuit theory. Their original derivation is summarized and discussed in Appendix III. In the present form (4.23), the relations have been derived from the coupled amplitude equations (4.19) - (4.21) without specification of the particular interaction. They are therefore valid for sum-frequency and difference-frequency generation. Considering first sum-frequency generation, with laser input beams at ω_1 and ω_2, one sees from equation (4.23) that both lasers will lose power, which is gained by the wave generated at the sum frequency $\omega_3 = \omega_1 + \omega_2$. For difference-frequency generation ($\omega_3 - \omega_2 = \omega_1$), one sees however that the source at ω_3 loses power not only to the beam at the generated frequency ω_1, but also to the source of ω_2. Thus both waves at ω_1 and ω_2 gain power if ω_1 is generated from sources of ω_3 and ω_2.

By division of (4.23) by $\hbar$ (the reduced Planck's constant), the above results may be restated in terms of the photon picture. In sum-frequency generation, a photon at ω_1 combines with a photon at ω_2 to give a photon at ω_3; in difference-frequency generation, a photon at ω_3 is split into a photon at ω_1 and a photon at ω_2. It must be emphasized, however, that these relations arise from conservation of energy and the non-linear response of the medium, and do not have special quantum-mechanical significance (see Appendix III).

One example of an application of these relations is the parametric oscillator {28}, in which a signal at the difference frequency ω_1 between a strong source at ω_3 (the pump) and a very weak source (e.g. noise in the system) is generated. This case corresponds to the inverting demodulator discussed in Appendix III.

(c) SECOND HARMONIC GENERATION

In the case where both input frequencies are equal, one has the interaction known as second harmonic generation, $\omega_2 = 2\omega_1$. Equations (4.4) - (4.6) are now replaced by the pair (note carefully {9} the factor of 2 in (4.24)):

$$P_1^{NL}(z,t) = 2d\,\mathcal{E}_1^*(z)\,\mathcal{E}_2(z)\exp i\left[(k_2 - k_1)z - (\omega_2 - \omega_1)t\right] \tag{4.24}$$

$$P_2^{NL}(z,t) = d\,\mathcal{E}_1^2(z)\exp i\left[2k_1 z - 2\omega_1 t\right] \tag{4.25}$$

whence (4.9) and (4.11) yield (still in the slowly varying amplitude and phase approximation):

$$\frac{d\mathcal{E}_1}{dz} = \frac{16\pi i\omega_1^2}{k_1 c^2}\, d\, \mathcal{E}_1^* \mathcal{E}_2 \exp[i(k_2 - 2k_1)z] \tag{4.26}$$

$$\frac{d\mathcal{E}_2}{dz} = \frac{8\pi i\omega_1^2}{k_2 c^2}\, d\, \mathcal{E}_1^2 \exp[i(2k_1 - k_2)z] \tag{4.27}$$

and again one can substitute:

$$\Delta k = 2k_1 - k_2 \,. \tag{4.28}$$

From equations (2.18), (4.26) and (4.27) one obtains for the Poynting vector corresponding to the second harmonic, for a homogeneous slab of thickness L,

$$S(2\omega) = \frac{512\,\pi^5 d^2 L^2 S^2(\omega)}{\mu(2\omega)\,\mu^2(\omega)\,\lambda_1^2 c}\left(\frac{\sin x}{x}\right)^2 \tag{4.29}$$

where λ_1 is the free-space wavelength corresponding to the fundamental frequency, and $x = \Delta k\, L/2$. This solution again applies only in the small-signal approximation.

The coupled equations (4.26) and (4.27) can, however, be solved without recourse to the small-signal approximation [9,14]. With the simplifying assumptions of phase matching ($\Delta k = 0$), and thus equal phase velocities for the fundamental and second harmonic [14], i.e. $k_2 = 2k_1$, the solution is particularly simple. In that case, one obtains from (4.26) and (4.27) for the real field amplitudes:

$$\frac{d\mathcal{E}_{10}}{dz} = -\frac{4\pi\omega_1^2}{k_1 c^2}\, d\, \mathcal{E}_{10}\mathcal{E}_{20} \sin(\phi_2 - 2\phi_1) \tag{4.30}$$

$$\frac{d\mathcal{E}_{20}}{dz} = +\frac{4\pi\omega_1^2}{k_1 c^2}\, d\, \mathcal{E}_{10}^2 \sin(\phi_2 - 2\phi_1) \tag{4.31}$$

where we have substituted:

$$\mathcal{E}_1 = \mathcal{E}_{10} \exp i\phi_1(z) \,, \quad \mathcal{E}_2 = \mathcal{E}_{20} \exp i\phi_2(z) \,.$$

In addition, the imaginary parts of these equations yield for the spatial variation of the phases:

$$\frac{d}{dz}(\phi_2 - 2\phi_1) = \frac{4\pi\omega_1^2 d}{k_1 c^2}\left[\frac{\mathcal{E}_{10}^2}{\mathcal{E}_{20}} - 2\mathcal{E}_{20}\right]\cos(\phi_2 - 2\phi_1). \tag{4.32}$$

Now making the substitutions:

$$\left.\begin{aligned} L_{SH} &= \left[\frac{4\pi\omega_1^2 d}{k_1 c^2}\,\mathcal{E}_{10}(0)\right]^{-1} \\ \theta &= \phi_2 - 2\phi_1 \\ u &= \mathcal{E}_{10}(z)/\mathcal{E}_{10}(0) \\ v &= \mathcal{E}_{20}(z)/\mathcal{E}_{10}(0) \\ \zeta &= z/L_{SH} \end{aligned}\right\} \tag{4.33}$$

one obtains the following:

$$\frac{du}{d\zeta} = -uv\sin\theta \tag{4.34}$$

$$\frac{dv}{d\zeta} = u^2\sin\theta \tag{4.35}$$

$$\frac{d\theta}{d\zeta} = \left(\frac{u^2}{v} - 2v\right)\cos\theta \;. \tag{4.36}$$

Now from (4.34) and (4.35), it follows that:

$$u^2 + v^2 = \text{constant} \tag{4.37}$$

where the constant will be unity if:

$$u(\zeta = 0) = 1 \;, \quad v(\zeta = 0) = 0 \,.$$

We shall see below that (4.37) is also an immediate consequence of the Manley-Rowe relation for second harmonic generation. In addition, (4.34) - (4.36) imply (by inspection) that:

$$u^2 v \sin\theta \frac{d\theta}{d\xi} = 2uv\cos\theta \frac{du}{d\xi} + u^2 \cos\theta \frac{dv}{d\xi} \tag{4.38}$$

whence,

$$\Gamma \equiv u^2 v \cos\theta = \text{constant.} \tag{4.39}$$

The initial conditions:

$$u(\xi = 0) = 1 \quad , \quad v(\xi = 0) = 0$$

therefore constrain θ to the values $\pm \pi/2$, whence one has from (4.33), (4.34) and (4.35) the solutions:

$$\mathcal{E}_{10}(z) = \mathcal{E}_{10}(0)\, \mathrm{sech}(z/L_{SH}) \tag{4.40}$$

$$\mathcal{E}_{20}(z) = \mathcal{E}_{10}(0)\, \tanh(z/L_{SH}) \tag{4.41}$$

which immediately satisfy (4.37) as well. These relations describe the growth of second-harmonic amplitude with perfect phase matching in the crystal.

For completeness, the Manley-Rowe relation is also obtained for second-harmonic generation. With $\Delta k = 0$, equations (4.26) and (4.27) yield:

$$\frac{\mu_1 c}{\omega_1} \mathcal{E}_1^* \frac{d\mathcal{E}_1}{dz} = 4\pi i d\, \mathcal{E}_1^* \mathcal{E}_1^* \mathcal{E}_2 \tag{4.42}$$

$$\frac{\mu_2 c}{\omega_2} \mathcal{E}_2^* \frac{d\mathcal{E}_2}{dz} = 2\pi i d\, \mathcal{E}_2^* \mathcal{E}_1^2 \tag{4.43}$$

whence one obtains for the Poynting vectors:

$$\frac{1}{\omega_1} \frac{d}{dz} S(z,\omega_1) = -\frac{2}{\omega_2} \frac{d}{dz} S(z,\omega_2) \tag{4.44}$$

which is a special case ($\omega_2 = 2\omega_1$) of the Manley-Rowe relations (4.23). Now with $k_2 = 2k_1$ (or $\mu_2 = \mu_1$), one has again $u^2 + v^2$ = constant as in (4.37).

(d) OUTPUT ANGLE

The question of the angle over which a crystal of a particular size will radiate as a result of the non-linear term in the polarization becomes particularly important in difference-frequency generation, where the output wavelength may be considerably larger than the input wavelengths {9,18}. Consider the phase-matched interaction between two waves with angular frequencies ω_2 and $\omega_3 > \omega_2$, generating a difference frequency $\omega_1 = \omega_3 - \omega_2$. For simplicity, it will be assumed that the interaction takes place in a cylinder with radius a and axial length L, and that in this cylinder the two waves at ω_2 and ω_3 are plane waves, parallel to each other, with uniform intensities out to the edge. As a result, the polarization wave at ω_1 will also be a plane wave with uniform intensity out to the edge of the cylinder.

A cylindrical coordinate system may be defined with the z axis parallel to the direction of propagation of the waves (i.e. along the axis of the cylinder). The field at an arbitrary point in space may now be determined by summation of the contributions to this field from each point within the interaction cylinder, for which the amplitude and phase of each contribution are required {29}. In terms of cylindrical coordinates, one can express the distance r between any source point (ρ, θ, z) within the cylinder and any field point (ρ_0, θ_0, z_0) outside. Letting the field point be a distance r_0 from the origin, we shall consider the simplified situation in which $^{r_0}/_{r}$ is sufficiently large for the field point to lie in the radiation zone {7,10,11} of the cylinder. In that case, it is sufficiently accurate to approximate r by:

$$r \simeq r_0 - s\cos\psi$$
$$= r_0 - (z\cos\varphi + \rho\sin\varphi\cos\theta'). \tag{4.45}$$

where s is the distance of the source point from the origin, ψ the angle between the vectors $\vec{s}$ and $\vec{r}_0$, $\theta' = \theta_0 - \theta$ and $\varphi = \arccos(^{z_0}/r_0)$. The radiation pattern in the generated field of frequency ω_1 is determined by the integral:

$$\mathcal{E}_1 = \mathcal{E}_0 \int \exp[-i k_1 (z + r)]\, dV \tag{4.46}$$

which is performed over the volume: $V = \pi a^2 L$

of the cylinder. On substituting from (4.45), and noting {30} that:

$$\int_0^a \int_0^{2\pi} \exp[i\, k_1 \rho' \sin\varphi \cos\theta']\, \rho'\, d\theta'\, d\rho'$$

$$= 2\pi a \; \frac{J_1(k_1 a \sin\varphi)}{k_1 \sin\varphi} \qquad (4.47)$$

in terms of the first-order Bessel function J_1, one obtains the <u>Fraunhofer diffraction pattern</u> {9,18}:

$$|\mathcal{E}_1|^2 = |\mathcal{E}_0|^2 V^2 \left[\frac{2 J_1(k_1 a \sin\varphi)}{k_1 a \sin\varphi}\right]^2 \left[\frac{\sin\{k_1 L(1-\cos\varphi)/2\}}{k_1 L(1-\cos\varphi)/2}\right]^2. \qquad (4.48)$$

The first term within parentheses on the right-hand side of (4.48) will be recognized as the term which determines the diffraction pattern of a circular aperture with radius a {7}. The second term within parentheses is reminiscent of the $(\sin x/x)^2$ term in equation (4.18), but with Δk replaced by $k_1(1-\cos\varphi)$. This result shows therefore that the angular distribution of the output radiation is determined by the Fraunhofer diffraction pattern of an aperture with the same radius as the interaction cylinder, multiplied by a term that depends upon the phase mismatch due to the angular deviation φ of the line joining the field point to the origin, from the axis of the cylinder.

We see, therefore, that the output angle in far-infrared difference-frequency generation can be much larger than the input angle. This fact allows the crystal to be placed inside the resonator of the laser which generates the input beams. The output beam can then be brought out of the resonator with only very small losses by a mirror which permits the input beams to pass through a hole in its centre {18}.

(e) PHASE MATCHING

As one sees from equations (4.18) and (4.29), "the term $(\sin x/x)^2$ is crucial to the success of any frequency-mixing experiment" {9}, since it effectively prevents the growth of significant amounts of power in the generated electromagnetic wave, unless the phase velocities are matched ($\Delta k = 0$) for the particular frequencies involved. Without phase matching, the generated signal will reach a maximum over a crystal length termed the <u>coherence length</u>, L_{coh} {6}. For example, from equation (4.29) one has for second-harmonic generation:

$$L_{coh} = \frac{\pi}{|\Delta k|} = \frac{\lambda_1}{4|\mu_1 - \mu_2|} \qquad (4.49)$$

where λ_1 is the free-space wavelength at the fundamental frequency. It corresponds to the order of twenty wavelengths for typical crystals investigated with ruby-laser radiation.

This effect, which is due to dispersion within the crystal (whereby a dephasing results between the second-harmonic and input fields), was first verified in an experiment by Maker et al. {31}, with crystalline quartz. A parallel ruby-laser beam was projected through a thin (0.8 mm) plate of quartz, and the production of the second harmonic observed as a function of the angle between the plate normal and the laser beam. The rotation axis was normal to the beam, and parallel to the crystal z axis, which was also perpendicular to the direction of polarization of the input laser beam. "This experiment not only demonstrates the effect of dispersion in a dramatic way but also provides the most useful method known today of obtaining quantitative measurements of the tensor elements describing the second-harmonic polarization. This is because the plane wave radiation problem from a flat plate can be accurately evaluated, whereas the situation with other geometries is exceedingly difficult and requires very precise information about the optical properties of the laser beam" {6}. In this experiment {31}, the distance between the successive maxima in the plot of output signal versus angular rotation corresponded to 14 μ, while the value of twice the coherence length calculated from refractive index data was 13.9 μ.

There are several different approaches to the problem of phase matching, which we briefly discuss next.

i) <u>Quasi-phase-matching methods</u>

A change in the phase difference between the generated second harmonic wave and the incoming electromagnetic wave by $\pi/2$ per coherence length, would produce a quasi-phase-matched condition {9}. The term "quasi" is used in the present context because of the factor

$(\sin x/_x)^2$ in equation (4.29), which ensures a reduction in output signal strength of $^4/_{\pi^2}$ per coherence length; hence the output from this crystal would still be reduced below that from a phase-matched crystal with the same tensor d and length L. One suggestion for achieving this is to manufacture thin plates of the particular crystal, each one coherence length in thickness, and to reverse the direction of the crystal axis in alternate platelets so that the polarization wave would undergo a phase change of π in going from one plate to the next {14}. A major drawback of this method is the thickness of the plates (e.g. 7 μ for quartz) and the requirement that each successive one would have to be in optical contact. A more recent proposal is to grow semiconductor layers epitaxially onto one another {32}.

A different method {14} that has been employed experimentally {33, 34} is to utilize the phase change produced by total internal reflection. Both the fundamental and second harmonic are reflected between the top and bottom surfaces of a slab of crystal, with the angle of reflection so chosen that the phase mismatch accumulated in every pass between the two reflecting sides is exactly cancelled by the differential phase change between the fundamental and the second-harmonic reflections {9}.

ii) <u>Angle phase-matching</u>

The solution to this problem for a uniaxial crystal was first given by Giordmaine {35}, and independently by Maker et al. {31}. Generalized for non-collinear beams, the condition for phase matching in second-harmonic production is:

$$\Delta \underset{\sim}{k} = \underset{\sim}{k}_1 + \underset{\sim}{k}_1' - \underset{\sim}{k}_2 = 0 \tag{4.50}$$

where the waves with wave numbers $\underset{\sim}{k}_1$ and $\underset{\sim}{k}_1'$ both have the same angular frequency (ω_1). In the <u>collinear</u> case, this would imply equal phase velocities v_1 and v_2 and thus equal refractive indices, since $\omega_2 = 2\omega_1$. In general this would not be possible, owing to the effect of dispersion; in most materials, the dispersion is normal in the optical region and therefore the generated second harmonic radiation, with electric field varying as $\exp i\,(k_2 z - 2\omega_1 t)$ will lag behind the polarization wave which varies as $\exp 2i\,(k_1 z - \omega_1 t)$, launched by the input beam.

The phase velocities v_1 and v_2 may, however, be equal for certain directions in an anisotropic crystal. We shall first consider the case of a <u>negative uniaxial crystal</u>, such as KDP, with ordinary and extraordinary refractive indices denoted (as in Chapter 2) by $\mu_o = c/v_o$, $\mu_e = c/v_e$, respectively, where $\mu_e < \mu_o$. Now, provided that $\mu_{1o} \geqslant \mu_{2e}$, where the subscripts 1o and 2e refer to an ordinary wave of angular frequency ω_1, and an extraordinary wave of angular frequency $\omega_2 = 2\omega_1$, respectively, phase matching may be achieved in this particular crystal {6,9,31,35}. (This process is also termed index matching in the older literature.) When the above condition is fulfilled, one readily shows from equation (2.12) that surfaces mapped out by polar plots of μ_{1o} and $\mu_{2e}(\theta)$ versus angle θ between the wave normal and the optic axis, intersect along the rim of a cone of half-angle:

$$\psi_o = \arcsin\left[\frac{\mu_{2e}}{\mu_{1o}}\left(\frac{\mu_{2o}^2 - \mu_{1o}^2}{\mu_{2o}^2 - \mu_{2e}^2}\right)^{\frac{1}{2}}\right] \tag{4.51}$$

centred on the optic axis. From a single plane wave of wave number $\underset{\sim}{k}_1$ completely in-phase harmonic radiation is produced in a uniaxial crystal only if the vector $\underset{\sim}{k}_1$ is inclined at an angle ψ_o to the optic axis. (Note from Chapter 2 that the ordinary fundamental wave may readily be prepared, since it must be polarized perpendicularly to the optic axis.)

Analogously, in the case of a positive uniaxial crystal, such as cinnabar (HgS), phase matching cannot be achieved unless $\mu_{1e} \geqslant \mu_{2o}$. In this case, one finds from equation (2.12) that surfaces mapped out by polar plots of $\mu_{1e}(\theta)$ and μ_{2o} versus angle θ between the wave normal and the optic axis, intersect along the rim of a cone of half-angle:

$$\psi_o = \arcsin\left[\frac{\mu_{1e}}{\mu_{2o}}\left(\frac{\mu_{2o}^2 - \mu_{1o}^2}{\mu_{1e}^2 - \mu_{1o}^2}\right)^{\frac{1}{2}}\right] \tag{4.52}$$

An example of a positive uniaxial crystal which does not satisfy the above condition for phase matching (i.e. is not sufficiently birefringent) is quartz {6}.

We now consider the more general situation of second harmonic production in which the two input waves are non-collinear {35}. When equation (4.50) is satisfied, the entire irradiated crystal volume can radiate coherently, electromagnetic momentum is conserved and optimum radiation efficiency is achieved. In the particular case in a negative uniaxial crystal of an ordinary wave ($\underset{\sim}{k}_1$) propagating at an angle ψ_1 with the optic axis, and mixing with a second ordinary wave ($\underset{\sim}{k}_1'$) to form the extraordinary second harmonic wave ($\underset{\sim}{k}_2$) emitted at an angle Θ

to $\underset{\sim}{k}_1$, we see that equation (4.50) is satisfied if:

$$\cos \Theta = \mu_{2e}(\theta)/\mu_{1o} \qquad (4.53)$$

where θ is the angle between $\underset{\sim}{k}_2$ and the optic axis. Referring to (2.12), we see that (4.53) is a transcendental equation; its solutions are, however, simple provided that:

$$\psi_1 - \psi_0 = \Delta\psi$$

is a small angle; i.e. provided that $\underset{\sim}{k}_1$ lies close to the direction for which $\mu_{2e}(\theta) = \mu_{1o}$. In that case, equation (4.53) can be approximated by:

$$\Theta^2 = 2\left(1 - \frac{\mu_{2e}(\theta)}{\mu_{1o}}\right)$$

and from (2.12) with the assumption that both Θ and $\Delta\psi$ are much smaller than ψ_0 and θ, one obtains {35}:

$$\Theta^2 = K(\Delta\psi - \Theta\cos\alpha) \qquad (4.54)$$

where the constant:

$$K = 2\,\frac{(\mu_{1o}^2 - \mu_{2e}^2)^{\frac{1}{2}}(\mu_{2o}^2 - \mu_{1o}^2)^{\frac{1}{2}}}{\mu_{2o}\,\mu_{2e}} \qquad (4.55)$$

and α is the included angle between vectors parallel to $\underset{\sim}{k}_1 \times (\underset{\sim}{k}_2 \times \underset{\sim}{k}_1)$ and $\underset{\sim}{k}_1 \times (\hat{\underset{\sim}{z}} \times \underset{\sim}{k}_1)$, originating at the same point on a line through the origin parallel to $\underset{\sim}{k}_1$.

The solution to the quadratic equation (4.54) represents a circular cone of wave vectors $\underset{\sim}{k}_2$ of half-angle $\frac{1}{2}K\,(1 + \frac{4\Delta\psi}{K})^{\frac{1}{2}}$ centred about the direction in the $\underset{\sim}{k}_1\hat{\underset{\sim}{z}}$ plane making an angle $\psi_1 + \frac{1}{2}K$ with the optic axis. One easily shows that no completely in-phase second harmonic generation is possible unless $\psi_1 \geq \psi_0 - \frac{1}{4}K$. As ψ_1 increases from this value, emission can occur on a cone of increasing radius which crosses the direction of the incident wave $\underset{\sim}{k}_1$ when $\psi_1 = \psi_0$ {35}.

For example, for a ruby laser (λ_r = 6943 Å) incident on a KDP crystal, appropriate values for the corresponding refractive indices are: $\mu_{1o} = 1.506$, $\mu_{1e} = 1.466$, $\mu_{2o} = 1.534$ and $\mu_{2e} = 1.487$. For these, one obtains from equations (4.51) and (4.55):.

$$\psi_0 = 49.9^\circ, \quad K = 0.061 \quad .$$

Experimental values obtained by Giordmaine {35} for these quantities from the features of the second harmonic emission were in good agreement with the above values; qualitative and quantitative agreement was obtained with the general behaviour of the emission as discussed above.

The type of phase matching discussed up to this point in which both of the fundamental waves have the <u>same</u> polarization, is termed <u>type I</u> phase matching; if the fundamental waves are orthogonally polarized, the phase matching is termed <u>type II</u> {9}. The solution to the problem of phase matching in second-harmonic generation for type II processes in uniaxial crystals, is reviewed by Hobden {36}, who shows that the phase match angle (ψ_0) is always greater for type II than for type I processes, and therefore that if the type II process is possible in a particular crystal, then so is the corresponding type I process. The more general problem of phase matching in a biaxial crystal is considerably more complicated than for a uniaxial crystal, and simple solutions in closed form are not in general obtained. For further details on this subject, Ref. {36} may be consulted.

One of the major problems associated with angle phase matching is the fact that for the extraordinary wave, the ray direction and the wave-normal direction are parallel only when $\theta = 0$ or 90° (see Chapter 2). Thus in a phase-matched interaction at an intermediate angle θ, the extraordinary beam (i.e. direction of energy propagation) cannot overlap an ordinary beam in the entire interaction length {9}. Thus the fundamental and second-harmonic beams physically separate from each other, since they are necessarily in orthogonal states of polarization, a phenomenon termed "<u>walk-off</u>" in the technical literature {9,36}. The angle (α) between the directions of $\underset{\sim}{S}$ and $\underset{\sim}{n}$ in equation (2.18) is referred to as the <u>"walk-off" angle</u>. "For a type I interaction this effect, although present, is not too serious. It only means that the generated beam does not totally overlap the polarization wave, and thus the integration in equation (4.18) becomes more complicated. The exact form of integration has to be worked out for each specific case. In general, we find that the output is proportional not to the square, but rather to a lower power of the length. For a type II interaction the effect is more serious, because here the two fundamental beams do not overlap completely, and thus after a certain crystal length the polarization wave vanishes completely and mixing no longer occurs" {9}. An expression for the "walk-off" angle has been derived in Chapter 2 (equation (2.31)).

Another limiting factor in angle phase matching is due to the divergence of a focussed beam {9}. For second-harmonic generation in a negative uniaxial crystal and a type I interaction, equation (4.50) yields for collinear beams:

$$\Delta k = \frac{2\omega_1}{c}\left[\mu_{1o} - \mu_{2e}(\theta)\right] \tag{4.56}$$

where $\theta_m = \psi_o$ (equation (4.51)) for $\Delta k = 0$. For a small deviation $\Delta\theta$ from this angle, equations (2.12) and (4.56) yield {36}:

$$\Delta k = \frac{\omega_1}{c}\mu_{1o}^3\left[\mu_{2e}^{-2} - \mu_{2o}^{-2}\right]\sin 2\theta_m \,\Delta\theta \;. \tag{4.57}$$

For second-harmonic generation in a positive uniaxial crystal and a type I interaction, equation (4.50) yields for collinear beams:

$$\Delta k = \frac{2\omega_1}{c}\left[\mu_{1e}(\theta) - \mu_{2o}\right] \tag{4.58}$$

where $\theta_m = \psi_o$ (equation (4.52)) for $\Delta k = 0$. For a small deviation $\Delta\theta$ from this angle, equations (2.12) and (4.56) yield {36}:

$$\Delta k = \frac{\omega_1}{c}\mu_{2o}^3\left[\mu_{1o}^{-2} - \mu_{1e}^{-2}\right]\sin 2\theta_m \,\Delta\theta \;. \tag{4.59}$$

In the case of type II processes, the mismatch Δk is reduced by a factor of 2 {36}. For crystals with small birefringence and dispersion, equations (4.57) and (4.59) reduce to:

$$\Delta k \simeq \beta\,\frac{2\omega_1}{c}\left(\mu_e - \mu_o\right)\sin 2\theta_m \,\Delta\theta \tag{4.60}$$

where β has the same sign as the crystal birefringence and has modulus 1 for type I and modulus ½ for type II processes, respectively. These relations show that the variation of Δk is linear with $\Delta\theta$, a fact which causes practical difficulties for phase matching at intermediate angles θ_m. "The output is proportional to the energy density at the fundamental frequency, and so, to achieve the maximum energy density, the beam is focussed on the crystal. However, the linear variation of Δk with $\Delta\theta$ means that, for a given convergence of the beam, efficient phase matching will be obtained over a restricted crystal length only" {9}.

If it is possible to adjust the refractive indices so that $\theta_m = 90^\circ$, by variation of the temperature {9} (temperature-dependent phase matching) or chemical composition of the crystal, then the linear change in Δk with $\Delta\theta$ is replaced by a quadratic dependence {36}, since in equations (4.57), (4.59) and (4.60) now:

$$\sin 2\theta_m \,\Delta\theta \simeq 2(\Delta\theta)^2 .$$

Thus for $\theta_m = 90^\circ$, the allowable beam divergence is much larger, and in addition the "walk-off" effect due to double refraction disappears, according to equation (2.31). For these reasons, phase matching at an angle $\theta_m = 90^\circ$ to the optic axis is termed non-critical phase matching (NCPM) {36}. By contrast, phase matching at $0 < \theta_m < 90^\circ$ is termed critical phase matching (CPM) {36}. As a numerical example of CPM, consider the case of a KDP crystal irradiated with the beam of a ruby laser, as in the experiment of Giordmaine {35}. For the given values of the refractive indices in this paper, one obtains a value of $\theta_m = \psi_0 = 49.9^\circ$. For a coherence length of 1 cm, equations (4.49) and (4.57) yield a value for the allowable beam divergence from the phase matched direction of:

$$\Delta\theta = 0.37 \text{ mrad} = 1.3'.$$

On the other hand with NCPM, values of $\Delta\theta$ some 30 times larger could be obtained {36}.

The topic of angle phase matching is resumed in the following chapter, with reference to the three-wave interaction.

iii) Other phase matching methods

A different technique whereby phase matching may be achieved in a non-linear medium, is to employ optical rotatory dispersion, the optical rotation arising either from natural optical activity {37,38} or from magneto-optic rotation as in the Faraday effect {39}. In addition, optical waveguides have been employed to reduce the degree of mismatch {40}, and phase matching has been induced acoustically {41}. For details on these techniques, references {37-41} should be consulted.

CHAPTER 5

PRACTICAL APPLICATIONS

In this chapter we consider various fields of application of the preceding theory, as well as limitations imposed by practical considerations.

(a) SECOND HARMONIC GENERATION

This important application is reviewed in detail in Ref. {9}, where the reader will also find numerical results for cases of practical interest. An important modification of our earlier equation (4.29) is to allow for the finite cross-section (A) of the incoming laser beam. With $W(2\omega)$ and $W(\omega)$ denoting the total second-harmonic and input powers respectively, this equation may be re-written as:

$$W(2\omega) = \frac{512\pi^5 d^2 L^2 W^2(\omega)}{A\mu(2\omega)\mu^2(\omega)\lambda_1^2 c} \qquad (5.1)$$

provided that the condition of phase matching has been satisfied (i.e. $\mu(2\omega) = \mu(\omega)$). (As before, λ_1 denotes the wavelength in vacuo of the fundamental.) An efficiency η_{SH} for the generation of second-harmonic power may now be defined by the ratio of $W(2\omega)$ to $W(\omega)$:

$$\eta_{SH} = \frac{512\pi^5 d^2 L^2 W(\omega)}{A\mu(2\omega)\mu^2(\omega)\lambda_1^2 c}. \qquad (5.2)$$

It should be recalled from the previous chapter that this result is strictly applicable only within the small-signal approximation. A more general result {14} follows from equations (4.33) and (4.41):

$$\eta_{SH} = \frac{\mu(2\omega)}{\mu(\omega)} \tanh^2\left(\frac{L}{L_{SH}}\right) \qquad (5.3)$$

where:

$$L_{SH} = \left[\frac{512\, \pi^5 d^2\, S(z=0,\omega)}{\mu^3(\omega)\, \lambda_1^2\, c} \right]^{-1/2} . \tag{5.4}$$

Noting the phase match condition, equations (5.3) and (5.4) immediately reduce to (5.2) in situations where the fractional power lost to the second harmonic by the fundamental is very small.

Now equation (5.2) suggests that the efficiency of the process of second harmonic generation may be improved by:

i) reduction of the beam cross-section (A);

ii) increasing the interaction length (L);

iii) judicious choice of effective susceptibility (d), i.e. crystal type;

iv) increasing the input power in the fundamental ($W(\omega)$).

Above, A has been assumed constant throughout the interaction region in the crystal. As A is reduced, diffraction effects (equation (4.48)) and hence beam divergence will become more important, with a consequent reduction in efficiency below that predicted by equation (5.2). Moreover, since the second-harmonic and fundamental beams are of opposite polarization, the effect of "walk-off" will tend to limit the effective interaction length (see Chapter 4). However, as long as the effects of beam divergence and "walk-off" do not predominate, improved focussing of the laser beam will necessarily improve the efficiency {9}. This suggests a need to investigate the optimum degree of focussing {42,43}.

In this study {42,43}, the input laser beam is considered to be Gaussian, while the second harmonic wave is ensured to be Gaussian in character by use of an optical resonator {44,45}. Gaussian beams are specified by the following parameters: direction of the beam axis, location of the focus, the confocal parameter, the frequency and the power, the first three of these being treated as optimizable parameters. In the confocal resonator {44}, identical spherical reflectors of radius b are employed, separated by a distance equal to b. With crystal length denoted by L, a <u>focussing parameter</u> is defined by:

$$\xi = L/b \tag{5.5}$$

and double refraction enters in the theory through the parameter:

$$B = \alpha (L k_1)^{1/2} / \alpha \tag{5.6}$$

where α denotes the <u>double refraction</u> ("walk-off") <u>angle</u>, which for an ordinary input beam subject to the phase matching condition, follows from equation (2.31) as:

$$\tan \alpha = \frac{1}{\mu_{20}\mu_{2e}} \left[(\mu_{20}^2 - \mu_{10}^2)(\mu_{10}^2 - \mu_{2e}^2) \right]^{1/2} . \tag{5.7}$$

(This expression is in agreement with the formula of Boyd et al. {46}, where the symbol ρ is used as in Refs. {42,43} to denote the double refraction angle.) For the case B = 0, Boyd and Kleinman {43} show that the <u>optimum focussing parameter</u>:

$$\xi_m = 2.84 .$$

Values for the optimum focussing parameter as a function of B may be obtained from figure 4 of Ref. {43}. Another characteristic length of importance in this study, called the <u>aperture length</u> L_a {46}, corresponds to the critical distance for which second-harmonic generation no longer increases with the square of the crystal length, owing to the "walk-off" effect. (The name originates from the earlier appellation of "walk-off" as the <u>aperture effect</u> {47}). From Ref. {46} one has the expression:

$$L_a = \sqrt{\pi} \, \frac{w_0}{\alpha} \tag{5.8}$$

where w_0 denotes the optimum spot size of a Gaussian beam, the spot size being defined {42} as the distance from the beam axis at which the amplitude of the field falls to $1/e$ of its axial value. It is simply given by {46}:

$$w_0 = (b/k_1)^{1/2} . \tag{5.9}$$

For crystals of length $L > L_a$, the dependence of second-harmonic generation power on length changes from quadratic to linear {46}.

A fourth characteristic length is the <u>effective length of the focus</u> {42}:

$$L_f = \pi b/2 \quad . \tag{5.10}$$

In summary, the study of optimization introduces four characteristic lengths (L, w_o, L_a and L_f), the last three of which may be expressed in terms of L, B and ξ by {43}:

$$\left.\begin{aligned} w_o &= \left(\frac{L}{k_1}\right)^{\frac{1}{2}} \xi^{-\frac{1}{2}} \\ L_a &= \left(\frac{\pi}{4}\right)^{\frac{1}{2}} \frac{L}{B} \xi^{-\frac{1}{2}} \\ L_f &= \frac{\pi}{2} \frac{L}{\xi} \end{aligned}\right\} \tag{5.11}$$

On defining the following grouping of factors in equation (5.2) by {43}:

$$K_o = \frac{512 \pi^4 d^2}{\mu(2\omega)\mu^2(\omega)\lambda_1^2 c} \tag{5.12}$$

the original expression for the efficiency may now be re-written in the ideal case as:

$$\eta_{SH} = \frac{K_o W(\omega)}{w_o^2} L^2 . \tag{5.13}$$

The study of Boyd and Kleinman {43} shows that owing to the effects of "walk-off", beam divergence and diffraction, this idealized formula is only valid subject to the conditions:

$$L_a , L_f \gg L .$$

Four other asymptotic representations may be obtained:

$$\eta_{SH} = \frac{K_o W(\omega)}{w_o^2} L L_a \qquad (L_f \gg L \gg L_a) \tag{5.14}$$

$$\eta_{SH} = \frac{k_o W(\omega)}{W_o^2} L_f L_a \quad (L \gg L_f \gg L_a) \tag{5.15}$$

$$\eta_{SH} = \frac{4 k_o W(\omega)}{W_o^2} L_f^2 \quad (L \gg L_a \gg L_f) \tag{5.16}$$

$$\eta_{SH} = \frac{4.75 k_o W(\omega)}{W_o^2} L_f^2 \quad (L_a \gg L \gg L_f). \tag{5.17}$$

It should be pointed out, however, that all five expressions (5.13) - (5.17) for η_{SH} assume that the point of focus of the Gaussian beam occurs mid-way between the entrance and exit faces of the crystal, and that the absorption coefficients are zero for both the fundamental and the second harmonic.

With regard to the third and fourth points stated above for improving η_{SH}, the list of suitable materials in Ref. {9} may be consulted. An upper limit to the input power for a crystal of a particular effective susceptibility is imposed by the onset of radiation damage. As a rough empirical rule, the threshold field-strength at which this occurs is such that the maximum allowed value of the factor $d^2W(\omega)/A$ in equation (5.2) is approximately constant from one medium to the next.

(b) PARAMETRIC UP-CONVERSION

The process of second-harmonic generation is in fact a special case of the situation in which electromagnetic waves at two different frequencies (ω_1, ω_2) interact to generate a wave at the sum frequency $\omega_3 = \omega_1 + \omega_2$. The coupled equations which describe this process have already been derived ((4.12)-(4.14)). In the present case, it is assumed that the input power at the frequency ω_2 is much larger than the power at ω_1, and that the electromagnetic field at ω_3 is absent in the absence of the driving fields at ω_1 and ω_2 {9}. With the subsidiary condition:

$$\frac{d\mathcal{E}_2}{dz} = 0 \tag{5.18}$$

in the coupled equations (4.12)-(4.14), together with the phase matching condition:

$$\Delta k = k_1 + k_2 - k_3 = 0, \tag{5.19}$$

one has for the real field amplitudes:

$$\frac{d\mathcal{E}_{10}}{dz} = \frac{4\pi\omega_1^2 d}{k_1 c^2}\,\mathcal{E}_{20}\mathcal{E}_{30}\sin(\phi_1-\phi_3) \tag{5.20}$$

$$\frac{d\mathcal{E}_{30}}{dz} = -\frac{4\pi\omega_3^2 d}{k_3 c^2}\,\mathcal{E}_{10}\mathcal{E}_{20}\sin(\phi_1-\phi_3) \tag{5.21}$$

where we have substituted:

$$\mathcal{E}_1 = \mathcal{E}_{10}(z)\exp i\phi_1(z) \quad , \quad \mathcal{E}_3 = \mathcal{E}_{30}(z)\exp i\phi_3(z) \; .$$

In addition, the imaginary parts of these equations yield for the spatial variation of the phases:

$$\frac{d}{dz}(\phi_1-\phi_3) = \frac{4\pi d}{c^2}\left[\frac{\omega_1^2}{k_1}\,\frac{\mathcal{E}_{20}\mathcal{E}_{30}}{\mathcal{E}_{10}} - \frac{\omega_3^2}{k_3}\,\frac{\mathcal{E}_{10}\mathcal{E}_{20}}{\mathcal{E}_{30}}\right]\cos(\phi_1-\phi_3) . \tag{5.22}$$

Now making the substitutions:

$$\left.\begin{aligned}
L_{uc} &= \left[\frac{4\pi d}{c^2}\left(\frac{\omega_1^2\,\omega_3^2}{k_1\,k_3}\right)^{1/2}\mathcal{E}_{20}\right]^{-1}\\
\theta &= \phi_1-\phi_3\\
u &= \mathcal{E}_{10}(z)/\mathcal{E}_{10}(0)\\
v &= \mathcal{E}_{30}(z)/\mathcal{E}_{10}(0)\\
\alpha &= \left[\frac{\omega_1^2\,k_3}{\omega_3^2\,k_1}\right]^{1/2}\\
\zeta &= z/L_{uc}
\end{aligned}\right\} \tag{5.23}$$

one obtains the following:

$$\frac{du}{d\xi} = \alpha v \sin\theta \tag{5.24}$$

$$\frac{dv}{d\xi} = -\frac{1}{\alpha} u \sin\theta \tag{5.25}$$

$$\frac{d\theta}{d\xi} = \left(\alpha \frac{v}{u} - \frac{u}{\alpha v}\right) \cos\theta \, . \tag{5.26}$$

Thus from (5.24) and (5.25), it follows that:

$$u^2 + \alpha^2 v^2 = \text{constant} \tag{5.27}$$

where the constant will be unity subject to the initial conditions:

$$u(\xi = 0) = 1 \, , \quad v(\xi = 0) = 0 \, . \tag{5.28}$$

In addition, (5.24)-(5.26) imply (by inspection) that:

$$u v \sin\theta \frac{d\theta}{d\xi} = v \cos\theta \frac{du}{d\xi} + u \cos\theta \frac{dv}{d\xi} \tag{5.29}$$

whence:

$$\Gamma \equiv u v \cos\theta = \text{constant} . \tag{5.30}$$

The initial conditions (5.28) therefore constrain θ to the values $\pm \frac{\pi}{2}$, whence one has from equations (5.24), (5.25) and (5.27) the result {9}:

$$\mathcal{E}_{10}(z) = \mathcal{E}_{10}(0) \cos(z/L_{uc}) \tag{5.31}$$

$$\mathcal{E}_{30}(z) = \left(\frac{\omega_3^2 k_1}{\omega_1^2 k_3}\right)^{1/2} \mathcal{E}_{10}(0) \sin(z/L_{uc}) \, . \tag{5.32}$$

Thus we see that the flux at the frequency ω_1 is totally transferred to the sum-frequency beam after a characteristic length $\pi L_{uc}/2$, where L_{uc} may be re-written from (5.23) as:

$$L_{uc} = \left[\frac{512\, \pi^5 d^2 S_2(z=0)}{\mu(\omega_1)\mu(\omega_2)\mu(\omega_3)\lambda_1\lambda_3 c} \right]^{-1/2} \tag{5.33}$$

It is interesting to compare this expression with that for L_{SH} (equation (5.4)); in spite of apparent similarities, the interpretation of the two lengths is completely different, in view of the dissimilar nature of the pairs of relations (4.40), (4.41) and (5.31), (5.32).

By a suitable choice of crystal length, one therefore has the means of achieving an efficient conversion of low-frequency (e.g. infrared) into higher frequency (visible) radiation. Such conversion has two major advantages: firstly, low-noise detectors of visible radiation are far more efficient than those operating in the infrared; secondly, the detection process can be carried out at room temperature after up-conversion, whereas direct measurement in the infrared would require detector cooling to between 4 and 77°K {9}.

From (5.31) and (5.32) one has for the Poynting vectors:

$$S_3(L) = \frac{\lambda_1}{\lambda_3} S_1(z=0) \sin^2(L/L_{uc}) \tag{5.34}$$

which becomes, in the limit $L \ll L_{uc}$:

$$S_3(L) = \frac{512\, \pi^5 d^2 L^2 S_1(z=0) S_2(z=0)}{\mu(\omega_1)\mu(\omega_2)\mu(\omega_3)\lambda_3^2 c} \tag{5.35}$$

in agreement with the small-signal result (4.18), subject to the phase match condition being met.

With reference to practical applications, one may introduce the subscripts s ("sum"), p ("pump") and ir ("infrared") in place of 3, 2 and 1, respectively {9}. For a beam cross-section A, the conversion efficiency η_{uc} may now be defined by the ratio of the photon fluxes corresponding to $S_s(L)$ and $S_{ir}(o)$, which follows from (5.35):

$$\eta_{uc} = \frac{512\, \pi^5 d^2 L^2 W_p(z=0)}{\mu_{ir}\mu_p\mu_s\lambda_{ir}\lambda_s c} . \tag{5.36}$$

A further application of up-conversion is parametric image conversion {9}, in which image information is transferred from the infrared input to the sum frequency output.

(c) PARAMETRIC DOWN-CONVERSION

This process, which is not of the same technological importance as parametric up-conversion, is a particular case of the inverse of sum-frequency generation ($\omega_3 = \omega_1 + \omega_2$), in which a signal ($\omega_1$) is generated by interaction of a wave of higher frequency (ω_3) with a laser beam of frequency ω_2. It is assumed in this case that the input power at the frequency ω_2 is much larger than the power at ω_3. We have already seen from the previous chapter that the Manley-Rowe relations for this case imply that the wave of frequency ω_3 loses power not only to the beam at the generated frequency ω_1, but also to the source of ω_2 (the laser beam).

It is interesting to note from the nature of the solutions (5.31) and (5.32) to the problem of parametric up-conversion, that down-conversion will be a necessary by-product of the latter process after depletion of the available field energy at ω_1 has occurred.

For further details on this process the reader is referred to the book by Bloembergen {24}.

(d) OPTICAL PARAMETRIC AMPLIFICATION

This is a particular case of difference-frequency generation ($\omega_3 - \omega_2 = \omega_1$), in which a weak signal (ω_2) is made to interact with a strong, higher-frequency pump (ω_3), and both the generated difference frequency (ω_1) and the original signal are amplified. The signal at the generated frequency difference is known as the idler, and this process corresponds to the inverting modulator discussed in Appendix III.

With the subsidiary condition:

$$\frac{d\mathcal{E}_3}{dz} = 0 \tag{5.37}$$

in the coupled equations (4.12)-(4.14), together with the phase matching condition (5.19), one obtains for the real field amplitudes:

$$\frac{d\mathcal{E}_{10}}{dz} = \frac{4\pi\omega_1^2 d}{k_1 c^2}\, \mathcal{E}_{20}\,\mathcal{E}_{30} \sin(\phi_1 + \phi_2) \tag{5.38}$$

$$\frac{d\mathcal{E}_{20}}{dz} = \frac{4\pi\omega_2^2 d}{k_2 c^2}\,\mathcal{E}_{10}\mathcal{E}_{30}\sin(\phi_1+\phi_2) \tag{5.39}$$

where we have substituted:

$$\mathcal{E}_1 = \mathcal{E}_{10}(z)\exp i\phi_1(z) \quad , \quad \mathcal{E}_2 = \mathcal{E}_{20}(z)\exp i\phi_2(z).$$

In addition, the imaginary parts of these equations yield for the spatial variation of the phases:

$$\frac{d}{dz}(\phi_1+\phi_2) = \frac{4\pi d}{c^2}\left[\frac{\omega_1^2}{k_1}\frac{\mathcal{E}_{20}\mathcal{E}_{30}}{\mathcal{E}_{10}} + \frac{\omega_2^2}{k_2}\frac{\mathcal{E}_{10}\mathcal{E}_{30}}{\mathcal{E}_{20}}\right]\cos(\phi_1+\phi_2). \tag{5.40}$$

On substituting as follows:

$$\left.\begin{aligned} L_{pa} &= \left[\frac{4\pi d}{c^2}\left(\frac{\omega_1^2\omega_2^2}{k_1 k_2}\right)^{1/2}\mathcal{E}_{30}\right]^{-1} \\ u &= \mathcal{E}_{10}(z)/\mathcal{E}_{20}(0) \\ v &= \mathcal{E}_{20}(z)/\mathcal{E}_{20}(0) \\ \alpha &= \left[\frac{\omega_1^2 k_2}{\omega_2^2 k_1}\right]^{1/2} \\ \zeta &= z/L_{pa} \end{aligned}\right\} \tag{5.41}$$

one obtains the relations:

$$\frac{du}{d\zeta} = \alpha v \sin\Theta \tag{5.42}$$

$$\frac{dv}{d\zeta} = \frac{1}{\alpha} u \sin\Theta \tag{5.43}$$

$$\frac{d\Theta}{d\zeta} = \left(\alpha\frac{v}{u} + \frac{1}{\alpha}\frac{u}{v}\right)\cos\Theta\,. \tag{5.44}$$

From (5.42) and (5.43) it immediately follows that:

$$u^2 - \alpha^2 v^2 = \text{constant} \tag{5.45}$$

where the constant will be equal to:

$$-\alpha^2 v^2(\xi = 0)$$

subject to the initial condition:

$$u(\xi = 0) = 0. \tag{5.46}$$

In addition, (5.42)-(5.44) imply (by inspection) that:

$$u v \sin\theta \frac{d\theta}{d\xi} = v \cos\theta \frac{du}{d\xi} + u \cos\theta \frac{dv}{d\xi} \tag{5.47}$$

whence:

$$\Gamma \equiv u v \cos\theta = \text{constant}. \tag{5.48}$$

The initial condition (5.46) therefore constrains θ to the value $\pm \frac{\pi}{2}$, whence one has from equations (5.42), (5.43) and (5.45) the results {9, 48}:

$$\mathcal{E}_{20}(L) = \mathcal{E}_{20}(0) \cosh(L/L_{pa}) \tag{5.49}$$

$$\mathcal{E}_{10}(L) = \left(\frac{\omega_1^2 k_2}{\omega_2^2 k_1}\right)^{1/2} \mathcal{E}_{20}(0) \sinh(L/L_{pa}). \tag{5.50}$$

For the usual situation $L/L_{pa} \ll 1$, the amplification of the signal (E_2) therefore depends upon L as:

$$1 + \frac{1}{2}\left(\frac{L}{L_{pa}}\right)^2$$

The disadvantage of the parametric amplifier in comparison with the parametric oscillator discussed below is this rather low gain {9}.

An important application of this principle is in the field of <u>real-time holography</u> {49}.

(e) OPTICAL PARAMETRIC OSCILLATION

With the addition of feedback (by means of a resonator), the weak signal at ω_2 may be built up by repeated passage through the non-linear crystal. As a result of this process, the weak signal at ω_1 can also be built up if the resonator is effective at this frequency as well. In that case, the terms "signal" and "idler" could apply to either of the two lower frequencies {9}. The pump may be assumed to make a single pass, without depletion {48}.

The solutions derived in the previous section are now inapplicable, because of the different boundary conditions, and instead assume the form {9,48}:

$$\mathcal{E}_{20}(L) = \mathcal{E}_{20}(0) \exp(L/L_{pa}) \tag{5.51}$$

$$\mathcal{E}_{10}(L) = \mathcal{E}_{20}(0)\left(\frac{\omega_1^2 k_2}{\omega_2^2 k_1}\right)^{1/2} \exp(L/L_{pa}). \tag{5.52}$$

The condition for oscillation is that the gain per pass should exceed the losses. With a high Q resonator used to provide the proper feedback by means of mirrors, the reflectivity of these mirrors becomes an important consideration, in addition to diffraction effects and bulk crystal losses, in determining the performance of the oscillator {48}.

The first successful demonstration of this effect was by Giordmaine and Miller {28}, who employed a Q-switched $CaWO_4:Nd^{3+}$ giant pulse laser producing 10580 Å radiation. By means of a lithium niobate crystal, the pump frequency was doubled ($\omega_p = 2\omega_0$) and then passed through an infrared absorbing filter into a second lithium niobate crystal, this one coated with dielectric films on the entrance and exit surfaces, designed for peak reflectivity at the original wavelength (10580 Å). The pump wave was removed after emerging from the second crystal by passage through a silicon filter, leaving only the signal and the idler (where $\lambda_s \approx \lambda_i \approx 10^4$ Å, from the phase match condition).

Whereas in this experiment the threshold power for oscillation was so high that only a pulsed pump source could be used to produce parametric oscillation, it has subsequently been demonstrated {50} that continuous parametric oscillation is feasible by suitable choice of non-linear material, such as barium sodium niobate ($Ba_2NaNb_5O_{15}$). This crystal, which has interesting optical and ferroelectric properties, has a filled tungsten bronze (orthorhombic) structure, and belongs to the C_{2v} class; its non-linear coefficients are three times those of $LiNbO_3$ and $LiIO_3$ {51}. Good crystals are, however, difficult to grow.

The major problem associated with the optical parametric oscillator is the requirement of double resonance for both signal and idler frequencies, together with the constraints of phase matching and a prescribed sum frequency $\omega_p = \omega_s + \omega_i$. In general, the cavity will be resonant at frequencies ω_{so} and ω_{io} such that $\omega_p = \omega_{so} + \omega_{io} + \Delta\omega$, while for the actual signal and idler frequencies, the Q-factor will be lower (i.e. higher losses and thus higher oscillation threshold). Since the decrease in Q owing to a phase mismatch at $\Delta\omega = 0$ is much less than the decrease owing to a finite $\Delta\omega$ under phase matched conditions, the actual operating point of the oscillator may be quite far removed from the operating point determined with the assumption of phase matching [9]. Slight variations in operating conditions then lead to sudden shifts in output, a phenomenon known as mode hopping or the cluster effect [9].

(f) PHASE MATCHING IN THE THREE-WAVE INTERACTION

Whereas phase matching is discussed in Chapter 4 with specific reference to second harmonic generation, we conclude this chapter by pointing out the required generalization to the three-wave interaction. Subject to the conditions:

$$\omega_3 = \omega_1 + \omega_2 \tag{5.53}$$

$$k_3 = k_1 + k_2 \tag{5.54}$$

applicable to collinear beams, the following alternatives arise in uniaxial crystals [9,52]:

$$\mu_{3e}^{(\theta)}\omega_3 = \mu_{1o}\omega_1 + \mu_{2o}\omega_2 \qquad \text{(type Ia)} \tag{5.55}$$

$$\mu_{3o}\omega_3 = \mu_{1e}^{(\theta)}\omega_1 + \mu_{2e}^{(\theta)}\omega_2 \qquad \text{(type Ib)} \tag{5.56}$$

$$\mu_{3e}^{(\theta)}\omega_3 = \mu_{1o}\omega_1 + \mu_{2e}^{(\theta)}\omega_2 \qquad \text{(type IIa)} \tag{5.57}$$

$$\mu_{3o}\omega_3 = \mu_{1o}\omega_1 + \mu_{2e}^{(\theta)}\omega_2 \, . \qquad \text{(type IIb)} \tag{5.58}$$

It has already been shown that for second-harmonic generation, type Ia is appropriate to negative uniaxial crystals, whereas type Ib is appropriate to positive uniaxial crystals (noting again that owing to insufficient birefringence [6] or the particular symmetry of the crystal [9], phase matching may not be possible in particular cases).

Now for the present equations ((5.55)-(5.58)), curves may be drawn of ω_3 versus the pair of variables ω_1, ω_2 for various phase matching angles θ_m to the optic axis. Some examples are shown in Ref. [52].

In the case of $AgGaS_2$, which is a crystal of chalcopyrite (tetragonal) structure (class D_{2d}), solutions are found to the phase matching problem for both type I and type II processes investigated. In the case of type Ia phase matching, the plot is found to have two branches, for $\theta_m = 90^\circ$, one in the high-frequency (band-gap) dispersive region, and one in the low-frequency (reststrahlen {9}) dispersive region. For the high-frequency branch, second-harmonic generation occurs at $\lambda_3 \simeq 8930$ Å, while for the low-frequency branch, it occurs at $\lambda_3 \simeq 56000$ Å. Between these two wavelength values, the birefringence is too large to allow phase-matched second-harmonic generation for $\theta_m = 90^\circ$. However, as θ_m is decreased, the birefringence is reduced and the extrema of the two branches (corresponding to $\omega_1 = \omega_2$) move closer together on the ω_3 axis, the two type Ia solutions coalescing to a single closed curve which shrinks to a single point as θ_m approaches 30°. For $\theta_m < 30^\circ$, no type I solution of any kind is possible for this crystal.

In the case of type IIa phase matching in $AgGaS_2$, two branches are again obtained and second-harmonic generation corresponds to the intersection of the two curves, no solution being possible for $\theta_m < 45^\circ$.

For two other semiconductor crystals ($CuGaS_2$ and $CuInS_2$) investigated by these authors {52}, no phase matched solutions are obtained.

While the above comments apply to collinear phase matching, it is clear that for wavelengths such that:

$$k_1 + k_2 > k_3$$

phase matching can only be achieved by employing non-collinear beams. An example of this situation, which arises in parametric amplification in KDP, is described in Ref. {53}.

For further details on the subject of collinear phase matching in the three-wave interaction, the reader is referred to Appendix IV.

CHAPTER 6

ADDITIONAL NON-LINEAR OPTICAL EFFECTS

(a) RESONANCE EFFECTS

A basic requirement of the power-series expansion approach (equation (1.13)) from which the above description of non-linear optics has been developed, namely convergence {54}, may be expressed in the following way, in the notation of equation (3.14):

$$\hbar^{-1}|x_j e \tilde{E}| \leq |\omega - \omega_{0j} + i\frac{\Gamma_j}{M_j}| . \tag{6.1}$$

This expression for semi-classical mode j may be re-expressed quantum-mechanically {6,24,54} in terms of the electric dipole moment μ_{ng} connecting the states $|n\rangle$ and $|g\rangle$, separated in energy by $\hbar\omega_{ng}$. The factor Γ_j/M_j is now interpreted {24} as the damping constant of the off-diagonal element of the density matrix ρ_{ng} corresponding to the homogeneous width of this one-photon transition. The interpretation {54} is then that the Rabi frequency, given by the left-hand side of inequality (6.1), should be small compared with the de-tuning, i.e. for non-resonant parametric processes. The perturbation approach will remain valid even on resonance provided that the Rabi frequency is small compared with the homogeneous line-width.

This inequality is violated in the coherent interaction between radiation and atomic systems known as <u>self-induced transparency</u> {55-58}. In this non-linear process, a short pulse of coherent light with energy above a certain critical value for the particular pulse width {57} passes through an optically resonant medium as though it were transparent, whereas pulses of energy below the critical threshold are absorbed. Although the width, energy and shape of the transmitted pulses are preserved (after some initial re-shaping), their speed is far below the ordinary phase velocity of light in the medium. A complete discussion of this process and related phenomena will be found in Refs. {55-58} and further references cited by Shen {55}.

(b)   CALCULATION OF NON-LINEAR SUSCEPTIBILITIES

The non-linearities of interest may be divided into two categories, viz. _fast_ and _slow_, according to the relevant time-scales involved. In the former case, the non-linearities are associated with electronic motion, and the relevant frequency responses greatly exceed optical frequencies; in the latter case, motions of the nuclei need to
be considered, relaxation times are finite and absorption of electromagnetic wave energy by the medium takes place to some extent. Slow
non-linearities are usually associated with infrared frequencies.

Models for the calculation of $\tilde{X}^{(n)}$ are discussed by Shen {55}.
Of particular interest is the bond model for molecular polarizabilities
$\alpha_{//}$ and $\alpha_{\perp}$ (components of α parallel and perpendicular to a cylindrically
symmetrical bond). According to the bond additivity rule for crystals
the induced polarization in a molecule (or crystal) is the vector sum
of the induced polarizations of all bonds between the atoms. Extending
this rule to the non-linear polarizabilities, we may write:

$$\tilde{X}^{(n)} = \sum_i \tilde{\beta}_i^{(n)} \tag{6.2}$$

where $\tilde{\beta}_i^{(n)}$ is the nth-order polarizability tensor of the ith bond, and
the summation is over all bonds in a unit volume. The $\tilde{\beta}^{(n)}$ are in turn
related to the polarizability α and electric field $\tilde{\underset{\sim}{E}}$ by the derivative:

$$\tilde{\beta}^{(n)} = \frac{\partial^{(n-1)}}{\partial \tilde{\underset{\sim}{E}}^{(n-1)}} \underset{\approx}{\alpha} . \tag{6.3}$$

In component form, for example:

$$\tilde{\beta}_{ijk}^{(2)} = \frac{\partial}{\partial \tilde{E}_i} \alpha_{jk} . \tag{6.4}$$

Further details are developed in Chapter 7, section (f).

(c)   LIQUID CRYSTALS

Of recent interest as well is the subject of non-linearities in
liquid crystalline materials {55,59}. These materials are composed of
long molecules with strong anisotropy, whose shapes and intermolecular
forces tend to produce parallel alignment against thermal agitation.
This results in the appearance of mesomorphic phases {60,61}, between
the phases of liquid and solid.   In these new phases, the molecules
are approximately aligned with possible long-range structural order,
but may still retain certain degrees of translational and rotational

freedom. Of the mesomorphic phases, the nematic phase {60,61} has the least molecular ordering, as the molecules are aligned in one direction but are free to translate and rotate about their long axes. In both the liquid and nematic phases the material has inversion symmetry and in consequence a vanishing second-order susceptibility.

On the other hand, the third-order susceptibility in the nematic phase is quite large. Because of the strong molecular anisotropy, even an optical field can induce appreciable molecular alignment, the induced alignment (and hence induced birefringence) being proportional to the square of the impressed optical electric field {55}. In the liquid phase, this effect is known as the optical (quadratic) Kerr effect {4}, and is a special case of the processes listed in section (c) of Chapter 3. By analogy with paramagnetic-ferromagnetic phase transitions (the light intensity playing the role of the magnetic field, and the induced birefringence taking the place of the induced magnetization), a Curie Law may be formulated for the liquid-nematic phase transition {62}:

$$\Delta\mu = \frac{C|E|^2}{T - T^*} . \tag{6.5}$$

(C is a constant). This expression for the induced birefringence ($\Delta\mu$) at temperatures $T > T^*$ (the second-order transition temperature) applies well except in the vicinity of the critical divergence, which is better described by first-order theory {55}.

The induced alignment also exhibits a critical slowing-down behaviour {55}, i.e. the relaxation (response) time τ of $\Delta\mu$ also diverges as T approaches T*, this behaviour being analogous to the critical slowing-down of spin alignment in a paramagnetic system. The relaxation time τ may in turn be described by {55,62}:

$$\tau = \frac{C' \nu}{T - T^*} \tag{6.6}$$

where C′ is a constant, and ν a viscosity coefficient. In the vicinity of T*, τ for liquid crystalline materials is typically several hundred nanoseconds, considerably longer than for ordinary liquids. In the case of CS_2, for example, τ is only 2 picoseconds. The large values of $\Delta\mu$ and τ make liquid crystals suitable materials for the study of self-focussing, stimulated Raman and Brillouin scattering (see below).

Third-harmonic generation in cholesteric liquid crystals {60,61} has also been a subject of recent study {55,59,63}. "A cholesteric liquid crystal can be considered as a nematic liquid crystal twisted around an axis perpendicular to the direction of molecular alignment. The molecules form layers; in each layer the molecules are aligned parallel to the layer, but as the layer advances, the direction of alignment gradually rotates. As a result, the material has an overall helical structure, and hence a one-dimensional periodic structure with the period equal to one-half of the helical pitch p. The pitch of a cholesteric liquid crystal can easily be varied from about 0.2 μ to several hundred microns by almost any external perturbation, such as temperature" {55}.

As a result of this one-dimensional periodic structure, optical Bragg reflection can take place in such a medium. Furthermore, in order to achieve phase matching in wave mixing processes, momentum need only be conserved modulo $\frac{4\pi}{p}\hbar$ in the case of propagation along the helical axis {54,55,63}. In the case of third-harmonic generation, we require for maximum efficiency {63}:

$$\pm k(\omega) \pm k(\omega) \pm k(\omega) = \pm k(3\omega) \pm \frac{4\pi m}{p} \qquad (6.7)$$

where the + and - signs indicate forward and backward propagation, respectively, and m is an integer. While phase-matching is achieved, the balance of momentum ($4\pi m\hbar/p$) is transferred to the periodic structure {32}. From the dispersion of the refractive index of the medium, the helical pitch required for phase matching can be calculated from equation (6.7), and hence the appropriate sample temperature chosen {63, 64}. Various phase-matching cases with fundamental beams propagating in the same or opposite directions, or with fundamental and third-harmonic beams in opposite directions, have been studied experimentally by Shelton and Shen {63-65}. By analogy with "Umklapp" processes of electrons and phonons in crystals, the term "coherent optical umklapp processes" has been introduced by these authors.

(d) OPTICAL PHASE CONJUGATION

The related processes of optical phase conjugation, phase-conjugate reflection, degenerate four-wave mixing and real-time holography may be studied in terms of the principle that for any electromagnetic wave that propagates through an inhomogeneous, non-absorbing medium, a time-reversed replica of this wave may be generated by non-linear effects which can be used to correct the distortion of the initial wave fronts by the inhomogeneous medium {54,66,67}. Consider an experimental arrangement in which a strong pump beam:

$$\underset{\sim}{\mathcal{E}}_1 \exp[i(\underset{\sim}{k}_1 \cdot \underset{\sim}{r} - \omega t + \phi_1)]$$

is reflected by a mirror in such a way as to set up a strong standing-wave pattern in a non-linear medium, by interference of beam $\underset{\sim}{k}_1$ with the reflected beam ($\underset{\sim}{k}_3 = -\underset{\sim}{k}_1$):

$$\underset{\sim}{\mathcal{E}}_3 \exp[i(\underset{\sim}{k}_3 \cdot \underset{\sim}{r} - \omega t + \phi_3)].$$

A weak signal beam:

$$\underset{\sim}{\mathcal{E}}_2 \exp[i(\underset{\sim}{k}_2 \cdot \underset{\sim}{r} - \omega t + \phi_2)]$$

where $\underset{\sim}{k}_2$ is in some arbitrary direction relative to $\underset{\sim}{k}_1$ is also incident, giving rise parametrically to a reflected wave with wave vector $\underset{\sim}{k}_4 = -\underset{\sim}{k}_2$, as a result of non-linear polarization terms of the type {68}:

$$\chi^{(3)}_{ijk\ell}(\omega;\omega,-\omega,\omega)\,\mathcal{E}_{1j}\,\mathcal{E}^{*}_{2k}\,\mathcal{E}_{3\ell}\exp i[(\underset{\sim}{k}_1 - \underset{\sim}{k}_2 + \underset{\sim}{k}_3)\cdot\underset{\sim}{r} + \phi_1 - \phi_2 + \phi_3 - \omega t].$$

The phase of this reflected wave $\underset{\sim}{k}_4$ is therefore given by:

$$\begin{aligned}\underset{\sim}{k}_4 \cdot \underset{\sim}{r} + \phi_4 - \omega t &= (\underset{\sim}{k}_1 - \underset{\sim}{k}_2 + \underset{\sim}{k}_3)\cdot\underset{\sim}{r} + \phi_1 - \phi_2 + \phi_3 - \omega t \\ &= -\underset{\sim}{k}_2 \cdot \underset{\sim}{r} - \phi_2 - \omega t + \text{constant} \qquad (6.8)\end{aligned}$$

i.e. is equal and opposite to the phase of wave $\underset{\sim}{k}_2$, apart from an arbitrary constant. Note that the generation of the fourth beam will be enhanced if ω is chosen in the vicinity of a sharp atomic resonance line (see also section (g) of this chapter), owing to a three-fold resonant denominator in the third-order susceptibility {14,68}.

Now consider the situation where the phase $\phi_2(x,y)$ has a transverse distribution because the signal $\underset{\sim}{k}_2$ has undergone some distortion from a plane wave caused by phase aberrations in its passage through an inhomogeneous optical medium {66}. The fourth wave will now have its phase aberrations corrected in its passage through the same inhomogeneous optical medium. This may be expressed in the following way {54, 66,67,69,70}: the phase conjugate reflected wave 4 is the time-reversed replica of wave 2, since:

$$\underset{\sim}{E}_4 = \underset{\sim}{\mathcal{E}}_4 \exp i(\phi_1 + \phi_3) \exp i[-\underset{\sim}{k}_2 \cdot \underset{\sim}{r} - \omega t - \phi_2]$$
$$\propto [\chi^{(3)} \underset{\sim}{\mathcal{E}}_1 \underset{\sim}{\mathcal{E}}_3 \exp i(\phi_1 + \phi_3)] \underset{\sim}{E}^*_2(\underset{\sim}{r}, -t) \quad . \tag{6.9}$$

This concept of a phase-conjugate mirror is of considerable technological importance {69,70}, since if the pump field and the non-linearity of the medium are sufficiently strong, the output field $\underset{\sim}{E}_4$ and signal $\underset{\sim}{D}_2$ can undergo considerable amplification. Moreover, the conjugate mirror may act as a very sharp filter, owing to the behaviour of $\tilde{\chi}^{(3)}$ in the vicinity of a narrow optical resonance line {54,68}.

The theory of phase conjugation by stimulated scattering in a wave-guide has been derived by Hellwarth {71}.

(e) REFLECTION AND REFRACTION

The theoretical analysis of the behaviour of light waves at the boundary of a non-linear medium is due to Bloembergen and Pershan {72}, who obtained the appropriate generalizations of the laws of reflection, refraction and Fresnel's equations. In linear optics, the directions of the reflected and refracted waves are derived from the condition that the tangential component of the wave vector be conserved at the boundary between the two media. In the non-linear case, one considers e.g. the effect of the generation of second-harmonic radiation ($\omega_2 = 2\omega_1$) when a wave of angular frequency ω_1 impinges on a non-linear surface. The requirement that the tangential components of $\underset{\sim}{E}$ and $\underset{\sim}{H}$ be continuous everywhere on the boundary at all times {7,10,11} imposes the condition that the individual frequency components, at ω_1 and $2\omega_1$, be separately continuous across the boundary. With superscripts i, R and T denoting the incident, reflected and transmitted waves, respectively, subscripts 1 and 2 denoting the fundamental and second-harmonic fields, and supposing planar boundary to be specified by z = 0 with y = 0 denoting the plane of incidence, we have for the fundamental frequency:

$$k^i_{1x} = k^R_{1x} = k^T_{1x} \quad . \tag{6.10}$$

In the non-linear medium, a non-linear source term {14} (superscript S), with wave vector $\underset{\sim}{k}^S{}_2 = 2\underset{\sim}{k}^T{}_1$ will be generated by the fundamental. This in turn means that both a <u>forced</u> polarization wave $\underset{\sim}{k}^S{}_2$ and a <u>free</u> harmonic wave $\underset{\sim}{k}^T{}_2$ will propagate in the second medium, corresponding respectively to the particular solution of the inhomogeneous equation (4.7) and the solution of the corresponding homogeneous equation {14,72}. For the second-harmonic frequency one now has the condition:

$$2k^{T}_{1x} = k^{S}_{2x} = k^{R}_{2x} = k^{T}_{2x} \; . \tag{6.11}$$

Relations (6.10) and (6.11) reflect the general requirement of conservation of the tangential component of momentum {54,72}, and immediately yield the angles of reflection and refraction for the second-harmonic waves:

$$\left.\begin{aligned} \sin\theta^{R}_{2} &= \frac{k^{R}_{2x}}{|k^{R}_{2}|} = \frac{k^{i}_{1x}}{|k^{R}_{1}|} = \sin\theta^{i}_{1} \\ \sin\theta^{T}_{2} &= \frac{k^{T}_{2x}}{|k^{T}_{2}|} = \frac{1}{\mu(2\omega)}\sin\theta^{i}_{1} \\ \sin\theta^{S}_{2} &= \frac{k^{S}_{2x}}{|k^{S}_{2}|} = \frac{1}{\mu(\omega)}\sin\theta^{i}_{1} \; . \end{aligned}\right\} \tag{6.12}$$

Here we have considered for simplicity the first medium to be vacuum, while only one refractive index (for a particular frequency) has been assumed for the second medium. The present treatment (which can be generalized where necessary to account for more complicated situations) is therefore applicable to a cubic crystal (e.g. ZnS), or to a uniaxial crystal (e.g. KDP) when the plane of incidence contains the optic axis and the incident wave is, e.g. polarized within this plane. Since the first medium (here vacuum) is non-dispersive, the reflected second harmonic travels in the same direction as the reflected fundamental wave. On the other hand, the forced polarization wave $\underset{\sim}{k}^{S}_{2}$ and the free harmonic wave $\underset{\sim}{k}^{T}_{2}$ will in general (as a result of dispersion) propagate in somewhat different directions; they will travel parallel in the limiting cases of exact phase matching, $\mu(2\omega) = \mu(\omega)$, or normal incidence. The total harmonic field in the non-linear medium is determined by the interference between the free and forced waves, which prevents the generation of a significant amount of second-harmonic power in the absence of phase matching (see Chapter 4), unless the waves are spatially separated by the use of oblique angles of incidence {54}. (It is of course clear from (6.12) that the forced polarization wave is confined to those regions of the crystal in which there is fundamental field intensity.)

For the present situation where the fundamental incident wave is E polarized in the plane of incidence, which contains the optic axis of the crystal, the transmitted fundamental wave is an extraordinary wave. In that case, crystal symmetry requires that the $\underset{\sim}{P}^{NL}$ field for the forced second-harmonic wave, and therefore (from the boundary conditions at the interface) $\underset{\sim}{E}$ for the free second-harmonic wave be perpendicular to the plane of incidence {72}. Both of these second-harmonic waves therefore propagate as ordinary waves in the second medium. The same boundary conditions will be satisfied with $\underset{\sim}{E}^R(2\omega)$ perpendicular to the plane of incidence as well.

The various fields can readily be derived from the Maxwell equations together with the appropriate boundary conditions, for the special case of second harmonic generation as well as the more general situation where two waves of frequencies ω_1 and ω_2, incident from a linear medium, impinge on the surface of a non-linear medium, thereby generating inter alia free and forced waves at the sum frequency $\omega_3 = \omega_1 + \omega_2$ {72}.

The non-linear analogues of the Fresnel formulae {72}, the Brewster angle {73} and total internal reflection {72} for the two different critical angles predicted by equation (6.12) have been studied experimentally {73,74}. In all cases, quantitative agreement has been obtained with the analysis of Bloembergen and Pershan {72}, in a manner which serves as a beautiful confirmation of the correctness of Maxwell's equations in describing both linear and non-linear optical propagation.

A special case of particular interest in this connection is the phenomenon of conical diffraction in second-harmonic generation {75}, which is the non-linear analogue of a refraction process (conical refraction) in biaxial crystals first predicted by Hamilton {76} in 1833 and observed by Lloyd {77} in the same year. As shown in Chapter 2, the plane of polarization is arbitrary when wave propagation takes place along the optic axis of a biaxial crystal. The Poynting vectors belonging to the wave-vector direction exactly parallel to the optic axis lie on a cone, with apex angle α given by:

$$\tan\alpha = \frac{1}{\mu_1\mu_3}\left[(\mu_2^2 - \mu_1^2)(\mu_3^2 - \mu_2^2)\right]^{\frac{1}{2}} \tag{6.13}$$

where we have assumed for the principal velocities $v_i = \frac{c}{\sqrt{\epsilon_i}} = \frac{c}{\mu_i}$, with $\mu_1 < \mu_2 < \mu_3$, and the optic axes lie in the xz plane, each making an angle:

$$\eta = \arctan\left[\frac{\mu_3}{\mu_1}\left(\frac{\mu_2^2-\mu_1^2}{\mu_3^2-\mu_2^2}\right)^{\frac{1}{2}}\right] \tag{6.14}$$

with the z axis (see Appendix IV).

In the non-linear case, two types of internal conical refraction arise, namely <u>forced</u> and <u>free second-harmonic conical refraction</u> {75, 78-80}. Firstly, if the fundamental wave vector is parallel to one of the fundamental optical axes, both the fundamental and forced second-harmonic intensities will be distributed in a conical pattern. If the dispersion in the direction of the optic axes is sufficiently large {9} the free second-harmonic wave vectors will not contain the direction of the second-harmonic optic axis. The energy associated with the free wave mode will therefore be refracted in a single spot {80}.

If the incident laser beam has a bundle of wave normals that contain the second-harmonic optic axis, the free wave at 2ω gives rise to a free second-harmonic conical intensity pattern. Provided that the fundamental optic axis is not contained in the wave-normal bundle, the fundamental intensity is now confined to a single ray and consequently the forced second-harmonic wave gives rise to a forced ray coinciding with the fundamental.

Both of these cases were observed by Schell and Bloembergen {78-80}, who employed crystals of aragonite ($CaCO_3$) and α-iodic acid (α-HIO_3), both of which are orthorhombic in structure, but with the additional complication in the latter case of natural optical activity (optical rotatory power) {17}. Good agreement between measurement and theoretical predictions was again obtained. The experimental conditions which must be fulfilled in order to observe this phenomenon are rather stringent {75,78}:

(1) The diffraction angle of the primary laser beam should be small compared with the cone apex α. This implies that the radius of the waist w_0 (called the optimum spot size in Chapter 5) of the Gaussian beam focussed at the <u>entrance surface</u> of the crystal, and the crystal length L, must satisfy the condition:

$$\begin{aligned} \tfrac{1}{2}L\alpha \gg w(L) &= w_0[1+\xi^2]^{\frac{1}{2}} \\ &= w_0\left[1+\left(\frac{\lambda_1 L}{\pi w_0^2 \mu(\omega)}\right)^2\right]^{\frac{1}{2}} \end{aligned} \tag{6.15}$$

where we have used equations (5.5), (5.9) and Ref. {42}, equation (4.4). The refractive index $\mu(\omega) \approx \mu_2(\omega)$ is an "average" of the principal refractive indices.

(2) The angular dispersion between the optic axes at the fundamental and the second-harmonic frequency should be larger than the diffraction angle.

(3) The thickness of the crystal L should be sufficiently large to obtain a dark centre of the base formed by the intersection of the cone with the exit surface. This again requires that condition (6.15) be fulfilled.

(4) The coherence length should be smaller than the distance over which the forced and free second-harmonic waves overlap. In that case, the two harmonic waves have equal intensity {72}. This requires that the coherence length should be much less than the aperture length:

$$L_{coh} \ll L_a$$

or, from (4.49) and (5.8):

$$\lambda_1 \ll 4\sqrt{\pi}\,\frac{W_0}{\alpha}\,|\Delta\mu| \,. \tag{6.16}$$

(Note {75} that for present purposes (biaxial crystals) the apex angle α in equation (6.13) effectively replaces the aperture ("walk-off") angle in equation (5.7), used in Chapter 5 in the discussion of uniaxial crystals.)

Some specific numerical examples are discussed in Ref. {75}, notably the case of α-iodic acid (α-HIO_3). The non-linear analogue of external conical refraction is also discussed in this paper.

(f) SELF-FOCUSSING

This phenomenon is found to occur in media (liquids and solids) subjected to a laser beam whose intensity exceeds a certain critical value: the beam diameter contracts as the electromagnetic wave propagates through the material, eventually producing a sharply focussed spot which can cause appreciable damage in solids {55,56,81-83}. Self-focussing is caused by a field-induced non-linearity in the refractive index:

$$\mu = \mu_0 + \Delta\mu(|\underset{\sim}{E}|^2) \quad , \tag{6.17}$$

the possible physical mechanisms to which $\Delta\mu$ may be ascribed being libration, reorientation and redistribution of the molecules, electrostriction, deformation of electronic clouds, and heating {4,55}. While the complete solution to the problem is rather complex, a simple model for the process may readily be described {55,56}. When a laser beam with an initially Gaussian transverse profile enters a medium whose refractive index may be described by (6.17), the central part of the beam, which encounters a region of higher refractive index, is retarded relative to the edge, with consequent distortion of the original plane wave front. This distortion continues to increase with distance travelled in the medium, while the corresponding rays therefore tend to contract to a focus. This effect is somewhat offset by the tendency of any beam of finite cross-section to diffract, as described in Chapters 4 and 5, and the final outcome is determined by the competition between the two processes {55}. When the effects of self-focussing and diffraction exactly cancel, the beam will propagate without any change in its transverse profile, a situation known as <u>self-trapping</u> {81}.

With the use of high-intensity pulsed lasers, two situations arise: firstly, if the pulse duration is much longer than the response time of $\Delta\mu$ (in particular the case of a fast non-linearity), the response of $\Delta\mu$ to the laser intensity variation can be considered to be instantaneous, i.e. the case of <u>quasi-steady-state self-focussing</u> {55}. Secondly, if the pulse duration is comparable with or shorter than the response time of $\Delta\mu$, the lagging part of the pulse will encounter a variation in refractive index brought about by the passage of the leading part. This is the case of <u>transient self-focussing</u> {55}.

Both quasi-steady-state and transient self-focussing can be described formally by the non-linear wave equation (see (1.17)-(1.19)):

$$-\nabla^2\underset{\sim}{E} + \frac{1}{c^2}\frac{\partial^2}{\partial t^2}\left[(\mu_0 + \Delta\mu)^2\underset{\sim}{E}\right] = 0 \tag{6.18}$$

where $\Delta\mu$ obeys the appropriate dynamical equation (depending upon the physical mechanism responsible for $\Delta\mu$). "For example, in liquids with strongly anisotropic molecules (<u>Kerr liquids</u>), $\Delta\mu$ induced by a Q-switched laser pulse is mainly due to field-induced orientation of molecules and hence should obey the Debye relaxation equation" {55}. However, a complete and fully rigorous analytical solution to equation (6.18) applicable to all cases, cannot readily be derived; a qualitative discussion of many aspects of self-focussing together with an extensive bibliography will be found in Ref. {55}.

(g) OPTICAL MIXING IN VAPOURS

Because of inversion symmetry, $\tilde{\chi}^{(2)}$ vanishes in a vapour system in the electric-dipole approximation {6}, and in view of the comparatively low atomic density, one would expect that non-linear effects arising from $\tilde{\chi}^{(3)}$ would normally not be significant. However, when the optical frequencies are close to strong resonances, inequality (6.1) will be violated {54} and because of resonant enhancement, $\tilde{\chi}^{(3)}$ can become so large that the third-order non-linear optical processes in a vapour can appear as strong as the second-order processes in a crystal {55}. For efficient third-harmonic generation, phase matching is again required, and in the present case this may be achieved by introduction of an appropriate buffer gas whose density may be treated as a variable {84}. A discussion of the limitations on the laser intensity and atomic density will be found in the paper by Miles and Harris {84}. Numerous other references on this subject are cited by Shen {55}.

(h) DISPERSIVE OPTICAL BISTABILITY

In the phenomenon of optical bistability, matter and light are coupled together in such a way that phase transitions occur far from thermodynamic equilibrium. Initially {85}, a Fabry-Perot interferometer filled with a non-linear vapour (sodium irradiated with light from a CW dye laser) was used to demonstrate the effects of hysteresis in the curve of transmission versus input intensity, and optical bistability (dispersive and absorptive). Thereafter, similar observations were made by means of etalons containing semiconductor (GaAs and InSb) material {86,87} and distributed feedback structures {88}. For a discussion of these effects, Refs. {85-89} may be consulted.

CHAPTER 7

SCATTERING BY NON-LINEAR MEDIA

INTRODUCTION

Ordinary Raman spectroscopy has long been a valuable means of studying the vibrational energy levels of molecules and of optical branch lattice vibrations in crystals {90,91}. When a parallel beam of light with a discrete line spectrum traverses a gas, liquid or transparent solid body, and the scattered light (Tyndall effect) is analysed, it is found to contain primarily the same frequencies as those in the incident beam (Rayleigh scattering), but also weaker additional lines which do not appear in the spectrum of the light source (Raman scattering). The frequency shifts (Raman displacements) are found to be independent of the frequencies of the exciting lines, but characteristic of the scattering substance under consideration. The Raman lines displaced towards lower frequencies (relative to the unperturbed lines) are called Stokes lines, those displaced towards higher frequencies called anti-Stokes lines, this nomenclature originating from a law for fluorescent light formulated by Sir George Stokes {90,91}. In conventional molecular spectroscopy, the intensity of anti-Stokes scattering is significantly lower than that of Stokes scattering, by the Boltzmann factor $\exp(-h\nu_{01}/kT)$, where ν_{01} corresponds to the frequency of the unperturbed molecular transition between an excited state (1) and the ground state (0). In the anti-Stokes case, an incoming photon (frequency ν) is scattered by an initially excited molecule, and its frequency increased to:

$$\nu_a = \nu + \nu_{01} \tag{7.1}$$

while the molecule is simultaneously de-excited. In the case of Stokes scattering, the incoming photon (frequency $\nu > \nu_{01}$) encounters a molecule in its ground state, imparts a fraction of its energy to the molecule which is excited in the process, and is scattered with reduced frequency:

$$\nu_S = \nu - \nu_{01} \quad . \tag{7.2}$$

Until the advent of the laser (maser), this type of spectroscopy was performed with intense incoherent light sources.

In 1922, Brillouin {92} predicted that a liquid traversed by compression waves of short wavelength (ultrasound), when irradiated by visible light, would give rise to a diffraction phenomenon similar to that produced by a ruled grating. A simple model for this process is discussed in Chapter 12 of Ref. {13}. Experimental confirmation of the idea that ultrasonic pressure waves in an elastic medium could simulate the action of a diffraction grating, was produced by Debye and Sears {93} who used a high-frequency quartz crystal oscillator ($10^6 - 10^7$ Hz) to drive the elastic waves in toluene and carbon tetrachloride, and observed the resultant diffraction of a mercury arc spectrum. Comparison between calculated (from the adiabatic compressibility of the fluid) and measured velocity, yielded agreement within a few percent, on the basis of an assumed formula:

$$\sin \theta_n = n\lambda/\Lambda \tag{7.3}$$

for the angular deviation of the nth order with respect to the central image, Λ being the wavelength of the ultrasound. A more detailed study by Lucas and Biquard {94}, with a careful theoretical analysis of their findings, followed in the same year. A full treatment of the diffraction of light by ultrasonic waves on the basis of Maxwell's equations was subsequently undertaken by Raman and Nath {95}. A résumé of their treatment, together with an extensive bibliography on the subject, will be found in Ref. {13}.

The term Brillouin scattering is now associated with situations where a spectrum of thermally excited acoustic waves is present in a liquid or crystal, producing partial scattering of an incident light beam as a result of an interchange of energy and momentum between the incident photons and acoustic phonons {56}.

(a) STIMULATED RAMAN SCATTERING

The first reported observation of this phenomenon was made by Woodbury and Ng {96}, who noted that a ruby laser switched with a nitrobenzene (a Raman-active liquid) Kerr cell, emitted a copious amount of light at 7670 Å in addition to the normal ruby laser light at 6943 Å. This was shortly thereafter explained by Eckhardt et al. {97} who demonstrated that when various Raman-active liquids are placed inside an optical Fabry-Perot cavity and illuminated by light of frequency ω_α and intensity greater than a certain threshold, then coherent light builds up in the cavity at a frequency ω_β which equals ω_α minus the frequency of a Raman-active vibration {24,98}. It was shown by these authors {97} that while negligible resonant absorption at ω_α takes place, the lasing action is caused by stimulated Raman scattering.

The Raman laser spectra of various liquids and solids are listed in Table 5-2 of Ref. {24}. Only the frequencies belonging to the sharpest and most intense spontaneous Raman line will be produced in the stimulated process, or occasionally two lines belonging to a symmetrical vibration {24,90}. The reason for this is provided by a phenomenological theory due to Hellwarth {98}, in which the process of stimulated Raman scattering is described in terms of tabulated or measurable material parameters, namely ordinary Raman scattering cross-sections. According to a simple model {24}, the process with the lowest threshold will tend to limit the laser power below higher threshold values: further increase in pump power will increase the intensity of this Stokes line, but not create other lines of the spontaneous Raman spectrum. However, when the Stokes line has attained a sufficient intensity, it can in turn create a second Stokes line of the first Stokes line, at $\omega_L - 2\omega_V$ (where ω_L is the pump frequency, and ω_V the frequency of the optical phonon or vibrational wave). This process can continue, with the production of higher-order Stokes lines at exact harmonics of the first vibrational transition, whose existence was reported in Ref. {97}.

It might be supposed that these higher-order lines are due to a Raman transition, induced by the laser pump at ω_L, while the molecule undergoes a transition with a change of two or more in the vibrational quantum number. As explained by Bloembergen {24}, this cannot be the case in view of the small matrix elements for such transitions and the anharmonicity of the vibrational potential energy well {90} which is sufficiently large to ensure, e.g. that the frequency of spontaneous Raman emission for the double vibrational quantum transition $\omega_L - \omega_{V(0\to2)}$ does not coincide with $\omega_L - 2\omega_{V(0\to1)}$.

(b) STIMULATED RAMAN SCATTERING (THEORY)

A theoretical description of the process of stimulated Raman scattering is provided by the papers of Bloembergen and Shen {24,99,100} and Loudon {101}. The first two authors base their treatment on a model {24} for the (third-order) complex Raman (or Stokes) susceptibility of a molecule exhibiting the Raman effect, $\chi^{(3)}_S$. With ω_L again denoting the laser frequency, and ω_V a vibrational resonant frequency of the molecular system, one finds that for photon emission (ω_S) exactly at resonance ($\omega_S = \omega_L - \omega_V$), $\chi^{(3)}_S$ is negative pure imaginary, corresponding to a negative absorption or positive gain at ω_S, proportional to the intensity of the laser beam. The non-linear polarization for N molecules per unit volume may be written:

$$\underset{\sim}{P}^{NL}(\omega_S) = \chi^{(3)}_S(\omega_S;\omega_L,-\omega_L,\omega_S)|\underset{\sim}{E}_L|^2\underset{\sim}{E}_S \tag{7.4}$$

where $\underset{\sim}{E}_L$ and $\underset{\sim}{E}_S$ denote the electric field-strengths at ω_L and ω_S, respectively. Similarly, the non-linear polarization at ω_L is given by:

$$\underset{\sim}{P}^{NL}(\omega_L) = \chi^{(3)*}_S(\omega_L;\omega_L,\omega_S,-\omega_S)\underset{\sim}{E}_L|\underset{\sim}{E}_S|^2. \tag{7.5}$$

Corresponding to the electromagnetic fields at ω_S and ω_L, one has in addition a non-linear polarization at the anti-Stokes frequency $2\omega_L - \omega_S$, given by:

$$\underset{\sim}{P}^{NL}(\omega_a) = \chi^{(3)}(\omega_a;\omega_L,\omega_L,-\omega_S)\underset{\sim}{E}_L^2\underset{\sim}{E}_S^* \tag{7.6}$$

where $\chi^{(3)}$ is related to the anti-Stokes susceptibility $\chi^{(3)}_a$ and the Stokes susceptibility $\chi^{(3)}_S$ by {24,100}:

$$\chi^{(3)} = (\chi^{(3)*}_S\chi^{(3)}_a)^{\frac{1}{2}} \tag{7.7}$$

which becomes simply $\chi^{(3)*}_S$ in the limit of negligible dispersion. An important <u>complex symmetry relationship</u> is derived by Bloembergen and Shen {24,100}:

$$(\chi^{(3)*}_S\chi^{(3)}_a)^{\frac{1}{2}} = \chi^{(3)}(\omega_a;\omega_L,\omega_L,-\omega_S) = \chi^{(3)*}(\omega_S;-\omega_a,\omega_L,\omega_L). \tag{7.8}$$

This is a generalization of equation (3.37). When absorption of energy by the medium must be considered, the susceptibility elements are complex quantities; in that case, a permutation of frequencies involving a change of sign of the frequency combination near resonance requires simultaneous complex conjugation of the relevant susceptibility elements. (Compare also equation (3.2).) It may be shown {24} that (7.8) is tantamount to the requirement that the Raman-type susceptibilities obey the same Kramers-Kronig relation {10,24} as in the case of linear susceptibilities.

In addition to the resonances described above at $\omega_S = \omega_L - \omega_V$ and $\omega_a = \omega_L + \omega_V$, one has other non-resonant contributions to the total non-linear susceptibility which are non-dispersive in character, and therefore expressible by a purely real tensor quantity $\chi^{(3)}_{NR}$. With $\underset{\sim}{E}_a$ denoting the electric field-strength at the anti-Stokes frequency ω_a, one has finally the following general expressions for the non-linear polarization at ω_S, ω_L and ω_a {24}:

$$\underset{\sim}{P}^{NL}(\omega_S) = (\chi_S + \chi_{NR})|\underset{\sim}{E}_L|^2 \underset{\sim}{E}_S + \{(\chi_S \chi_a^*)^{\frac{1}{2}} + \chi_{NR}\} \underset{\sim}{E}_L^2 \underset{\sim}{E}_a^* \tag{7.9}$$

$$\underset{\sim}{P}^{NL}(\omega_L) = (\chi_S^* + \chi_{NR}) \underset{\sim}{E}_L |\underset{\sim}{E}_S|^2 + (\chi_a^* + \chi_{NR}) \underset{\sim}{E}_L |\underset{\sim}{E}_a|^2 + \{(\chi_S \chi_a^*)^{\frac{1}{2}} + (\chi_S^* \chi_a)^{\frac{1}{2}} + 2\chi_{NR}\} \underset{\sim}{E}_L^* \underset{\sim}{E}_S \underset{\sim}{E}_a \tag{7.10}$$

$$\underset{\sim}{P}^{NL}(\omega_a) = (\chi_a + \chi_{NR})|\underset{\sim}{E}_L|^2 \underset{\sim}{E}_a + \{(\chi_S^* \chi_a)^{\frac{1}{2}} + \chi_{NR}\} \underset{\sim}{E}_L^2 \underset{\sim}{E}_S^* \quad . \tag{7.11}$$

(For brevity, we have omitted the superscript 3 on the third-order Raman susceptibilities.)

The following physical interpretation may be given for the various terms in (7.9)-(7.11). The imaginary part of the terms with $\chi^{(3)}_S$ and $\chi^{(3)}_a$ describes the emission of Stokes and anti-Stokes photons resonantly in the interaction between the electromagnetic field and the molecular systems. The real part of $\chi^{(3)}_S$ and $\chi^{(3)}_a$ describes a parametric process, the simultaneous scattering of quanta at ω_L and ω_S, and ω_L and ω_a, respectively. The interference of all these higher scattering processes in the homogeneous medium leads to a change in index of refraction at ω_S and ω_a proportional to $|\underset{\sim}{E}_L|^2$, and at ω_L proportional to $|\underset{\sim}{E}_S|$ and to $|\underset{\sim}{E}_a|^2$. The real part of the terms in $(\chi^{(3)}_S \chi^{(3)*}_a)^{\frac{1}{2}}$ corresponds to a scattering process in which two quanta at ω_L scatter into a quantum at ω_S and one at $\omega_a = 2\omega_L - \omega_S$, or in which a pair of quanta at ω_S and ω_a combine to yield two quanta at ω_L. The interference of all scattering processes in the homogeneous medium leads to a parametric generation of ω_S and ω_a by the laser beam, or vice versa {24}.

The imaginary part of the terms in $(\chi_S \chi_a^*)^{\frac{1}{2}}$ and $(\chi_S^* \chi_a)^{\frac{1}{2}}$ can be described as the interference of two Raman processes operating between the same initial and final state, which may either enhance or decrease the total transition rate between the vibrational levels. In the former case, the generation of ω_S and the absorption of ω_a is enhanced; in the latter case the generation of ω_S and the absorption of ω_a is decreased by the interference. The relative phases of $\underset{\sim}{E}_L$, $\underset{\sim}{E}_S$ and $\underset{\sim}{E}_a$ determine which situation applies: this in fact suggests that a better description of these processes would be in terms of coupled coherent wave packets of electromagnetic oscillator states {24}.

The concepts outlined above can now be applied to the description of the stimulated Raman effect as a parametric process {99,100}. The expressions for $P^{NL}(\omega_S)$ and $P^{NL}(\omega_a)$ in equations (7.9) and (7.11) are substituted in the differential equations for the corresponding field-strengths (cf. (4.7)), here written for simplicity for an isotropic medium:

$$\nabla^2 \underset{\sim}{E}_s - \frac{\epsilon_s}{c^2} \frac{\partial^2 \underset{\sim}{E}_s}{\partial t^2} = \frac{4\pi}{c^2} \frac{\partial^2}{\partial t^2} \underset{\sim}{P}^{NL}(\omega_s) \tag{7.12}$$

$$\nabla^2 \underset{\sim}{E}_a - \frac{\epsilon_a}{c^2} \frac{\partial^2 \underset{\sim}{E}_a}{\partial t^2} = \frac{4\pi}{c^2} \frac{\partial^2}{\partial t^2} \underset{\sim}{P}^{NL}(\omega_a) . \tag{7.13}$$

Along the lines of analogous coupled amplitude equations in Chapters 4 and 5, these equations for the field-strengths may first be simplified by adopting the slowly varying amplitude and phase approximation. The solutions for the Stokes and anti-Stokes intensities are outlined and discussed in Refs. {24,99} as a function of the real and imaginary parts of the complex Raman susceptibility, and the wave vector (momentum) mismatch:

$$\Delta \underset{\sim}{k} = 2 \underset{\sim}{k}_L - \underset{\sim}{k}_s - \underset{\sim}{k}_a . \tag{7.14}$$

Loudon {101} considers the process where an incident photon of frequency ω_1 is destroyed in a crystal, accompanied by the creation of a scattered photon of frequency ω_2 and an optical phonon (lattice vibration quantum) of frequency ω. The photons ω_1 and ω_2 have wave vectors $\underset{\sim}{k}_1$ and $\underset{\sim}{k}_2$ directed perpendicular to the end mirrors of the laser cavity {97}; wave vector conservation requires that the phonon wave vector $\underset{\sim}{k}$ equal $\underset{\sim}{k}_1 - \underset{\sim}{k}_2$. For forward scattering (as considered here), where $\underset{\sim}{k}_1$ and $\underset{\sim}{k}_2$ are collinear, $\underset{\sim}{k}$ is of the order of 0 to 10^3 cm^{-1}, i.e. we are dealing with the creation of long wavelength phonons. The dominant scattering interaction {102} is one in which the photons and phonons are coupled indirectly through electron-photon and electron-phonon interactions with the electrons in the crystal.

The theory of stimulated Raman scattering by lattice vibrations is developed by Loudon {101} by analogy with the treatment of spontaneous Raman scattering from phonons {102}. It is important to note the restrictions imposed by crystal symmetry in the present case as well. Since the transverse (optical) phonons under consideration have the same symmetry character as photons, the crystal symmetry restrictions are identical to those which apply to collinear three-photon processes {103,104}. An analysis of the phonon modes that satisfy conservation of energy and momentum and hence can participate in forward Raman scattering, requires a detailed knowledge of the phonon dispersion curves {101,102} for the particular crystal. A detailed discussion for the case of uniaxial crystals in terms of ordinary and extraordinary electromagnetic and lattice waves will be found in Ref. {101}, together with an analysis of the stimulated Raman scattering threshold.

(c) HIGHER-ORDER STOKES AND ANTI-STOKES RADIATION

Already by 1963, power flux densities in excess of 100 megawatts/ cm^2 were readily obtainable in the focus of an external laser beam, and these giant pulses could be utilised to generate copious Stokes and anti-Stokes radiation in various orders, from liquids such as benzene, liquid nitrogen, and nitrobenzene {105-107}. By means of a frosted glass plate placed immediately behind the Raman cell {97} and imaged on the slit of a high-resolution spectrograph, the frequency spectrum integrated over all directions could be measured. Several interesting features of the Stokes and anti-Stokes lines were noted by Stoicheff {106}, in particular the widths of the lines as well as the effect on the Raman spectra of mixing liquids such as benzene and carbon disulphide.

The angular dependence of laser-stimulated Raman radiation in calcite was measured by Chiao and Stoicheff {108}. Four orders of anti-Stokes emission were observed in well-defined cones, as well as diffuse first-order Stokes emission with cones of absorption in this diffuse radiation, and a well-defined cone of second-order Stokes emission, all in the forward direction. Excellent agreement was found between the measured and calculated half apex angles of the various cones, on the basis of the precisely known refractive index of calcite and the wave vector (momentum matching) relations for the laser and the various Stokes and anti-Stokes fields {24,108,109}. The calculation of these angles is discussed in detail by Bloembergen {24}.

(d) STIMULATED BRILLOUIN SCATTERING

As explained above, Brillouin scattering {92} involves a coupling between acoustic waves and electromagnetic waves. The effect may be described either as the diffraction of a light wave by a variable index of refraction grating {13}, set up by the acoustic vibration, or as a collision which occurs between an incident light wave ($\underset{\sim}{k}_1$, ω_1) and an acoustic wave ($\underset{\sim}{q}_2$, ω_2) to produce a light wave ($\underset{\sim}{k}_3$, ω_3). The conditions of conservation of energy and quasi-momentum require:

$$\omega_3 = \omega_1 + \omega_2 \tag{7.15}$$

$$\underset{\sim}{k}_3 = \underset{\sim}{k}_1 + \underset{\sim}{q}_2 . \tag{7.16}$$

For acoustical phonons with frequencies below 10^{10} Hz [10 GHz], these conditions can always be satisfied {24}. Although the momentum q_2 can be comparable in magnitude to k_1 and k_2, the acoustical frequency is very small compared with the light frequency, and consequently the momentum triangle corresponding to (7.16) may be considered isosceles, with $|\underset{\sim}{k}_1| \simeq |\underset{\sim}{k}_3|$. The Brillouin relation {92} for the angle of scattering of the light now follows immediately as:

$$q = 2k \sin\left(\frac{\theta}{2}\right) \tag{7.17}$$

where θ is the angle between the scattered and incident electromagnetic wave vectors.

As written, equation (7.15) represents the absorption of a phonon by an electromagnetic wave. It is also possible for a vibrational quantum of frequency ω_2 to be emitted, with the attendant scattering of a photon of frequency $\omega_1 - \omega_2$.

The phenomenon of stimulated Brillouin scattering, in which the acoustic wave that scatters the incident light is produced by the optical beam itself, was discovered by Chiao, Townes and Stoicheff {110} in

1964. In comparing this process with that of stimulated Raman scattering by lattice vibrations {101}, it is important to note {24} the very different dispersion law for acoustical phonons from that for optical phonons.

In this experiment {110}, intense radiation from a giant-pulse ruby laser was focussed inside a crystal (quartz or sapphire), and the radiation scattered in the backward direction detected and resolved with the aid of two Fabry-Perot interferometers, whose interferograms were compared to distinguish between the ruby radiation and that scattered directly backward from the sample. The Brillouin scattering was found to be very intense, and hence much amplified over that expected from normal Brillouin scattering. Excellent agreement was obtained between the observed frequency shifts and those calculated from the known elastic constants of the particular material, on the basis of a model for electrostrictively-driven compressional waves. The intense hypersonic waves generated in these crystals were found to have frequencies exceeding 10^{10} Hz {110}.

For a discussion of more recent developments such as phase conjugation in stimulated Brillouin {111} and stimulated Raman scattering {71}, the reader is referred to the review articles, Refs. {54,55,69, 70}.

(e) THE RAMAN LASER

While the inclusion of the rather small non-resonant contributions χ_{NR} in equations (7.9)-(7.11) is important for certain problems like Raman induced Kerr effect spectroscopy (RIKES) {112}, it is often a safe procedure to omit them entirely. In that case one has from equations (7.9)-(7.14) in the slowly-varying amplitude and phase approximation (compare equation (4.11)):

$$\frac{d\mathcal{E}_s}{dz} = \frac{2\pi i\omega_s}{\mu_s c}\left[\chi_s|\mathcal{E}_L|^2\mathcal{E}_s + (\chi_s\chi_a^*)^{1/2}\mathcal{E}_L^2\mathcal{E}_a^* e^{i\Delta k z}\right] - \alpha_s\mathcal{E}_s \tag{7.18}$$

$$\frac{d\mathcal{E}_a}{dz} = \frac{2\pi i\omega_a}{\mu_a c}\left[\chi_a|\mathcal{E}_L|^2\mathcal{E}_a + (\chi_s^*\chi_a)^{1/2}\mathcal{E}_L^2\mathcal{E}_s^* e^{i\Delta k z}\right] - \alpha_a\mathcal{E}_a \quad . \tag{7.19}$$

Again, co-linear wave propagation has been assumed, and in addition linear attenuation in the medium has been allowed for by writing {13,24}:

$$\frac{\epsilon_s \omega_s^2}{c^2} = k_s^2 + 2i\alpha_s k_s \tag{7.20}$$

$$\frac{\epsilon_a \omega_a^2}{c^2} = k_a^2 + 2i\alpha_a k_a \tag{7.21}$$

with $\mu_S = (\mathrm{Re}\ \epsilon_S)^{\frac{1}{2}}$, $\mu_a = (\mathrm{Re}\ \epsilon_a)^{\frac{1}{2}}$. In cases where the wave vector mismatch:

$$\Delta k = 2k_L - k_s - k_a \tag{7.22}$$

is not small, the terms in (7.18) and (7.19) which couple the Stokes and anti-Stokes waves will not contribute significantly to the generation of power at the Stokes and anti-Stokes frequencies (compare for example equations ((4.16)-(4.18)). Neglecting these terms in this case, and making the further assumption that the laser beam intensity is not significantly altered as a result of the scattering processes, i.e. treating ε_L as constant, one may integrate (7.18) directly to obtain:

$$\mathcal{E}_s(L) = \mathcal{E}_s(0)\ \exp\left[\left(\frac{2\pi i\,\omega_s}{\mu_s c}\chi_s|\mathcal{E}_L|^2 - \alpha_s\right)L\right]. \tag{7.23}$$

As before, L is the thickness of the medium in the beam direction.

Noting {24} that the imaginary part of the complex Raman susceptibility is negative, one may define the Stokes power gain per unit length of the scattering medium as {24,99}:

$$g_s = -\frac{4\pi\,\omega_s|\mathcal{E}_L|^2}{\mu_s c}\,\mathrm{Im}\,\chi_s. \tag{7.24}$$

Provided that this quantity exceeds $2\alpha_S$, exponential growth of the Stokes scattered wave with distance, takes place. This is the basis of the Raman laser {112}, which is the most common application of stimulated Raman scattering. It is easily shown that if the scattering occurs from a single vibrational line, then Im χ_a = - Im χ_S for the anti-Stokes frequency ($\omega_a = 2\omega_L - \omega_S$), so that for anti-Stokes light the gain is always negative. It follows from equation (7.19) that anti-Stokes radiation can be generated once the intensity of the Stokes wave has attained a large value, provided that Δk is non-zero. This is, however, a parasitic effect which reduces the gain substantially, and in conventional Raman laser design, an attempt is usually made to eliminate it.

The more general cases of Stokes and anti-Stokes wave generation in non-colinear wave propagation, are treated in Refs. (24), (99) and (109).

The theory of stimulated Brillouin scattering is not very different from that of stimulated Raman scattering. The relevant equations containing $\chi^{(3)}$ follow from a wave equation for the sound wave, and are difficult to solve except when the sound wave is highly damped. In this case, however, the Stokes and anti-Stokes light waves are produced by sound waves which move in opposite directions. These two waves are related to different normal coordinates of the system, and hence the Stokes and anti-Stokes processes are uncoupled. In the case of stimulated Raman scattering, however, the Stokes and anti-Stokes waves arise from a driven molecular vibration described by a single vibrational coordinate, and are coupled except when the anti-Stokes wave is suppressed by a phase mismatch.

(f) CALCULATION OF NON-LINEAR SUSCEPTIBILITIES (Continued)

In the calculation of third-order non-linear susceptibilities, there are two categories which can be labelled as fast and slow, as stated in Chapter VI, section (b). Fast non-linearities are related to non-resonant electronic motion as described in Chapter 3, and the response times are shorter than optical periods. Slow non-linearities may be further subdivided into: (i) near-resonant electronic interactions such as those used in sodium vapour to demonstrate optical bistability {85} and four-wave mixing {113}, and (ii) all interactions, whether resonant or not, which are related to nuclear motion. In the latter case, the relevant frequency response of the mechanism can range from the near-infrared down to virtual frequency independence. In the slow non-linearities, some interchange of energy with the light wave takes place and the non-linearity has a finite response time; in the case of fast non-linearities, the role of the medium is strictly catalytic.

The treatment of a slow non-linearity depends upon whether one is dealing with transitions which are optically allowed (single-photon) or optically forbidden (two-photon). In the former case one can use a model such as that outlined in Chapter 3, provided it is borne in mind that one is infringing on a domain in which a quantum-mechanical treatment of matter is needed. In the case of optically forbidden transitions, a different approach is required, since the non-linearity enters to lowest order in the interaction. The medium then responds as a damped harmonic oscillator to a force that is quadratic in the field.

The susceptibility is developed as follows {24}. Let x_k denote a small displacement of a normal coordinate of a molecule from its equilibrium position. The polarizability of the molecule is expanded as:

$$\alpha_{ij} = \alpha_{ij}^{(0)} + \sum_k \alpha_{ijk}^{(1)} x_k \tag{7.25}$$

where the corresponding induced dipole moment is given by:

$$d_i = \sum_j \alpha_{ij} \tilde{E}_j \tag{7.26}$$

The component of the driving force acting on the molecule which corresponds to x_k, is:

$$\frac{\partial}{\partial x_k}\left(\sum_i d_i \tilde{E}_i\right) = \sum_{ij} \alpha_{ijk}^{(1)} \tilde{E}_i \tilde{E}_j \tag{7.27}$$

and therefore the differential equation obeyed by x_k may be written as for a damped harmonic oscillator in the form:

$$\frac{d^2 x_k}{dt^2} + \frac{\Gamma_k}{M_k}\frac{dx_k}{dt} + \omega_k^2 x_k = \frac{\alpha_{ijk}^{(1)}}{M_k} \tilde{E}_i \tilde{E}_j \tag{7.28}$$

which may be compared with equation (3.14). In equation (7.28), summation over repeated indices (ij) is implied.

There are, in general, many driving terms in the product $\tilde{E}_i\tilde{E}_j$, and the coordinate responds independently to each of these. The dyadic product of the electric fields may be written as:

$$\tilde{E}_i \tilde{E}_j = \sum_{K\Omega} G_{ij} \exp[i(\underset{\sim}{K}\cdot\underset{\sim}{r} - \Omega t)] \tag{7.29}$$

where $\underset{\sim}{K} = \underset{\sim}{k}_\beta \pm \underset{\sim}{k}_\gamma$ and $\Omega = \omega_\beta \pm \omega_\gamma$, and the symbols β, γ run over all the frequency components of the total electric field. The amplitude G_{ij} is then the appropriate dyadic product of the field amplitudes. Similarly, with:

$$x_k = \sum_{K\Omega} X_k(K,\Omega) \exp[i(\underset{\sim}{K}\cdot\underset{\sim}{r} - \Omega t)] \tag{7.30}$$

one has from (7.28):

$$X_k(K,\Omega) = \frac{1}{\left[\omega_k^2 - \Omega^2 - i\dfrac{\Omega\Gamma_k}{M_k}\right]} \frac{\alpha^{(1)}_{\ell mk} G_{\ell m}}{M_k} . \tag{7.31}$$

The non-linear polarization is constructed by summing the second term in equation (7.25) over all atoms. In the case of scattering from sound waves (Brillouin scattering), it is more convenient to start with x_k denoting a collective normal coordinate, which requires suitable generalization {56} of equation (7.28).

With the effective non-linear polarizability defined as the sum over coordinates and atoms per unit volume of the $\alpha^{(1)}_{ijk}$ term in (7.25), one has from equations (7.30) and (7.31):

$$\alpha^{NL}_{ij} = \sum_{K\Omega} \sum_{ka} \frac{\alpha^{(1)a}_{ijk} \; \alpha^{(1)a}_{\ell mk}}{\left[\omega_k^2 - \Omega^2 - i\dfrac{\Omega\Gamma_k}{M_k}\right]} \left(\frac{G_{\ell m}}{M_k}\right) \exp[i(\underset{\sim}{K}\cdot\underset{\sim}{r} - \Omega t)] \tag{7.32}$$

where the superscript a is used to denote the atoms. With the $G_{\ell m}$ written as explicit functions of the field amplitudes, equation (6.3) can be used to generate the coefficients $\chi^{(3)}_{\ell mij}$:

$$\tilde{\chi}^{(3)}_{\ell mij} = \frac{\partial}{\partial \tilde{E}_\ell} \frac{\partial}{\partial \tilde{E}_m} \alpha^{NL}_{ij} . \tag{7.33}$$

The sums in equation (7.32) have been written explicitly in order to emphasize the fact that α^{NL}_{ij} is a sum of phased structures impressed on the medium by the light fields. When these scatter light waves, the process is termed <u>stimulated scattering</u>; whereas when the normal coordinates x_k have a random structure arising from thermal or quantum excitation, the process is termed spontaneous scattering.

The non-linear processes are classified according to the spontaneous processes with which they are associated, and are identified by the values of Ω and K that make the resonance denominators in equation (7.32) large enough for the effect to be observable.

In Rayleigh scattering, ω_k is zero and the value of K is arbitrary insofar as the denominators do not depend upon it. Resonance occurs at $\Omega = 0$, and therefore the phased arrays in (7.32) are independent of time. Thus the scattered waves always have the same frequency as the incident waves. When $K \neq 0$ the susceptibility is in the form of a static Bragg grating, which leads to the non-linear process called real-time holography or degenerate four-wave mixing {66,70}. In the case of $K = 0$, the susceptibility is independent of space, and describes a non-linear index of refraction which gives rise to self-focussing {55,56,81-83}.

Raman scattering is characterized by a non-zero ω_k and arbitrary K. In this case $\Omega \approx \omega_k \neq 0$, so that the susceptibility is time-dependent. The scattered frequencies are shifted by $\pm\,\Omega$ from the incident frequencies. There are two cases that arise within this category, depending upon whether the optical frequencies ω_β, ω_γ are larger or smaller than the phonon frequency ω_k. In the first case, we have:

$$\omega_\beta - \omega_\gamma = \Omega \approx \omega_k \tag{7.34}$$

which describes coherent Raman scattering. As we have already seen, this comprises several possibilities: the frequency of the scattered light may be greater (anti-Stokes scattering) or less (Stokes scattering) than that of the incident light, and multiple scattering may take place (higher-order Stokes and anti-Stokes scattering). These processes are also reviewed in Ref. {112}.

The case where both frequencies are smaller than ω_k leads to a resonance condition:

$$\omega_\beta + \omega_\gamma = \Omega \approx \omega_k . \tag{7.35}$$

Then, energy from both light waves is absorbed by the medium, in the process known as two-photon absorption {3,24}. A special case of this occurs when:

$$\omega_\beta = \omega_\gamma = \omega = \Omega/2 . \tag{7.36}$$

This two-photon resonance can then be exploited either in the sum process:

$$\Omega + \omega = 3\omega \qquad (7.37)$$

which leads to two-photon resonantly enhanced tripling {114}, or to the difference process:

$$\Omega - \omega = \omega \qquad (7.38)$$

which gives a two-photon resonant enhancement of real-time holography, a subject of current investigation.

In Brillouin scattering, ω_k is non-zero, and $\Omega/|\tilde{K}|$ must be approximately equal to the speed of sound in the medium. After suitable generalization of equation (7.28) as in Ref.{56}, an expression is obtained for the non-linearity which gives rise to stimulated Brillouin scattering, in which the scattered fields are shifted in frequency with respect to the incident field, and the shift varies with scattering angle. In both Raman and Brillouin scattering processes, where energy is exchanged with the medium, the Manley-Rowe relations should be generalized to allow for the coherently driven vibrations in the medium.

The new developments termed coherent anti-Stokes (CARS) and coherent Stokes (CSRS) Raman spectroscopy, are treated in Ref.{112}.

ACKNOWLEDGEMENTS

One of us (JDH) wishes to thank the librarian-in-charge of the Engineering and Science Library, U.C.T., Miss Ann Borland and her staff, for their kind assistance with this project. We are indebted to Miss Lesley Jennings of the Physics Department, U.C.T., for her painstaking work of typing the manuscript, and to Mrs. Joan Parsons for typing the corrections to the first draft.

APPENDIX I

MACROSCOPIC AND LOCAL QUADRATIC SUSCEPTIBILITIES IN ANISOTROPIC CRYSTALS

The local field at the site of the ith atom in the unit cell may be written {14} as:

$$\tilde{\underset{\sim}{E}}^{(i)} = \underset{\sim}{E} + \sum_j \underset{\approx}{L}^{(ij)} \cdot \left(\underset{\sim}{P}^{L(j)} + \underset{\sim}{P}^{NL(j)} \right) \tag{A1}$$

and the polarizations of the ith atom at angular frequency $\omega_3 = \omega_1 + \omega_2$ are related to the local fields at ω_1 and ω_2 by:

$$\underset{\sim}{P}^{L(i)}(\omega_3) = \tilde{\underset{\approx}{X}}^{(1)}(\omega_3) \cdot \tilde{\underset{\sim}{E}}^{(i)}(\omega_3) \tag{A2}$$

$$\underset{\sim}{P}^{NL(i)}(\omega_3) = \tilde{\underset{\approx}{X}}^{(2)}(\omega_3\,;\,\omega_1, \omega_2) : \tilde{\underset{\sim}{E}}^{(i)}(\omega_1)\, \tilde{\underset{\sim}{E}}^{(i)}(\omega_2)\,. \tag{A3}$$

Note that $\underset{\approx}{L}^{(ij)}$, like $\tilde{\underset{\approx}{X}}^{(1)}$, is a 3×3 matrix.

From equations (A1) and (A2), it follows that:

$$\underset{\sim}{P}^{L(i)} = \tilde{\underset{\approx}{X}}^{(1)} \cdot \underset{\sim}{E} + \sum_j \tilde{\underset{\approx}{X}}^{(1)} \cdot \underset{\approx}{L}^{(ij)} \cdot \left(\underset{\sim}{P}^{L(j)} + \underset{\sim}{P}^{NL(j)} \right). \tag{A4}$$

Now with the definition :

$$\underset{\approx}{M}^{(ij)} = \delta_{ij}\, \underset{\approx}{I} - \tilde{\underset{\approx}{X}}^{(1)} \cdot \underset{\approx}{L}^{(ij)} \tag{A5}$$

it follows that :

$$\sum_j \underset{\approx}{M}^{(ij)} \cdot \underset{\sim}{P}^{L(j)} = \tilde{\underset{\approx}{X}}^{(1)} \cdot \underset{\sim}{E} + \underset{\sim}{P}^{NL(i)} - \sum_j \underset{\approx}{M}^{(ij)} \cdot \underset{\sim}{P}^{NL(j)}\,. \tag{A6}$$

Now $\underset{\approx}{M}^{(ij)}$ in component form would be a $3N \times 3N$ matrix, where N is the number of atomic sites per unit cell. This "supermatrix" clearly has an inverse, which enables a new set of 3×3 matrices $\underset{\approx}{R}^{(ij)}$ to be defined such that:

$$\sum_j \underset{\approx}{M}^{(ij)} \cdot \underset{\approx}{R}^{(jk)} = \sum_j \underset{\approx}{R}^{(ij)} \cdot \underset{\approx}{M}^{(jk)} = \delta_{ik} \underset{\approx}{I} . \tag{A7}$$

From equation (A6) one finds, on applying the inverse matrix:

$$\underset{\sim}{P}^{L(k)} + \underset{\sim}{P}^{NL(k)} = \sum_i \underset{\approx}{R}^{(ki)} \cdot \underset{\approx}{\tilde{\chi}}^{(1)} \cdot \underset{\sim}{E} + \sum_i \underset{\approx}{R}^{(ki)} \cdot \underset{\sim}{P}^{NL(i)} . \tag{A8}$$

On summing over all atoms in the unit cell, one therefore obtains:

$$\underset{\sim}{P} = \underset{\sim}{P}^{L} + \underset{\sim}{P}^{NL} = \sum_{ik} \underset{\approx}{R}^{(ki)} \cdot \underset{\approx}{\tilde{\chi}}^{(1)} \cdot \underset{\sim}{E} + \sum_{ik} \underset{\approx}{R}^{(ki)} \cdot \underset{\sim}{P}^{NL(i)} \tag{A9}$$

whence by inspection:

$$\frac{(\underset{\approx}{\epsilon} - \underset{\approx}{I})}{4\pi} = \sum_{ik} \underset{\approx}{R}^{(ki)} \cdot \underset{\approx}{\chi}^{(1)} \tag{A10}$$

and

$$\underset{\sim}{D}(\omega_3) = \underset{\approx}{\epsilon}(\omega_3) \cdot \underset{\sim}{E}(\omega_3) + 4\pi \sum_{ik} \underset{\approx}{R}^{(ki)}(\omega_3) \cdot \underset{\sim}{P}^{NL(i)}(\omega_3) \quad . \tag{A11}$$

We now write the last term in (A11), which is the effective macroscopic non-linear polarization, in terms of the microscopic non-linear polarizability tensors. Provided that $P^{NL} \ll P^{L}$, equations (A1) and (A2) yield:

$$\sum_j (\delta_{ij} \underset{\approx}{I} - \underset{\approx}{L}^{(ij)} \cdot \underset{\approx}{\tilde{\chi}}^{(1)}) \cdot \underset{\sim}{\tilde{E}}^{(j)} = \underset{\sim}{E} . \tag{A12}$$

Now, energy considerations require the symmetry of polarizability ($\tilde{\underset{\approx}{\chi}}^{(1)}$) and Lorentz ($\underset{\approx}{L}$) tensors. With T ≡ transpose,

$$\tilde{\underset{\approx}{\chi}}^{(i)} = (\tilde{\underset{\approx}{\chi}}^{(i)})^T$$

$$\underset{\approx}{L}^{(ij)} = \underset{\approx}{L}^{(ji)} = (\underset{\approx}{L}^{(ji)})^T$$

equation (A12) can be written as:

$$\sum_j (\underset{\approx}{M}^{(ji)})^T \cdot \tilde{\underset{\sim}{E}}^{(j)} = \underset{\sim}{E} . \tag{A13}$$

By means of the transpose of (A7), this can be written as:

$$\tilde{\underset{\sim}{E}}^{(k)} = \sum_i (\underset{\approx}{R}^{(ik)})^T \cdot \underset{\sim}{E} . \tag{A14}$$

On combining (A14) with (A11) and (A3), the effective macroscopic non-linear polarization is found to be:

$$\sum_{ik} \underset{\approx}{R}^{(ki)}(\omega_3) \cdot \tilde{\underset{\approx}{\chi}}^{(2)}(\omega_3; \omega_1, \omega_2) : \left[\sum_j (\underset{\approx}{R}^{(ji)}(\omega_1))^T \cdot \underset{\sim}{E}(\omega_1)\right]$$

$$\times \left[\sum_l (\underset{\approx}{R}^{(li)}(\omega_2))^T \cdot E(\omega_2)\right] .$$

This last expression may be put equal to $\underset{\approx}{\chi}^{(2)} : \underset{\sim}{E}(\omega_1)\underset{\sim}{E}(\omega_2)$, and on defining:

$$\underset{\approx}{N}^{(i)}(\omega) = \sum_l (\underset{\approx}{R}^{li}(\omega))^T , \tag{A15}$$

one has for the third-rank macroscopic polarizability tensor:

$$\underset{\approx}{\chi}^{(2)}(\omega_3 ; \omega_1, \omega_2) = \sum_i \tilde{\underset{\approx}{\chi}}^{(2)}(\omega_3; \omega_1, \omega_2) \vdots \underset{\approx}{N}^{(i)}(\omega_3) \underset{\approx}{N}^{(i)}(\omega_1) \underset{\approx}{N}^{(i)}(\omega_2) . \tag{A16}$$

In component form (a b c),

$$\chi^{(2)}_{abc}(\omega_3 ; \omega_1, \omega_2) = \sum_i \sum_{def} \tilde{\chi}^{(2)}_{def}(\omega_3; \omega_1, \omega_2) N^{(i)}_{da}(\omega_3) N^{(i)}_{eb}(\omega_1) N^{(i)}_{fc}(\omega_2) . \tag{A17}$$

Now it has been shown in the text that the $\tilde{\chi}^{(2)}_{ijk}$ satisfy the set of permutation symmetry relations {14}:

$$\tilde{\chi}^{(2)}_{ijk}(\omega_3;\omega_1,\omega_2) = \tilde{\chi}^{(2)}_{jki}(\omega_1;-\omega_2,\omega_3) = \tilde{\chi}^{(2)}_{kij}(\omega_2;\omega_3,-\omega_1) \qquad \text{(A18)}$$

or, in other words, the frequencies may be permuted at will provided the Cartesian indices are simultaneously permuted so that a given frequency is always associated with the same index. Clearly, the macroscopic non-linear susceptibility $\underset{\approx}{\chi}^{(2)}$ satisfies the same permutation symmetry relationships as in (A18). In addition, $\underset{\approx}{\chi}^{(2)}$ has the point symmetry properties of the crystal lattice as a whole, whereas the individual non-linear polarizabilities $\underset{\approx}{\tilde{\chi}}^{(2)}$ have the symmetry properties of the individual lattice sites {14}.

APPENDIX II

CRYSTAL CLASSES EXHIBITING QUADRATIC SUSCEPTIBILITY

There are seven crystal systems comprising thirty-two crystal classes {17}, but of these only twenty lack a centre of inversion and thus possess quadratic susceptibility. These are listed in order of increasing symmetry, and for each class (in the Schönfliess description {16}) are indicated the non-zero tensor elements {4}. The relationships between the components are stated with the assumption of the Kleinman symmetry relation (3.30) in addition to the general symmetry relation (3.28). For brevity, non-zero components which may be obtained from those already stated merely by permutation of the last two indices (as in (3.29)), are omitted.

1. Triclinic System (Class C_1)

The crystals of this class lack symmetry elements, and thus all 18 (10 independent) tensor elements χ_{ijk} are non-zero.

2. Monoclinic System (Class C_2)

These crystals have a twofold symmetry axis (z); rotation through 180° about this axis leaves the crystal invariant. The non-zero elements are $\chi_{113} = \chi_{311}$; $\chi_{223} = \chi_{322}$; $\chi_{123} = \chi_{321} = \chi_{213}$, and χ_{333} (4 independent components).

3. Monoclinic System (Class C_s)

These crystals possess a plane of symmetry (xy), reflection in which leaves the crystal invariant. The non-zero elements are those in which the index 3 appears only in pairs or not at all. χ_{111}; $\chi_{122} = \chi_{212}$; $\chi_{133} = \chi_{313}$; $\chi_{112} = \chi_{211}$; χ_{222}; $\chi_{233} = \chi_{323}$ (6 independent components).

4. Orthorhombic System (Class D_2)

Crystals of this class have three mutually perpendicular twofold axes (x,y,z). Rotation through 180° about any of these axes reverses the direction of the two remaining axes. The only non-zero components of χ_{ijk} are those in which all three indices are different.
$\chi_{123} = \chi_{312} = \chi_{213}$ (1 independent component).

5. Orthorhombic System (Class C_{2v})

These crystals have one twofold axis (z) parallel to two planes of symmetry, xz and yz. The components of χ must be invariant under rotations about the z axis and reflection in the xz and/or yz plane. The non-zero components χ_{ijk} are $\chi_{311} = \chi_{113}$; $\chi_{322} = \chi_{223}$; χ_{333} (3 independent components).

6. Tetragonal System (Class S_4)

Crystals of this class are invariant under 90° rotation about the z axis followed by reflection in the xy plane. The non-zero components χ_{ijk} are $\chi_{312} = \chi_{123} = \chi_{231}$; $\chi_{311} = \chi_{131} = -\chi_{232} = -\chi_{322}$ (2 independent components).

7. Tetragonal System (Class C_4)

These crystals are invariant under 90° rotations about the fourfold z axis. The non-zero elements χ_{ijk} are χ_{333}; $\chi_{311} = \chi_{113} = \chi_{223} = \chi_{322}$ (2 independent components).

8. Tetragonal System (Class D_{2d})

This class is characterized by three mutually perpendicular twofold axes (x,y,z) and two planes of symmetry through the z axis that bisect the angles between the x and y axes. Thus, in addition to the symmetry elements in class D_2, the components χ_{ijk} must also be invariant under permutation of the indices 1 and 2 (independent of Kleinman symmetry). Therefore the only non-zero components of χ_{ijk} are $\chi_{123} = \chi_{213} = \chi_{321}$ (1 independent component).

9. Tetragonal System (Class D_4)

Crystals of this class have a fourfold axis of symmetry (z) and four twofold axes in the xy plane. In addition to the symmetry elements of class D_2, only 90° rotation about the z axis need be considered. Because of this transformation the component χ_{312} must vanish, the Kleinman symmetry condition therefore requiring that all components χ_{ijk} be zero. Without the Kleinman condition one would have $\chi_{123} = -\chi_{213}$. However, the fact that no crystal of this class possessing non-linear properties is known, is a strong indication of the practical usefulness of Kleinman's conjecture. (0 independent components)

10. Tetragonal System (Class C_{4v})

Crystals of this class have a fourfold axis of symmetry z parallel to four planes of symmetry, one of which is the plane xz. In addition to the symmetry elements of class C_4, the components must be invariant under reflection in these planes. The non-zero components are χ_{333}; $\chi_{311} = \chi_{113} = \chi_{223} = \chi_{322}$ (2 independent components).

11. Trigonal or Rhombohedral System (Class C_3)

These crystals are invariant under rotations of 120° about the threefold symmetry axis z. The non-zero components are χ_{333}; $\chi_{113} = \chi_{223} = \chi_{311} = \chi_{322}$; $\chi_{111} = -\chi_{122} = -\chi_{212}$; $\chi_{211} = -\chi_{222} = \chi_{112}$ (4 independent components).

12. Trigonal System (Class D_3)

Crystals of this class have a threefold axis of symmetry (z axis) and three twofold axes of symmetry in the xy plane, one of the latter being the x axis. The non-zero components are $\chi_{111} = -\chi_{122} = -\chi_{212}$ (1 independent component).

13. Trigonal System (Class C_{3v})

These crystals have a threefold symmetry axis z parallel to three planes of symmetry, one of these being the plane xz. The non-zero components are χ_{333}; $\chi_{113} = \chi_{223} = \chi_{311} = \chi_{322}$; $\chi_{111} = -\chi_{122} = -\chi_{212}$ (3 independent components).

14. Hexagonal System (Class C_{3h})

Crystals of this class have a threefold axis of symmetry (z) and a plane of symmetry (xy). The non-zero components are $\chi_{122} = \chi_{212} = -\chi_{111}$; $\chi_{112} = \chi_{211} = -\chi_{222}$ (2 independent components).

15. Hexagonal System (Class D_{3h})

These crystals have a threefold axis of symmetry z and three twofold axes, one of which coincides with the x axis, as well as a plane of symmetry (xy). The non-zero components are $\chi_{122} = \chi_{212} = -\chi_{111}$ (1 independent component).

16. Hexagonal System (Class C_6)

Crystals of this class are invariant under 60° rotations about a sixfold symmetry axis (z). The non-zero components are χ_{333}; $\chi_{113} = \chi_{223} = \chi_{311} = \chi_{322}$ (2 independent components).

17. Hexagonal System (Class D_6)

These crystals have a sixfold symmetry axis (z) and six twofold axes lying in the plane xy. As in the case of the class D_4, non-zero components can only exist in violation of Kleinman's conjecture. These are $\chi_{123} = -\chi_{213}$ (0 independent components).

18. Hexagonal System (Class C_{6v})

Crystals of this class have a sixfold symmetry axis (z) parallel to six symmetry planes, one of which is the xz plane. The non-zero components are χ_{333}; $\chi_{113} = \chi_{223} = \chi_{311} = \chi_{322}$ (2 independent components).

19. Cubic System (Class T)

These crystals have three orthogonal twofold symmetry axes x,y,z and four diagonal threefold axes. The non-zero components are $\chi_{123} = \chi_{312} = \chi_{231}$ (1 independent component).

20. Cubic System (Class T_d)

In addition to the symmetry elements of class T, crystals of this class have six symmetry planes, each of which contains two of the four diagonal threefold axes. The non-zero components are the same as for class T: $\chi_{123} = \chi_{312} = \chi_{231}$ (1 independent component).

The above classes are summarized in Table II.1, in which the reader will also find listed some typical crystals of importance in non-linear optics. Crystals with point symmetry D_4 and D_6 are forbidden by the Kleinman condition {15} from exhibiting second-harmonic generation and related effects.

TABLE II.1

Crystal Class {16}	Type {9}	(non-zero) Independent Elements $\chi^{(2)}_{ijk}$ {4}	Typical Examples {4,16}
Triclinic (C_1)	Biaxial	10	KIO_3 (potassium iodate)
Monoclinic (C_2)	Biaxial	4	$(CH_2NH_2COOH)_3H_2SO_4$) (triglycine sulphate)
Monoclinic (C_s)	Biaxial	6	
Orthorhombic (D_2)	Biaxial	1	Rochelle salt
Orthorhombic (C_{2v})	Biaxial	3	$NaNO_2$ (sodium nitrite)
Tetragonal (S_4)	Uniaxial	2	$C(CH_2OH)_4$
Tetragonal (C_4)	Uniaxial	2	
Tetragonal (D_{2d})	Uniaxial	1	KDP (KH_2PO_4) ADP ($NH_4H_2PO_4$)
Tetragonal (D_4)	Uniaxial	0	
Tetragonal (C_{4v})	Uniaxial	2	$BaTiO_3$ (barium titinate)
Trigonal (C_3) (Rhombohedral)	Uniaxial	4	
Trigonal (D_3)	Uniaxial	1	SiO_2 (α-quartz); HgS (cinnabar)
Trigonal (C_{3v})	Uniaxial	3	$LiNbO_3$ (lithium niobate)
Hexagonal (C_{3h})	Uniaxial	2	
Hexagonal (D_{3h})	Uniaxial	1	
Hexagonal (C_6)	Uniaxial	2	$LiIO_3$ (lithium iodate)
Hexagonal (D_6)	Uniaxial	0	
Hexagonal (C_{6v})	Uniaxial	2	CdS (cadmium sulphide)
Cubic (T)	Isotropic	1	$NaC\ell O_3$ (sodium chlorate)
Cubic (T_d)	Isotropic	1	ZnS (zinc sulphide) GaAs (gallium arsenide)

APPENDIX III

THE MANLEY-ROWE RELATIONS

Because of their importance in sum- and difference-frequency generation in a non-linear medium, the original derivation of the Manley-Rowe relations [25,26] is summarized here, with the aim both of clarifying the underlying assumptions upon which they are based and of illustrating the strong analogy that can be drawn between electrical circuit theory and laser physics.

Based upon unpublished work by Manley some twenty years earlier, these relations were derived [25] with reference to a hysteresisless and loss-free non-linear reactance (specifically a non-linear capacitor). The characteristic of the non-linear capacitor is given by specifying the voltage as some arbitrary function of the charge:

$$V = f(q) \tag{B1}$$

where f(q) is assumed to be single-valued. The situation is considered where two generators of fundamental (incommensurable) frequencies $f_1 = \omega_1/2\pi$ and $f_0 = \omega_0/2\pi$ are connected to the non-linear capacitor; as a result, all of the frequencies:

$$f_{m,n} = mf_1 + nf_0$$

will be present in the circuit, where m and n take on all integral values, positive, negative and zero. The charge q flowing into the non-linear capacitor may be written as the double Fourier series:

$$q = \sum_{m=-\infty}^{+\infty} \sum_{n=-\infty}^{+\infty} Q_{m,n} \exp[i(mx + ny)] \tag{B2}$$

where $x = \omega_1 t$, $y = \omega_0 t$. Since q is real,

$$Q_{m,n} = Q^*_{-m,-n} \, . \tag{B3}$$

The variables x and y are initially considered as independent, taking on any values in the xy plane, and are subsequently replaced by the above substitution in terms of angular frequency, so that only values along a straight line in the xy plane appear in the final results (method of calculating modulation products {27}). The current flowing into the non-linear capacitor is therefore given by:

$$i = \sum_{m=-\infty}^{+\infty} \sum_{n=-\infty}^{+\infty} I_{m,n} \exp[i(mx+ny)] \tag{B4}$$

where

$$I_{m,n} = i(m\omega_1 + n\omega_0) Q_{m,n} \tag{B5}$$

and

$$I_{m,n} = I^*_{-m,-n} . \tag{B6}$$

Now

$$v = f(q) = f[q(x,y)] = F(x,y) \tag{B7}$$

and since the voltage is now a single-valued function, periodic in x and y, it may also be expanded in a double Fourier series:

$$v = \sum_{m=-\infty}^{+\infty} \sum_{n=-\infty}^{+\infty} V_{m,n} \exp[i(mx+ny)] \tag{B8}$$

$$V_{m,n} = V^*_{-m,-n} \tag{B9}$$

where the coefficients are given by:

$$V_{m,n} = \frac{1}{4\pi^2} \int_0^{2\pi} dy \int_0^{2\pi} dx \; F(x,y) \exp[-i(mx+ny)] . \tag{B10}$$

On multiplying both sides of (B10) by im $Q^*_{m,n}$ and summing over the full range of m and n, one obtains:

$$\sum_{m=-\infty}^{+\infty}\sum_{n=-\infty}^{+\infty} im\, Q^*_{m,n} V_{m,n} = \frac{1}{4\pi^2}\int_0^{2\pi} dy \int_0^{2\pi} dx\, F(x,y)$$

$$\times \sum_{m=-\infty}^{+\infty}\sum_{n=-\infty}^{+\infty} im\, Q^*_{m,n} \exp[-i(mx+ny)] \,. \tag{B11}$$

From the derivative of (B2) with respect to x, and using (B3) and (B5), one finds that

$$\sum_{m=-\infty}^{+\infty}\sum_{n=-\infty}^{+\infty} \frac{m V_{m,n} I^*_{m,n}}{m\omega_1 + n\omega_0} = \frac{1}{4\pi^2}\int_0^{2\pi} dy \int_0^{2\pi} dx\, \frac{\partial q}{\partial x} F(x,y)$$

$$= \frac{1}{4\pi^2}\int_0^{2\pi} dy \int_{q(0,y)}^{q(2\pi,y)} f(q)\, dq. \tag{B12}$$

The limits of the second integral indicate that the variation of q is determined by allowing x to vary from 0 to 2π, holding y constant.

After a similar analysis with the roles of x and y interchanged, the following result is readily obtained:

$$\sum_{m=-\infty}^{+\infty}\sum_{n=-\infty}^{+\infty} \frac{n V_{m,n} I^*_{m,n}}{m\omega_1 + n\omega_0} = \frac{1}{4\pi^2}\int_0^{2\pi} dx \int_{q(x,0)}^{q(x,2\pi)} f(q)\, dq \tag{B13}$$

where the limits of the second integral indicate that the variation of q is determined by allowing y to vary from 0 to 2π with x constant.

While all of the above equations have included both positive and negative frequencies, there can be no physical distinction between a positive and a negative frequency of the same magnitude. In order to relate the quantities $V_{m,n}I^*_{m,n}$ to the average powers associated with the various frequencies, pairs of terms on the left-hand sides of (B12) and (B13) should be appropriately combined. This is achieved by defining the time-averaged vector power:

$$S_{m,n} = P_{m,n} + iX_{m,n} = 2V_{m,n} I^*_{m,n} \tag{B14}$$

in terms of the real ($P_{m,n}$) and reactive ($X_{m,n}$) powers flowing into the non-linear element at a particular frequency. This frequency includes both the positive and negative components $\pm|mf_1 + nf_0|$. Thus with

$$S_{m,n} = S^*_{-m,-n} \tag{B15}$$

we have

$$P_{m,n} = V_{m,n} I^*_{m,n} + V^*_{m,n} I_{m,n} = P_{-m,-n} \,. \tag{B16}$$

We note in addition that since q is periodic in both x and y, and f(q) is single-valued, the integrals on q on the right-hand sides of equations (B12) and (B13) must be identically zero for all values of y and x, respectively. From (B12) and (B13), one therefore has (grouping together corresponding positive and negative frequency components and taking care not to count particular contributions twice):

$$\sum_{m=0}^{+\infty} \sum_{n=-\infty}^{+\infty} \frac{m P_{m,n}}{m\omega_1 + n\omega_0} = 0 \tag{B17}$$

$$\sum_{m=-\infty}^{+\infty} \sum_{n=0}^{+\infty} \frac{n P_{m,n}}{m\omega_1 + n\omega_0} = 0 \tag{B18}$$

These are the <u>Manley-Rowe relations</u>, which must be satisfied independently by the powers flowing into the non-linear capacitor at the various frequencies. A later derivation by the same authors {26} is even simpler mathematically than the approach {25} presented above. These relations clearly apply equally well in the case of a non-linear inductor.

With hysteresis present, corresponding relations may be derived provided that there are no departures from the principal single-frequency hysteresis loop established by the local oscillator (frequency f_1) alone. Thus, the path traversed in the q-v plane is assumed to be, at most, double-valued. Hence, the level of the local oscillator at the frequency f_1 is assumed to be large enough to drive the non-linear element well into saturation, so that the q-v characteristic will be single-valued over appreciable regions near its ends. The signal levels at the other frequencies are assumed to be small compared with the local oscillator level.

With these restrictions in mind, we now return to our original analysis, and regard f(q) as a double-valued function of q. In the following, the lower branch of this function must be chosen when the charge at the local oscillator frequency f_1 is increasing, the upper branch when it is decreasing. Thus with $Q_{1,0} = Q_{-1,0}$ so that both are real (from (B3)) and the local oscillator charge at frequency f_1 a cosine function, then for $0 < x < \pi$ the upper branch of the q-v characteristic must be chosen, while for $\pi < x < 2\pi$ the lower branch is selected. In (B7), while v is a double-valued function of q, it is a single-valued function of x and y, and therefore satisfies the requirement for expansion into a double Fourier series. Therefore, hysteresis does not alter the analysis leading up to (B12) and (B13).

The integrals on the right-hand sides of these equations must now be re-considered. With y constant in (B12), q will travel entirely around the hysteresis loop as x goes from 0 to 2π, and therefore the integral on q yields a constant, independent of y, and equal to the area of the hysteresis loop

$$h = \oint f(q)\,dq \; . \tag{B19}$$

This is also equal to the energy dissipated in the non-linear reactor in one transit around the hysteresis loop, corresponding to an average power dissipation

$$P = f_1 h \; . \tag{B20}$$

On the other hand in (B13), with x constant, q will travel back and forth along the same branch of the hysteresis loop, returning to its initial value without enclosing any area as y goes from 0 to 2π. Thus, this integral on q is identically equal to zero for all x, and the right-hand side of (B13) remains equal to zero, as in the case of no hysteresis. Equations (B12) and (B13) therefore yield the <u>modified Manley-Rowe relations</u>:

$$\sum_{m=0}^{+\infty} \sum_{n=-\infty}^{+\infty} \frac{m P_{m,n}}{m\omega_1 + n\omega_0} = \frac{P}{\omega} \tag{B21}$$

$$\sum_{m=-\infty}^{+\infty} \sum_{n=0}^{+\infty} \frac{n P_{m,n}}{m\omega_1 + n\omega_0} = 0 . \tag{B22}$$

In (B21), the hysteresis term may be transferred to the left-hand side and combined with the local oscillator term $P_{1,0}/\omega_1$ to give $(P_{1,0} - P)/\omega_1$. Therefore, the only difference from the results for the hysteresisless case is that the power lost in hysteresis is subtracted from the input power from the local oscillator. Thus, the power lost in hysteresis may be considered to come only from the local oscillator circuit, subject to the restriction that the hysteresis loop is no more than double-valued.

As an example of the foregoing analysis, we consider the simple case of non-linear modulators and demodulators in which only one of the principal sidebands of the signal about the carrier is allowed to carry a significant amount of power {25}. All other sidebands are assumed to be reactively terminated. Let f_1 again denote the local oscillator frequency, and f_2 a second applied frequency small compared with f_1. The corresponding sum and difference frequencies are denoted by

$$f_{\pm} = f_1 \pm f_2 \tag{B23}$$

the plus and minus signs corresponding to the terms <u>non-inverting modulator</u> (<u>demodulator</u>) and <u>inverting modulator</u> (<u>demodulator</u>), respectively, used in electrical engineering. For modulators the signal input is at f_2, the signal output at either f_+ or f_-; for demodulators the signal input is at either f_+ or f_-, the signal output at f_2.

For the non-inverting modulator and demodulator, and assuming that the powers at all but the three frequencies shown are equal to zero, the Manley-Rowe relations (B17) and (B18) yield:

$$\frac{P_1}{\omega_1} = \frac{P_2}{\omega_2} = -\frac{P_+}{\omega_+}, \tag{B24}$$

where P_1, P_2 and P_+ represent the powers flowing into the non-linear reactor at frequencies f_1, f_2 and f_+, respectively. For the modulator P_1 and P_2 are positive, representing power flowing into the non-linear reactor, while P_+ is negative, representing the useful power output of the device. For the demodulator, the reverse signs apply to the various powers.

For the inverting modulator and demodulator, the Manley-Rowe relations in turn yield:

$$\frac{P_1}{\omega_1} = -\frac{P_2}{\omega_2} = -\frac{P_-}{\omega_-}, \tag{B25}$$

where P_- represents the power flowing into the non-linear reactor at the frequency f_-. For both the inverting modulator and demodulator, P_1 is positive, P_2 and P_- negative; the input power from the local oscillator at the frequency f_1 flows out of the non-linear reactor at the two signal frequencies f_2 and f_-.

Now a striking analogy exists between equations (B24) and (B25) and the corresponding relations applicable to sum- and difference-frequency generation in a crystal with non-linear susceptibility owing to the interaction of wave fields of frequencies f_1 and f_2. After division of the corresponding equations (see Chapter 4) by the reduced Planck's constant ($\hbar$), relations are obtained connecting the changes in photon density (photon flux) at the various frequencies. Although the interpretation of these relations may then readily be given using the photon concept, it is clear from the foregoing that the relations arise from the two basic premises, viz. conservation of energy and the non-linear response of the medium {26}, and do not have special quantum-mechanical significance.

Consultation of Ref. {9} will show the analogies between the electrical circuit theory discussed in this section of the Appendix and non-linear optics applications, as listed in Table III.1.

TABLE III.1

ELECTRICAL CIRCUIT	ANALOGOUS LASER APPLICATION
non-inverting modulator	parametric up-conversion
non-inverting demodulator	parametric down-conversion
inverting modulator	parametric amplification
inverting demodulator	parametric oscillation

APPENDIX IV

THE INDEX ELLIPSOID

Many of the important results for extraordinary wave propagation in a uniaxial crystal follow immediately from the properties of the index ellipsoid {12,13}, i.e. the polar plot of $\mu_e(\theta)$ as given by equation (2.12). In the case of a negative uniaxial crystal, a plane section through this ellipsoid of revolution which includes the optic axis will be an ellipse of semi-major axis μ_o and semi-minor axis μ_e; in the case of the positive uniaxial crystal, the semi-major axis will be μ_e and the semi-minor axis μ_o. This is easily seen from consideration of the ellipse:

$$\frac{x^2}{\mu_o^2} + \frac{y^2}{\mu_e^2} = 1 \tag{C1}$$

where:

$$\left.\begin{aligned} x &= r\cos\theta \\ y &= r\sin\theta \end{aligned}\right\} \tag{C2}$$

θ being the angle between $\underset{\sim}{r}$ and the optic axis. Substitution from (C2) into (C1) yields:

$$r = \frac{\mu_o\mu_e}{\sqrt{\mu_o^2\sin^2\theta + \mu_e^2\cos^2\theta}} \tag{C3}$$

which immediately leads to the identification (from equation (2.12)):

$$r = \mu_e(\theta) \,. \tag{C4}$$

Next, we show that the outward drawn normal to this ellipsoid of revolution at any point $(\mu_e(\theta), \theta)$ is the ray direction (direction of energy propagation) for the particular wave normal at an angle θ to the optic axis. Differentiation of (C1) yields for the slope of the tangent to the index ellipsoid:

$$m_1 = -\left(\frac{\mu_e}{\mu_o}\right)^2 \frac{x}{y} \tag{C5}$$

whence the slope of the outward normal at the point $(\mu_e(\theta), \theta)$ equals:

$$\begin{aligned} m_2 \equiv \tan\varphi &= \left(\frac{\mu_o}{\mu_e}\right)^2 \frac{y}{x} \\ &= \left(\frac{\mu_o}{\mu_e}\right)^2 \tan\theta \end{aligned} \tag{C6}$$

in agreement with equation (2.30). (An elegant proof of this proposition will be found in the appendix to Ref. {46}.)

Moreover, from (C2)-(C5), it immediately follows that:

$$m_1 \equiv \tan\beta = \left(\frac{\mu^2 - \mu_e^2}{\mu^2 - \mu_o^2}\right) \tan\theta \tag{C7}$$

in agreement with equation (2.29). The electric field associated with the extraordinary ray is therefore tangential to the index ellipsoid, and lies in the plane containing $n = \underset{\sim}{r}/r$ and the optic axis. Its direction is in the sense of increasing θ. Therefore, the unit vectors $\left(\frac{1}{|\underset{\sim}{S}|}\underset{\sim}{S}, \frac{1}{|\underset{\sim}{E}|}\underset{\sim}{E}, \frac{1}{|\underset{\sim}{H}|}\underset{\sim}{H}\right)$ form a right-handed orthogonal set identical to the "usual" spherical polar unit vectors $(\underset{\sim}{i}_r, \underset{\sim}{i}_\theta, \underset{\sim}{i}_\phi)$.

The <u>double refraction</u> (aperture or "walk-off") <u>angle</u> $\alpha = |\theta-\varphi|$ now follows from (C3)-(C6):

$$\tan\alpha = \frac{1}{\mu_o\mu_e}\left[(\mu_o^2 - \mu^2)(\mu^2 - \mu_e^2)\right]^{\frac{1}{2}} \tag{C8}$$

in agreement with equation (2.31).

Next, we consider the general problem of collinear phase matching as discussed in section (f) of Chapter 5.

<u>Type Ia phase matching</u>:

The two conditions (5.53) and (5.54) yield for this case:

$$\mu_{3e}(\Theta) = \frac{1}{\omega_1 + \omega_2}\left[\mu_{1o}\omega_1 + \mu_{2o}\omega_2\right] \equiv \mu'_o \ . \tag{C9}$$

Now provided that:

$$\mu'_o \geq \mu_{3e} \tag{C10}$$

phase matching will be achieved for the angle:

$$\psi_o^{Ia} = \arc\sin\left[\frac{\mu_{3e}}{\mu'_o}\left(\frac{\mu_{3o}^2 - \mu_o'^2}{\mu_{3o}^2 - \mu_{3e}^2}\right)^{\frac{1}{2}}\right] . \tag{C11}$$

This type of phase matching is therefore suitable in the case of <u>negative uniaxial</u> crystals.

<u>Type Ib phase matching</u>:

The two conditions (5.53) and (5.54) yield for this case:

$$\mu_{3o} = \frac{1}{\omega_1 + \omega_2}\left[\mu_{1e}(\Theta)\,\omega_1 + \mu_{2e}(\Theta)\,\omega_2\right] . \tag{C12}$$

The curve described by the right-hand side of (C12) is, however, not in general expressible in the form (C3), and the calculation of the phase match angle therefore more cumbersome than in the previous case. However, provided that:

$$\mu'_e \equiv \frac{1}{\omega_1 + \omega_2}\left[\mu_{1e}\omega_1 + \mu_{2e}\omega_2\right] \geq \mu_{3o} \tag{C13}$$

a phase match angle ψ_o^{Ib} can be obtained from (C12). This type of phase matching is therefore suitable in the case of <u>positive uniaxial</u> crystals.

<u>Type IIa phase matching</u>:

The two conditions (5.53) and (5.54) yield for this case:

$$\mu_{3e}(\Theta) = \frac{1}{\omega_1+\omega_2}\left[\mu_{10}\omega_1 + \mu_{2e}(\Theta)\,\omega_2\right]. \tag{C14}$$

The solution for the phase match angle is again more cumbersome than (C11). However, provided that:

$$\frac{1}{\omega_1+\omega_2}\left[\mu_{10}\omega_1 + \mu_{2e}\omega_2\right] \geq \mu_{3e} \tag{C15}$$

a solution will exist for ψ_o^{IIa}. Now clearly in order for (C15) to hold in the case of <u>negative uniaxial</u> crystals, one has the necessary conditions:

i) $\omega_1 \geq \omega_2$, if $\mu_{20} \leq \mu_{3e}$;

ii) $\mu_{10} > \mu_{3e}$.

Conditions i) and ii) will be seen to hold in the solutions displayed in figure 2 of Ref. {52}.

<u>Type IIb phase matching</u>:

The two conditions (5.53) and (5.54) yield for this case:

$$\mu_{30} = \frac{1}{\omega_1+\omega_2}\left[\mu_{10}\omega_1 + \mu_{2e}(\Theta)\,\omega_2\right], \tag{C16}$$

which again lacks a simple solution of the type (C11). However, provided that:

$$\frac{1}{\omega_1+\omega_2}\left[\mu_{10}\omega_1 + \mu_{2e}\omega_2\right] \geq \mu_{30} \tag{C17}$$

a solution will be found for ψ_o^{IIb}. In order for (C17) to hold in the case of positive uniaxial crystals, one has the necessary conditions:

i) $\omega_2 \geq \omega_1$, if $\mu_{1e} \leq \mu_{30}$;

ii) $\mu_{2e} > \mu_{30}$.

It is clear from a comparison of (C10) with (C15) for the case of negative uniaxial crystals, and (C13) with (C17) in the case of positive uniaxial crystals, that for given frequencies ω_1 and ω_2, the phase match angle (where it exists) for type II processes is always greater than for type I processes, as noted for second harmonic generation by Hobden {36}.

BIAXIAL CRYSTALS

The Fresnel equation (2.8), with:

$$v_j = \frac{c}{\sqrt{\epsilon_j}} = \frac{c}{\mu_j} \tag{C18}$$

may be re-written as:

$$\mu_1^2(\mu_2^2 - \mu^2)(\mu_3^2 - \mu^2)x^2 + \mu_2^2(\mu_1^2 - \mu^2)(\mu_3^2 - \mu^2)y^2 + \mu_3^2(\mu_1^2 - \mu^2)(\mu_2^2 - \mu^2)z^2 = 0 \tag{C19}$$

where:

$$x_j = \mu n_j \tag{C20}$$

in terms of the direction cosines (n_j) of the wave normal.

Equation (C19), which applies to the triclinic, monoclinic and orthorhombic crystal systems, represents a two-sheeted surface with four points of intersection. When joined to the origin, these points yield the directions of the optic axes of the biaxial crystal. A cross section through the centre, perpendicular to one of the principal axes (i.e. a principal section) yields both an ellipse and a circle in each case. It is readily shown from (C19) that the principal section $x_j = 0$ consists of the circle:

$$\mu = \mu_j \tag{C21}$$

as well as the ellipse:

$$\frac{x_i^2}{\mu_k^2} + \frac{x_k^2}{\mu_i^2} = 1 . \tag{C22}$$

For the ordering $\mu_1 < \mu_2 < \mu_3$ chosen in the text, the following situations will be seen to apply: (a) In the xy plane, the ellipse lies within the circle. (b) In the yz plane, the circle lies within the ellipse. (c) In the xz plane, the circle and the ellipse intersect in four points. Opposite points, when connected, describe the two (primary) optic axes of the biaxial crystal, which are therefore perpendicular to the y axis.

From equations (C21) and (C22), it is readily found that the optic axes are each inclined at an angle:

$$\eta = \arctan\left[\frac{\mu_3}{\mu_1}\left(\frac{\mu_2^2 - \mu_1^2}{\mu_3^2 - \mu_2^2}\right)^{\frac{1}{2}}\right] \tag{C23}$$

to the z axis, in agreement with equation (6.14). For a wave normal directed along one of the optic axes, there are therefore two possible directions lying in the xz plane, of the corresponding Poynting vector. These two ray directions are given by the normal to the circle (C21) with j = 2, and the normal to ellipse (C22) with j = 2, calculated at one of the points of intersection. By differentiating these two equations, one determines the angle α between the two ray directions (i.e. the angle between the corresponding tangents) to be given by:

$$\tan\alpha = \frac{1}{\mu_1\mu_3}\left[(\mu_2^2 - \mu_1^2)(\mu_3^2 - \mu_2^2)\right]^{\frac{1}{2}}. \tag{C24}$$

This is the apex angle of the cone of ray directions corresponding to wave normals along an optic axis, which characterizes the process of <u>internal conical refraction</u> (equation (6.13)). It bears a striking resemblance to the double refraction angle (C8) for extraordinary wave propagation in a uniaxial crystal.

REFERENCES

1. P.A. FRANKEN, A.E. HILL, C.W. PETERS and G. WEINREICH, Phys. Rev. Lett. 7, 118 (1961).

2. R.W. TERHUNE, P.D. MAKER and C.M. SAVAGE, Phys. Rev. Lett. 8, 404 (1962).

3. W. KAISER and C.G.B. GARRETT, Phys. Rev. Lett. 7, 229 (1961).

4. S.A. AKHMANOV and R.V. KHOKHLOV, "Problems of Non-Linear Optics", Gordon and Breach, New York (1972).

5. D.A. KLEINMAN, Phys. Rev. 125, 87 (1962).

6. P.A. FRANKEN and J.F. WARD, Rev. Mod. Phys. 35, 23 (1963).

7. J.D. JACKSON, "Classical Electrodynamics", second edition, John Wiley and Sons, New York (1975).

8. F.N.H. ROBINSON, "Macroscopic Electromagnetism", Pergamon Press, Oxford (1973).

9. F. ZERNIKE and J.E. MIDWINTER, "Applied Non-Linear Optics", John Wiley and Sons, New York (1973).

10. L.D. LANDAU and E.M. LIFSHITZ, "Electrodynamics of Continuous Media", Pergamon Press, Oxford (1975).

11. D.S. JONES, "The Theory of Electromagnetism", Pergamon Press, Oxford (1964).

12. A.J.W. SOMMERFELD, "Optics", Academic Press, New York (1967).

13. M. BORN and E. WOLF, "Principles of Optics", Pergamon Press, Oxford (1965).

14. J.A. ARMSTRONG, N. BLOEMBERGEN, J. DUCUING and P.S. PERSHAN, Phys. Rev. 127, 1918 (1962).

15. D.A. KLEINMAN, Phys. Rev. 126, 1977 (1962).

16. M. SCHUBERT and B. WILHELMI, "Einführung in die Nichtlineare Optik", Teil I, B.G. Teubner, Leipzig (1971).

17. J.F. NYE, "Physical Properties of Crystals", Oxford University Press (1957).

18. F. ZERNIKE and P.R. BERMAN, Phys. Rev. Lett. 15, 999 (1965); 16, 117 (1966).

19. M. OKADA and S. IEIRI, Phys. Lett. 34A, 63 (1971).

20. R.C. MILLER, Appl. Phys. Lett. 5, 17 (1964).

21. C.G.B. GARRETT and F.N.H. ROBINSON, IEEE J. Quantum Electronics QE-2, 328 (1966).

22. C.K.N. PATEL, Phys. Rev. Lett. 15, 1027 (1965).

23. C.K.N. PATEL, Phys. Rev. Lett. 16, 613 (1966).

24. N. BLOEMBERGEN, "Non-Linear Optics", W.A. Benjamin, New York (1965).

25. J.M. MANLEY and H.E. ROWE, Proc. IRE 44, 904 (1956).

26. J.M. MANLEY and H.E. ROWE, Proc. IRE 47, 2115 (1959).

27. W.R. BENNETT, Bell System Tech. J. 12, 228 (1933).

28. J.A. GIORDMAINE and R.C. MILLER, Phys. Rev. Lett. 14, 973 (1965).

29. B.B. BAKER and E.T. COPSON, "The Mathematical Theory of Huygens' Principle", Clarendon Press, Oxford (1939).

30. G.N. WATSON, "A Treatise on the Theory of Bessel Functions", Cambridge University Press (1948).

31. P.D. MAKER, R.W. TERHUNE, M. NISENOFF and C.M. SAVAGE, Phys. Rev. Lett. 8, 21 (1962).

32. N. BLOEMBERGEN and A.J. SIEVERS, Appl. Phys. Lett. 17, 483 (1970).

33. A. ASHKIN, G.D. BOYD and D.A. KLEINMAN, Appl. Phys. Lett. 6, 179 (1965).

34. G.D. BOYD and C.K.N. PATEL, Appl. Phys. Lett. 8, 313 (1966).

35. J.A. GIORDMAINE, Phys. Rev. Lett. 8, 19 (1962).

36. M.V. HOBDEN, J. Appl. Phys. 38, 4365 (1967).

37. H. RABIN and P.P. BEY, Phys. Rev. 156, 1010 (1967).

38. P.P. BEY and H. RABIN, Phys. Rev. 162, 794 (1967).

39. C.K.N. PATEL and N. VAN TRAN, Appl. Phys. Lett. 15, 189 (1969).

40. P.K. TIEN, R. ULRICH and R.J. MARTIN, Appl. Phys. Lett. 17, 447 (1970).

41. D.F. NELSON and M. LAX, Appl. Phys. Lett. 18, 10 (1971).

42. D.A. KLEINMAN, A. ASHKIN and G.D. BOYD, Phys. Rev. 145, 338 (1966).

43. G.D. BOYD and D.A. KLEINMAN, J. Appl. Phys. 39, 3597 (1968).

44. G.D. BOYD and J.P. GORDON, Bell System Tech. J. 40, 489 (1961).

45. G.D. BOYD and H. KOGELNIK, Bell System Tech. J. 41, 1347 (1962).

46. G.D. BOYD, A. ASHKIN, J.M. DZIEDZIC and D.A. KLEINMAN, Phys. Rev. 137, A 1305 (1965).

47. D.A. KLEINMAN, Phys. Rev. 128, 1761 (1962).

48. G.D. BOYD and A. ASHKIN, Phys. Rev. 146, 187 (1966).

49. F.T.S. YU, "Introduction to Diffraction, Information Processing and Holography", MIT Press, Cambridge, Massachusetts (1973).

50. R.G. SMITH, J.E. GEUSIC, H.J. LEVINSTEIN, J.J. RUBIN, S. SINGH and L.G. VAN UITERT, Appl. Phys. Lett. 12, 308 (1968).

51. S. SINGH, D.A. DRAEGERT and J.E. GEUSIC, Phys. Rev. B2, 2709 (1970).

52. G.D. BOYD, H. KASPER and J.H. McFEE, IEEE J. Quantum Electronics QE-7, 563 (1971).

53. N. VAN TRAN, J. SPALTER, J. HANUS, J. ERNEST and D. KEHL, Phys. Lett. 19, 285 (1965).

54. N. BLOEMBERGEN, J. Opt. Soc. Am. 70, 1429 (1980).

55. Y.R. SHEN, Rev. Mod. Phys. 48, 1 (1976).

56. A. YARIV, "Quantum Electronics", second edition, John Wiley and Sons, New York (1975).

57. S.L. McCALL and E.L. HAHN, Phys. Rev. Lett. 18, 908 (1967).

58. S.L. McCALL and E.L. HAHN, Phys. Rev. 183, 457 (1969).

59. S.M. ARAKELYAN, G.A. LYAKHOV and Yu.S. CHILINGARYAN, Sov. Phys. Uspekhi 23, 245 (1980).

60. P.G. DE GENNES, "The Physics of Liquid Crystals", Clarendon Press, Oxford (1974).

61. S. CHANDRASEKHAR, "Liquid Crystals", Cambridge University Press (1977).

62. G.K.L. WONG and Y.R. SHEN, Phys. Rev. Lett. 30, 895 (1973).

63. J.W. SHELTON and Y.R. SHEN, Phys. Rev. A5, 1867 (1972).

64. J.W. SHELTON and Y.R. SHEN, Phys. Rev. Lett. 25, 23 (1970).

65. J.W. SHELTON and Y.R. SHEN, Phys. Rev. Lett. 26, 538 (1971).

66. R.W. HELLWARTH, J. Opt. Soc. Am. 67, 1 (1977); J.P. WOERDMAN, Optics Comm. 2, 212 (1970).

67. A. YARIV, Optics Commun. 25, 23 (1978); F.A. HOPF, A. TOMITA, K.H. WOMACK and J.L. JEWELL, J. Opt. Soc. Am. 69, 968 (1979).

68. P.D. MAKER and R.W. TERHUNE, Phys. Rev. 137, A801 (1965).

69. A. YARIV, IEEE J. Quantum Electronics QE-14, 650 (1978).

70. C.R. GIULIANO, Physics Today 34, 27 (1981).

71. R.W. HELLWARTH, J. Opt. Soc. Am. 68, 1050 (1978).

72. N. BLOEMBERGEN and P.S. PERSHAN, Phys. Rev. 128, 606 (1962).

73. N. BLOEMBERGEN, Opt. Acta 13, 311 (1966).

74. N. BLOEMBERGEN, H.J. SIMON and C.H. LEE, Phys. Rev. 181, 1261 (1969).

75. H. SHIH and N. BLOEMBERGEN, Phys. Rev. 184, 895 (1969).

76. W.R. HAMILTON, Trans. R. Irish Acad. 17, 1 (1833).

77. H. LLOYD, Trans. R. Irish Acad. 17, 145 (1833).

78. A.J. SCHELL and N. BLOEMBERGEN, J. Opt. Soc. Am. 68, 1093 (1978).

79. A.J. SCHELL and N. BLOEMBERGEN, J. Opt. Soc. Am. 68, 1098 (1978).

80. A.J. SCHELL and N. BLOEMBERGEN, Phys. Rev. A18, 2592 (1978).

81. R.Y. CHIAO, E. GARMIRE and C.H. TOWNES, Phys. Rev. Lett. 13, 479 (1964); 14, 1056 (1965).

82. P.L. KELLEY, Phys. Rev. Lett. 15, 1005 (1965); 16, 384 (1966).

83. C.R. GIULIANO and J.H. MARBURGER, Phys. Rev. Lett. 27, 905 (1971).

84. R.B. MILES and S.E. HARRIS, IEEE J. Quantum Electronics QE-9, 470 (1973).

85. H.M. GIBBS, S.L. McCALL and T.N.C. VENKATESAN, Phys. Rev. Lett. 36, 1135 (1976).

86. H.M. GIBBS, S.L. McCALL, T.N.C. VENKATESAN, A.C. GOSSARD, A. PASSNER and W. WIEGMANN, Appl. Phys. Lett. 35, 451 (1979).

87. D.A.B. MILLER, M.H. MOZOLOWSKI, A. MILLER and S.D. SMITH, Optics Commun. 27, 133 (1978).

88. H.G. WINFUL, J.H. MARBURGER and E. GARMIRE, Appl. Phys. Lett. 35, 379 (1979).

89. D.A.B. MILLER and S.D. SMITH, Optics Commun. 31, 101 (1979).

90. G. HERZBERG, "Molecular Spectra and Molecular Structure", Vols. I and II, D. Van Nostrand, Princeton (1962).

91. R.W. WOOD, "Physical Optics", Dover Publications, New York (1961).

92. L. BRILLOUIN, Ann. de Physique 17, 88 (1922).

93. P. DEBYE and F.W. SEARS, Proc. Nat. Acad. Sci. 18, 409 (1932).

94. R. LUCAS and P. BIQUARD, J. Phys. Radium 3, 464 (1932).

95. C.V. RAMAN and N.S.N. NATH, Proc. Ind. Acad. Sci. A2, 406 (1935); A3, 75 (1936).

96. E.J. WOODBURY and W.K. NG, Proc. IRE 50, 2367 (1962).

97. G. ECKHARDT, R.W. HELLWARTH, F.J. McCLUNG, S.E. SCHWARZ, D. WEINER and E.J. WOODBURY, Phys. Rev. Lett. 9, 455 (1962).

98. R.W. HELLWARTH, Phys. Rev. 130, 1850 (1963).

99. N. BLOEMBERGEN and Y.R. SHEN, Phys. Rev. Lett. 12, 504 (1964).

100. N. BLOEMBERGEN and Y.R. SHEN, Phys. Rev. 133, A37 (1964).

101. R. LOUDON, Proc. Phys. Soc. 82, 393 (1963).

102. R. LOUDON, Proc. Roy. Soc. (London) A275, 218 (1963); J.M. ZIMAN, "Electrons and Phonons", Clarendon Press, Oxford (1960).

103. R. LOUDON, Proc. Phys. Soc. 80, 952 (1962).

104. P.N. BUTCHER and T.P. McLEAN, Proc. Phys. Soc. 81, 219 (1963).

105. R.W. TERHUNE, Solid State Design 4, 38 (1963).

106. B.P. STOICHEFF, Phys. Lett. 7, 186 (1963).

107. H.J. ZEIGER, P.E. TANNENWALD, S. KERN and R. HERENDEEN, Phys. Rev. Lett. 11, 419 (1963).

108. R. CHIAO and B.P. STOICHEFF, Phys. Rev. Lett. 12, 290 (1964).

109. E. GARMIRE, F. PANDARESE and C.H. TOWNES, Phys. Rev. Lett. 11, 160 (1963).

110. R.Y. CHIAO, C.H. TOWNES and B.P. STOICHEFF, Phys. Rev. Lett. 12, 592 (1964).

111. B.Y. ZEL'DOVICH, V.I. POPOVICHEV, V.V. RAGUL'SKII and F.S. FAIZULLOV, Sov. Phys. JETP Lett. 15, 109 (1972).

112. M.D. LEVENSON and J.J. SONG, "Coherent Raman Spectroscopy", in "Coherent Non-Linear Optics, Recent Advances", edited by M.S. Feld and V.S. Letokhov, Topics in Current Physics series, Springer-Verlag, Berlin (1980).

113. D.M. BLOOM, P.F. LIAO and N.P. ECONOMOU, Optics Lett. 2, 58 (1978).

114. D.M. BLOOM, J.T. YARDLEY, J.F. YOUNG and S.E. HARRIS, Appl. Phys. Lett. 24, 427 (1974).

Lecture Notes in Physics

Vol. 114: Stellar Turbulence. Proceedings, 1979. Edited by D. F. Gray and J. L. Linsky. IX, 308 pages. 1980.

Vol. 115: Modern Trends in the Theory of Condensed Matter. Proceedings, 1979. Edited by A. Pekalski and J. A. Przystawa. IX, 597 pages. 1980.

Vol. 116: Mathematical Problems in Theoretical Physics. Proceedings, 1979. Edited by K. Osterwalder. VIII, 412 pages. 1980.

Vol. 117: Deep-Inelastic and Fusion Reactions with Heavy Ions. Proceedings, 1979. Edited by W. von Oertzen. XIII, 394 pages. 1980.

Vol. 118: Quantum Chromodynamics. Proceedings, 1979. Edited by J. L. Alonso and R. Tarrach. IX, 424 pages. 1980.

Vol. 119: Nuclear Spectroscopy. Proceedings, 1979. Edited by G. F. Bertsch and D. Kurath. VII, 250 pages. 1980.

Vol. 120: Nonlinear Evolution Equations and Dynamical Systems. Proceedings, 1979. Edited by M. Boiti, F. Pempinelli and G. Soliani. VI, 368 pages. 1980.

Vol. 121: F. W. Wiegel, Fluid Flow Through Porous Macromolecular Systems. V, 102 pages. 1980.

Vol. 122: New Developments in Semiconductor Physics. Proceedings, 1979. Edited by F. Beleznay et al. V, 276 pages. 1980.

Vol. 123: D. H. Mayer, The Ruelle-Araki Transfer Operator in Classical Statistical Mechanics. VIII, 154 pages. 1980.

Vol. 124: Gravitational Radiation, Collapsed Objects and Exact Solutions. Proceedings, 1979. Edited by C. Edwards. VI, 487 pages. 1980.

Vol. 125: Nonradial and Nonlinear Stellar Pulsation. Proceedings, 1980. Edited by H. A. Hill and W. A. Dziembowski. VIII, 497 pages. 1980.

Vol. 126: Complex Analysis, Microlocal Calculus and Relativistic Quantum Theory. Proceedings, 1979. Edited by D. Iagolnitzer. VIII, 502 pages. 1980.

Vol. 127: E. Sanchez-Palencia, Non-Homogeneous Media and Vibration Theory. IX, 398 pages. 1980.

Vol. 128: Neutron Spin Echo. Proceedings, 1979. Edited by F. Mezei. VI, 253 pages. 1980.

Vol. 129: Geometrical and Topological Methods in Gauge Theories. Proceedings, 1979. Edited by J. Harnad and S. Shnider. VIII, 155 pages. 1980.

Vol. 130: Mathematical Methods and Applications of Scattering Theory. Proceedings, 1979. Edited by J. A. DeSanto, A. W. Sáenz and W. W. Zachary. XIII, 331 pages. 1980.

Vol. 131: H. C. Fogedby, Theoretical Aspects of Mainly Low Dimensional Magnetic Systems. XI, 163 pages. 1980.

Vol. 132: Systems Far from Equilibrium. Proceedings, 1980. Edited by L. Garrido. XV, 403 pages. 1980.

Vol. 133: Narrow Gap Semiconductors Physics and Applications. Proceedings, 1979. Edited by W. Zawadzki. X, 572 pages. 1980.

Vol. 134: γγ Collisions. Proceedings, 1980. Edited by G. Cochard and P. Kessler. XIII, 400 pages. 1980.

Vol. 135: Group Theoretical Methods in Physics. Proceedings, 1980. Edited by K. B. Wolf. XXVI, 629 pages. 1980.

Vol. 136: The Role of Coherent Structures in Modelling Turbulence and Mixing. Proceedings 1980. Edited by J. Jimenez. XIII, 393 pages. 1981.

Vol. 137: From Collective States to Quarks in Nuclei. Edited by H. Arenhövel and A. M. Saruis. VII, 414 pages. 1981.

Vol. 138: The Many-Body Problem. Proceedings 1980. Edited by R. Guardiola and J. Ros. V, 374 pages. 1981.

Vol. 139: H. D. Doebner, Differential Geometric Methods in Mathematical Physics. Proceedings 1981. VII, 329 pages. 1981.

Vol. 140: P. Kramer, M. Saraceno, Geometry of the Time-Dependent Variational Principle in Quantum Mechanics. IV, 98 pages. 1981.

Vol. 141: Seventh International Conference on Numerical Methods in Fluid Dynamics. Proceedings. Edited by W. C. Reynolds and R. W. MacCormack. VIII, 485 pages. 1981.

Vol. 142: Recent Progress in Many-Body Theories. Proceedings. Edited by J. G. Zabolitzky, M. de Llano, M. Fortes and J. W. Clark. VIII, 479 pages. 1981.

Vol. 143: Present Status and Aims of Quantum Electrodynamics. Proceedings, 1980. Edited by G. Gräff, E. Klempt and G. Werth. VI, 302 pages. 1981.

Vol. 144: Topics in Nuclear Physics I. A Comprehensive Review of Recent Developments. Edited by T.T.S. Kuo and S.S.M. Wong. XX, 567 pages. 1981.

Vol. 145: Topics in Nuclear Physics II. A Comprehensive Review of Recent Developments. Proceedings 1980/81. Edited by T. T. S. Kuo and S. S. M. Wong. VIII, 571-1.082 pages. 1981.

Vol. 146: B. J. West, On the Simpler Aspects of Nonlinear Fluctuating. Deep Gravity Waves. VI, 341 pages. 1981.

Vol. 147: J. Messer, Temperature Dependent Thomas-Fermi Theory. IX, 131 pages. 1981.

Vol. 148: Advances in Fluid Mechanics. Proceedings, 1980. Edited by E. Krause. VII, 361 pages. 1981.

Vol. 149: Disordered Systems and Localization. Proceedings, 1981. Edited by C. Castellani, C. Castro, and L. Peliti. XII, 308 pages. 1981.

Vol. 150: N. Straumann, Allgemeine Relativitätstheorie und relativistische Astrophysik. VII, 418 Seiten. 1981.

Vol. 151: Integrable Quantum Field Theory. Proceedings, 1981. Edited by J. Hietarinta and C. Montonen. V, 251 pages. 1982.

Vol. 152: Physics of Narrow Gap Semiconductors. Proceedings, 1981. Edited by E. Gornik, H. Heinrich and L. Palmetshofer. XIII, 485 pages. 1982.

Vol. 153: Mathematical Problems in Theoretical Physics. Proceedings, 1981. Edited by R. Schrader, R. Seiler, and D.A. Uhlenbrock. XII, 429 pages. 1982.

Vol. 154: Macroscopic Properties of Disordered Media. Proceedings, 1981. Edited by R. Burridge, S. Childress, and G. Papanicolaou. VII, 307 pages. 1982.

Vol. 155: Quantum Optics. Proceedings, 1981. Edited by C.A. Engelbrecht. VIII, 329 pages. 1982.